Computational Fluid Dynamics
A Practical Approach

Second Edition

Jiyuan Tu
RMIT University, Australia

Guan-Heng Yeoh
Australian Nuclear Science
and Technology Organisation
University of New South Wales, Australia

Chaoqun Liu
University of Texas at Arlington

AMSTERDAM • BOSTON • HEIDELBERG • LONDON
NEW YORK • OXFORD • PARIS • SAN DIEGO
SAN FRANCISCO • SINGAPORE • SYDNEY • TOKYO
Butterworth-Heinemann is an imprint of Elsevier

ELSEVIER

Butterworth-Heinemann is an imprint of Elsevier
225 Wyman Street, Waltham, MA 02451, USA
The Boulevard, Langford Lane, Kidlington, Oxford, OX5 1 GB, UK

Library of Congress Cataloging-in-Publication Data
Application submitted.

British Library Cataloguing-in-Publication Data
A catalogue record for this book is available from the British Library.

ISBN: 978-0-08-098243-4

For information on all Butterworth-Heinemann publications,
visit our website: http://store.elsevier.com

Printed in the United Kingdom
12   13   14   15   16   10 9 8 7 6 5 4 3 2 1

Contents

Contents

Preface to the First Edition

Computational fluid dynamics (CFD), once the domain of academics, postdoctoral researchers, or trained specialists, is now progressively becoming more accessible to graduate engineers for research and development as well as design-oriented tasks in industry. Mastery of CFD in handling complex flow and heat industrial problems is becoming ever more important. Competency in such a skill certainly brings about a steep learning curve for practicing engineers, who constantly face extreme challenges to come up with solutions to fluid flow and heat transfer problems without *a priori* knowledge of the basic concepts and fundamental understanding of fluid mechanics and heat transfer.

Today's engineers are almost certainly geared more toward the use of commercial CFD codes, such as ANSYS-CFX, ANSYS-FLUENT, or STAR-CD. Without proper guidance, the use of these software packages poses risks likened to placing potent weaponry in the hands of poorly trained soldiers. There is every possibility of users with inadequate training causing more harm than good through flawed interpretations of results produced through such packages. This makes it ever more important that a sound knowledge of CFD be acquired. Furthermore, a changing workplace environment has imposed constraints on users in discerning the pitfalls of CFD by osmosis and through frequent failures. The number of users who have had the luxury of being fully equipped and who are conscious of the limitations of CFD from their own experiences is fast dwindling.

The purpose of this book is to offer CFD users a *suitable text* pitched at *the right level* of assumed knowledge. CFD is a mathematically sophisticated discipline and the authors' aim in this book has been to provide simple-to-understand descriptions of fundamental CFD theories, basic CFD techniques, and practical guidelines. It has never been our aim to overwhelm the reader with excessive mathematical and theoretical illustrations of computational techniques. Every effort has been made to discuss the material in a style to capture the reader's attention. The dominant feature of the present book is to maintain practicality in understanding CFD. In our lecturing experience on CFD, we have identified what it takes to present elementary concepts to initiate the student. This book incorporates specially designed *intuitive* and *systematic* worked-out CFD examples to enhance the learning process, and it provides students with examples for practice to better comprehend the basic principles. It is hoped that this approach will accomplish the purpose of offering techniques to *beginners* who are more focused on the *engineering practice* of CFD.

The basic structure of this book is as follows:

Chapter 1 presents an introduction to computational fluid dynamics and is specifically designed to provide the reader with an overview of CFD and its entailed advantages, the range of applications as a research tool on various facets of industrial problems, and the future use of CFD.

Chapter 2 aims to cultivate a sense of curiosity for the first-time user on how a CFD problem is currently handled and solved. The reader will benefit through guidance of these basic processes using any commercial, shareware and in-house CFD codes. More importantly, Chapter 2 serves as a guidepost for the reader to other chapters relating to fundamental knowledge of CFD. Chapter 2 has a unique design compared with many traditional CFD presentations.

The basic thoughts and philosophy associated with CFD, along with an extensive discussion of the governing equations of fluid dynamics and heat transfer, are treated in Chapter 3. It is vitally important that the reader can fully appreciate, understand, and feel comfortable with the basic physical equations and underlying principles of this discipline, as they are its lifeblood. By working through the worked-out examples, the reader will have a better understanding of the equations governing the conservation of mass, momentum, and energy.

Computational solutions are obtained in two stages. The first stage deals with numerical discretization, which is examined in Chapter 4. Here, the basic numerics are illustrated with popular discretization techniques, such as the finite-difference and finite-volume methods (adopted in the majority of commercial codes) for solving flow problems. The second stage deals with the specific techniques for solving algebraic equations. The pressure–velocity coupling scheme (SIMPLE and its derivatives) in this chapter forms the information core of the book. This scheme invariably constitutes the basis of most commercial CFD codes through which simulations of complex industrial problems have been successfully made.

The numerical concepts of stability, convergence, consistency, and accuracy are discussed in Chapter 5. As an understanding of the fundamental equations of fluid flow and heat transfer is the essence of CFD, it follows that the understanding of the techniques of achieving a CFD solution is the resultant substantive. This chapter will enable the reader to better assess the results produced when different numerical methodologies are applied.

The authors have included turbulence modeling in CFD, a subject not ordinarily treated in a book of this nature, but after careful consideration, we have felt it imperative to include it since *real-world applications of CFD* are turbulent in nature after all. In Chapter 6, the authors have therefore devised some practical guidelines for the reader to better comprehend turbulence modeling and other models commonly applied. The authors have also carefully designed worked-out examples that will assist students in the understanding of the complex modeling concept.

An increasing number of books and journals covering different aspects of CFD in mathematically abstruse terms are readily available, mainly for specialists associated with industry. It follows that it is more helpful to include in Chapter 7 illustrations of the power of CFD through a set of industrially relevant applications on a significant range of engineering disciplines. Special efforts have been made in this chapter to stimulate the inquisitive minds of the reader through exposition of some pioneering applications.

Although detailed treatment of advanced CFD techniques is usually outside the scope of a book of this nature, we have offered a general introduction to the basic concepts in Chapter 8, hoping, in the process, to reap the benefits of whetting the readers' appetite for more to come in the evolutionary use of CFD in any new emerging areas of science and engineering.

Jiyuan Tu
Guan-Heng Yeoh
Chaoqun Liu

Preface to the Second Edition

The acceptance of the first edition of our book by the CFD community has certainly been overwhelming and most welcomed. We were extremely pleased by the positive feedback received even within the short period since its publication. In responding to the numerous comments, the second edition aims to further enhance and update the fast-growing subject of CFD, including significant developments and important applications. In order not to stray away from our primary focus of offering CFD users a *suitable text* that is pitched at *the right level* of assumed knowledge, the structure and systematic approach of the first edition have been retained.

In the treatment of the fundamental physics of fluid flows, we have added the generic form of the equations pertaining to compressible flows, which are in retrospect an extension to the incompressible form of the equations for CFD.

At the time of the writing of the first edition, we focused predominantly on the description of the most popular discretization approaches in CFD, the finite-difference and finite-volume methods. Recognizing that other discretization methods, such as the finite-element and spectral methods, are still available in the mainstream of CFD, we have provided a summary of the basic ideas underpinning the use of these methods to solve the fluid-flow equations. To reflect the iterative approach that is also commonly being adopted to solve systems of discretized equations in commercial CFD codes, we have written a section in Chapter 4 dedicated to the multi-grid method.

In the first edition, we also identified a number of key sectors where CFD has been firmly established. They are aerospace, biomedical science and engineering, chemical and mineral processing, civil and environmental engineering, power generation, and sports. In this second edition, we add the application of CFD in the areas of metallurgy and nuclear safety. Considering the wide spread of CFD throughout a number of significant engineering areas, the proper handling of complex geometries becomes ever more important in view of different types of meshing approaches that can be employed. Discussions on key aspects of structured, body-fitted, and unstructured meshes have been provided within the practical guidelines of grid generation.

Finally, an alternative numerical approach based on the discrete element method has been added to the growing list of advanced topics in CFD. For instructors adopting this text for use in their courses, solutions to end-of-chapter problems and a set of PowerPoint slides are available by registering at www.textbooks.elsevier.com.

Jiyuan Tu
Guan-Heng Yeoh
Chaoqun Liu

Acknowledgments

The material presented in this book has been partly accumulated from teaching the course *Introduction to Computational Fluid Dynamics* for senior undergraduate students at the School of Aerospace, Mechanical and Manufacturing Engineering at the Royal Melbourne Institute Technology University, Australia, as well as the course *Computational Engineering* at the School of Mechanical and Manufacturing Engineering, University of New South Wales, Australia. We thank the students who have taken the course for providing us with feedback in designing particular project topics and in aiding understanding of the subject. The authors also thank many research students and colleagues who have generously assisted us in various ways.

For the first edition, we thank Jonathan Simpson, senior commissioning editor, and his colleagues at Elsevier Science & Technology who have offered us immense help in both their academic elucidation and their professional skills in the publication process. Our special thanks are also given to Dr. Risa Robinson, Associate Professor of Mechanical Engineering at the Rochester Institute of Technology, for reviewing the whole text within a very short period of time and giving us invaluable suggestions and comments that we have incorporated in this book.

We thank Joe Hayton and Fiona Geraghty from Elsevier Science and Technology for initiating and managing the project for the second edition of our book.

Dr. Tu expresses his deep gratitude to his wife, Xue, and his son, Tian, who have provided their unflinching support in the preparation and writing of the text.

Dr. Yeoh acknowledges the untiring support of his wife, Natalie, and his daughters, Genevieve, Ellana, and Clarissa, and thanks them for their understanding and encouragement during the seemingly unending hours spent in preparing and writing.

Dr. Liu acknowledges the strong support and encouragement received from his wife, Weilan, his daughter, Haiyan, and his son, Haifeng, during the preparation of the text.

To all who have been involved, we extend our deepest, heartfelt appreciation.

Introduction

1.1 WHAT IS COMPUTATIONAL FLUID DYNAMICS?

Computational fluid dynamics has certainly come of age in industrial applications and academic research. In the beginning, this popular field of study, usually referred to by its acronym CFD, was only known in the high-technology engineering areas of aeronautics and astronautics, but now it is becoming a rapidly adopted methodology for solving complex problems in modern engineering practice. CFD, which is derived from the disciplines of fluid mechanics and heat transfer, is also finding its way into important uncharted areas, especially in process, chemical, civil, and environmental engineering. Construction of new and improved system designs and optimization carried out on existing equipment through computational simulations are resulting in enhanced efficiency and lower operating costs. With the concerns about global warming and the world's increasing population, engineers in power-generation industries are heavily relying on CFD to reduce development and retrofitting costs. These computational studies are currently being performed to address pertinent issues relating to technologies for clean and renewable power as well as for meeting strict regulatory challenges, such as emissions control and substantial reduction of environmental pollutants.

Nevertheless, the basic question remains: What is *computational fluid dynamics*? In retrospect, it has certainly evolved, integrating not only the disciplines of fluid mechanics with mathematics but also computer science, as illustrated in Figure 1.1. Let's briefly discuss each of these individual disciplines. Fluid mechanics is essentially the study of fluids, either in motion (*fluid in dynamic mode*) or at rest (*fluid in stationary mode*). CFD is particularly dedicated to the former, fluids that are in motion, and how the fluid-flow behavior influences processes that may include heat transfer and possibly chemical reactions in combusting flows. This directly applies to the "*fluid dynamics*" description appearing in the terminology. Additionally, the physical characteristics of the fluid motion can usually be described through fundamental *mathematical* equations, usually in partial differential form, which govern a process of interest and are often called governing equations in CFD (see Chapter 3 for more insights). In order to solve these *mathematical* equations, *computer scientists* using high-level computer programming languages convert the equations into computer programs or software packages. The "*computational*" part simply means the study of the fluid flow

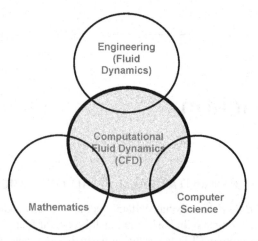

FIGURE 1.1   The different disciplines involved in computational fluid dynamics.

using numerical simulations, which involves employing computer programs or software packages performed on high-speed digital computers to attain the numerical solutions. Another question arises: Do we actually require the expertise of three specific people from each discipline—fluids engineering, mathematics, and computer science—to come together for the development of CFD programs or even to conduct CFD simulations? The answer is that it is more likely that a person who proficiently obtains more or less some subsets of the knowledge from each discipline can meet the demands of CFD.

CFD has also become one of the three basic methods or approaches that can be employed to solve problems in fluid dynamics and heat transfer. As demonstrated in Figure 1.2, the approaches that are strongly interlinked do not work in

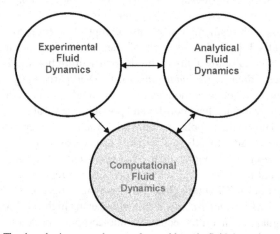

FIGURE 1.2   The three basic approaches to solve problems in fluid dynamics and heat transfer.

isolation. Traditionally, both experimental and analytical methods have been used to study the various aspects of fluid dynamics and to assist engineers in the design of equipment and industrial processes involving fluid flow and heat transfer. With the advent of digital computers, the computational (numerical) aspect has emerged as another viable approach. Although the analytical method is still practiced by many, and experiments will continue to be significantly performed, the trend is clearly toward greater reliance on the computational approach for industrial designs, particularly when the fluid flows are very complex.

In the past, potential or novice users would probably have learned CFD by investing a substantial amount of time in writing their own computer programs. With the increasing demands from industries or even within academia to acquire knowledge of CFD in a much shorter time, it is not surprising that interest in writing original computer programs is waning, in favor of using more commercially available software packages. Multi-purpose CFD programs are gradually earning approval, and with the advancement of models to better encapsulate the flow physics, these software packages are also gaining wide acceptance. There are numerous advantages in applying these computer programs. Since the mundane groundwork of writing and testing the computer codes has been thoroughly carried out by the "developers" in the respective software companies, today's potential or novice CFD users are relieved of dealing with these issues. Commercial programs can be readily employed to solve numerous fluid-flow problems.

Despite well-developed methodologies within the computational codes, CFD is certainly more than just being proficient in operating software packages. Bearing this in mind, the primary focus of this book is, therefore, to educate potential or novice users in how to employ CFD judiciously, so the text aims equally at supplementing the understanding of underlying basic concepts and at the technical know-how in better tackling fluid-flow problems. Users who are inclined to pursue postgraduate research or are currently undertaking research through the development of new mathematical models to solve more complex flow problems should consult other CFD books (e.g., Fletcher, 1991, Anderson, 1995, Versteeg and Malalasekera, 1995) and our future book, in which we intend to concentrate on presenting a step-by-step procedure for initially understanding the physics of new fluid dynamics problems at hand, developing new mathematical models to represent the flow physics, and implementing appropriate numerical techniques or methods to test the models in a CFD program.

CFD has indeed become a powerful tool that can be employed either for pure/applied research or for industrial applications. Computational simulations and analyses are increasingly performed in many fluid-engineering applications, which include aerospace engineering (airplanes, rocket engines...), automotive engineering (reducing drag coefficients for cars and trucks, improving air intake in engines...), biomedical engineering (blood flow in artificial hearts and, through scent flow, air flow in breathing...), chemical engineering (fluid flow through pumps and pipes...), civil and environmental engineering (river restoration, pollutant dispersion…), power engineering (improving turbine efficiency, wind farm siting and performance prediction…), and sports engineering (swimming equipment, golf swing mechanics, reducing drag in biking…). Through CFD, one can gain an increased knowledge of how system components are expected to perform, so as to make the required improvements for design and optimization studies. CFD actually asks the question "What if …?" before a commitment is undertaken to execute any design alteration. When one ponders the planet we live on, almost everything revolves in one way or another around a fluid or moves within a fluid.

CFD is also revolutionizing the teaching and learning of fluid mechanics and thermal science in higher-education institutions through visualization of complex fluid flows. Development of CFD-based software packages to be more user-friendly is allowing students to visually reinforce the concepts of fluid flow and heat transfer. The software allows teachers to create their own examples or to customize predefined existing ones. With carefully constructed examples, students are introduced to the effective use of CFD for solving fluid-flow problems, and they can develop an intuitive feel for flow physics. In the next section, we discuss some important advantages of CFD and further expound on how CFD has evolved and is applied in practice.

1.2 ADVANTAGES OF COMPUTATIONAL FLUID DYNAMICS

With the rapid advancement of digital computers, CFD is poised to remain at the forefront of cutting-edge research in the sciences of fluid dynamics and heat transfer. Also, the emergence of CFD as a practical tool in modern engineering practice is steadily attracting much interest.

There are many advantages of computational fluid dynamics. First, the theoretical development of the computational sciences focuses on the construction and solution of the governing equations and the study of various approximations to these equations. CFD presents the perfect opportunity to study specific terms in the governing equations in a more detailed fashion. New paths of theoretical development are realized which could not have been possible without the introduction of this computational approach. Second, CFD complements experimental and analytical approaches by providing an alternative cost-effective means of simulating real fluid flows. Particularly, CFD substantially reduces lead times and costs in design and production compared with experimentally based approaches and offers the ability to solve a range of complicated flow problems where the analytical approach is lacking. These advantages are realized through the increasing performance power of computer hardware and its declining costs. Third, CFD has the capacity to simulate flow conditions that are not reproducible in experimental tests found in geophysical and biological fluid dynamics, such as nuclear accident scenarios or scenarios that are too huge or too remote to be simulated experimentally (e.g., the Indonesian Tsunami of 2004). Fourth, CFD can provide

detailed visualization and comprehensive information when compared to analytical and experimental fluid dynamics.

In practice, CFD permits alternative designs to be evaluated over a range of
dimensionless parameters that may include the Reynolds number, Mach number, Rayleigh number, and flow orientation. The utilization of such an approach
is usually very effective in the early stages of development for fluid-system
designs. It may also prove to be significantly less expensive than the ever-
increasing spiraling cost of performing experiments. In many cases, where
details of the fluid flow are important, CFD can provide detailed information
and understanding of the flow processes to be obtained, such as the occurrence
of flow separation or whether the wall temperature exceeds some maximum
limit. With technological improvements and competition requiring a higher
degree of optimal designs, and as new high-technology applications demand
precise prediction of flow behaviors, experimental development may eventually
be too costly to initiate. CFD presents an alternative.

Nevertheless, the favorable appraisal of CFD thus far does not suggest that it
will soon replace experimental testing as a means for gathering information for
design purposes. Rather, it is considered to be complementary in solving fluid-
mechanics problems. For example, wind-tunnel testing is a typical experimental
approach that still provides invaluable information for the simulation of real
flows at reduced scale. For the design of engineering components, especially
for aircraft, which depend critically on the flow behavior, carrying out wind-
tunnel experiments remains an economically viable alternative to full-scale
measurement. Wind tunnels are very effective for obtaining global information
about the complete lift and drag on a body and the surface distributions at key
locations. In other applications where CFD still remains in a relatively primitive
state of development, experimental approaches are still the primary source of
information, especially when complex flows, such as multi-phase flows, boiling,
or condensation, are involved.

In spite of CFD's advantages, the reader must also be fully aware of some
inherent limitations of applying CFD. Numerical errors exist in computations;
therefore, there will be differences between computed results and reality. Visualization of numerical solutions using vectors, contours, or animated movies of
unsteady flows is by far the most effective way of interpreting the huge amount
of data generated from the numerical calculation. However, there is a danger that
an erroneous solution, which may look good, will not correspond to the expected
flow behavior! The authors have encountered numerous incorrect numerically
produced flow characteristics that could have been interpreted as acceptable
physical phenomena. Wonderfully bright color pictures may imply a realistic por-
trayal of the actual fluid mechanics inside the flow system, but they are worthless
if they are not quantitatively correct. Any numerical results obtained must always
be thoroughly examined before they are believed. Hence, a CFD user needs to
learn how to properly analyze and make critical judgments about the computed
results. This is one of the important aims of this book.

1.3   APPLICATION OF COMPUTATIONAL FLUID DYNAMICS

1.3.1   As a Research Tool

CFD can be employed to better understand the physical events or processes that occur in the flow of fluids around and within the designated objects. These events are closely related to the action and interaction of phenomena associated with dissipation, diffusion, convection, boundary layers, and turbulence. Whether the flows are incompressible or compressible, many of the most important aspects of these types of flows are non-linear and, as a consequence, often do not have any analytic solution. This motivates the search for numerical solutions for the partial differential equations, and it would seem, in hindsight, to invalidate the use of linear algebra for the classification of the numerical methods. Our experiences have nevertheless demonstrated that such is not the case.

CFD, analogous to wind-tunnel tests, can be employed as a *research tool* to perform *numerical experiments*. We examine one of these *numerical experiments*, garnered from our research in Li, Tu and Yeoh (2004), in order to demonstrate the feasible use of CFD as a *research tool* and to impart some understanding of this philosophy. Figure 1.3(a) represents a "snapshot" taken for an unsteady flow past two side-by-side cylinders at a given instant of time. In Figure 1.3(b), the comparative visualization of the numerical calculations based on a large eddy simulation (LES) model attests to the power of CFD modeling for capturing the complex flow characteristics. This example clearly illustrates how CFD can be utilized to better understand the observed flow structures and some important physical aspects of a flow field, similar to a real laboratory experiment. For the case of three side-by-side cylinders shown in Figure 1.4, here again is another example of how CFD simulations can work harmoniously with experiments, providing not only qualitative comparison but also a means to interpret some basic phenomenological aspects of the experimental condition. More important, *numerical experiments* can provide more comprehensive information and details of the flow visualized in three

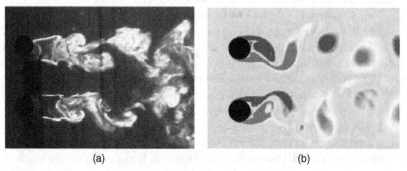

(a)   (b)

FIGURE 1.3   Example of a CFD numerical experiment for a flow past two side-by-side cylinders: (a) experimental observation and (b) numerical simulation.

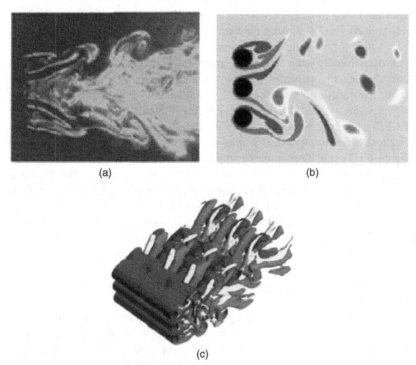

(a)　　　　　　　　　　　　　(b)

(c)

FIGURE 1.4　Example of CFD numerical experiment for flow past three side-by-side cylinders: (a) experimental observation, (b) numerical simulation on a two-dimensional cross-sectional plane, and (c) three-dimensional representation of the fluid flow through numerical simulation.

dimensions, as shown in Figure 1.4(c), when compared to laboratory experiments. These graphical examples clearly affirm the value of *numerical experiments* carried out within the framework of CFD.

1.3.2　As an Educational Tool in Basic Thermal-Fluid Science

While CFD is typically studied at the graduate level, the ease of use and broad capability of commercial CFD software packages have enabled CFD to be brought into the undergraduate classroom. The authors believe there are two prime benefits in exposing undergraduates to CFD. First, experience with a hands-on approach, as adopted in this book, leads to better understanding of the concepts of fluid flow and heat transfer and greatly enriches students' understanding of fluid-flow phenomena. In particular, the visualization capability enhances the students' intuition of flow behavior. Second, the approach opens the door to new classes of problems that can be solved by undergraduates, who are no longer limited by the narrow range of classical flow solutions.

For an introductory engineering curriculum, a CFD-based educational software package allows students to readily solve fluid-dynamics problems without requiring a long training period. The mission is to expose students to essential CFD concepts and to expand the learning experience with real-world applications, which is becoming an increasingly important skill in today's job market. With user-friendly student-specific graphical user interfaces guiding the students through the stages of geometry creation and mesh generation, which are further exemplified in Chapter 2, computational simulations and viewing the results by means of vectors, contours, or animated movies, the teaching of CFD has never been so visually exhilarating and welcoming. Within the graphical user interfaces, line graphs are also provided to assist users in assessing the CFD simulations, either by tracking the convergence history or by monitoring the surface distribution of certain fluid forces, such as the lift force through the lift coefficient. The prime derivative of these graphical user interfaces is certainly more than just introducing CFD technology to undergraduates. They are intended not only to arouse students' interest in basic fluid dynamics but also to entice them to further extend their learning experience to other transport phenomena of all kinds that may exist in practice or in nature.

1.3.3 As a Design Tool

Similarly, CFD is becoming an integral part of the engineering design and analysis environment in prominent industries. Companies are progressively seeking industrial solutions through the extensive use of CFD for the optimization of product development and processes and/or to predict the performance of new designs before they are manufactured or implemented. Software applications can now provide numerical analyses and solutions to pertinent flow problems through the employment of common desktop computers. As a viable design tool, CFD has assisted by providing significant and substantial insights into the flow characteristics within the equipment and processes required to increase production, improve longevity, and decrease waste. Increasing computer processing power is certainly revolutionizing the use of CFD in new and existing industries. These industrial solutions are expounded upon in succeeding sections of this book to further demonstrate the wide application of this specific technology in practice.

1.3.4 Aerospace

Computational fluid dynamics has certainly enjoyed a long and illustrious history of development and application in the aerospace and defense industries. To maintain an edge in a very competitive environment, CFD is playing a crucial role in overcoming many challenges faced by these industries in improving flight and in solving a diverse array of designs. Indeed, many engineers associate CFD with its well-known application to aerodynamics in the calculation of the lift force on an aircraft wingspan. Nevertheless, as methods and resources have augmented CFD's power and ease of use, practitioners have expanded the scope of application beyond the calculation of lift. Today, CFD is being applied

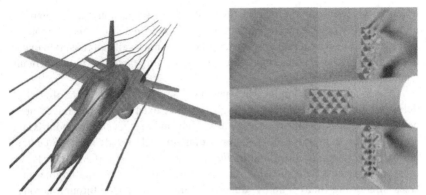

FIGURE 1.5 Example of CFD results for applications in the aerospace and defense industries. *(Courtesy of ANSYS-FLUENT and U. S. Army Research Laboratory)*

to many more-difficult operational problems that were too unwieldy to analyze or to solve with computational tools in the past.

Figure 1.5 illustrates the simulation of fluid path lines in the vicinity of an F18 jet (left) and prediction of pressure coefficient contours at a $10°$ angle of attack around a supersonic missile system with grid fins (right). These are just a small sample of the numerous applications of CFD in aerodynamic design and military applications. CFD has also been employed in resolving a number of complex operational problems in aircraft design, such as studying the impact of trailing vortices on the safe operation of successive aircraft taking off and landing on a runway, as well as in enhancing passenger and crew comfort by improving cabin ventilation, heating, and cooling. Efforts to better understand and suppress the noise produced by heavy artillery and the safe operation of a military helicopter upon firing a missile, which could affect the airframe or the tail rotor, are some other operational problems that CFD is increasingly being used to address in military applications. As a versatile, robust, and powerful tool, CFD is rigorously meeting the broad physical modeling demands of investigating relevant complex phenomena in aerospace and defense-related designs. CFD is undoubtedly becoming a household simulation tool within these industries, since the need to save time and money by reducing development costs and accelerating time-to-market and by improving the overall performance of system configurations is becoming more common.

1.3.5 Automotive Engineering

Automobile engineers are increasingly relying on more simulation techniques to bring new vehicle design concepts to fruition. Computer-aided engineering has been at the forefront of creating innovative internal systems that will enhance the overall driving experience, improve driver and passenger comfort and safety, and advance fuel economy. CFD has long been an essential element in automotive design and manufacture. Like the aerospace and aerodynamics industries, this branch of engineering has embraced much of this technology in research as well as in practice; the use of CFD is thus well entrenched in many disciplines as the engineering simulation tool for even the most difficult challenges.

CFD in automotive engineering has many advantages. The technology has delivered the ability to shorten cycles, to optimize existing engineering components and systems to improve energy efficiency and to meet strict standards and specifications, to improve the in-car environment, and to study important external aerodynamics, as illustrated in Figure 1.6. Specifically, CFD has shown measurable results in decreasing emissions with powertrain and engine analyses; increasing fuel economy, durability, and performance through aerodynamic investigations; and increasing reliability of brake components. As in the aerospace industry, CFD has been used to determine the effects of local geometry changes on the aerodynamic forces and provides a significant capability for directly comparing a multitude of vehicle designs. This reduces the dependence on time-consuming expensive clay models and wind-tunnel experiments and delivers quicker design turn-around.

More important, CFD modeling has provided insights into features of in-cylinder flows that would otherwise be too difficult and expensive to obtain experimentally. Numerical simulations allow the ease of investigating different valve and port designs that can lead to improved engine performance through better breathing and more induction change distribution. Within the cylinder itself, moving and deforming grids permit the means of simulating the piston and valve motion, such as described by the example in Figure 1.7 for a diesel internal combustion engine. In the figure, the intake runner is on the left and the

FIGURE 1.6   Examples of automotive aerodynamics. *(Both courtesy of ANSYS-FLUENT)*

FIGURE 1.7 Example of CFD applications in a diesel internal combustion engine. *(Courtesy of ANSYS-FLUENT, Internal Combustion Engine Design)*

exhaust runner is on the right. Here, CFD allows engineers to understand how changes in port and combustion chamber design affect engine performance, such as volume efficiency or swirl and tumble characteristics. The flow physics described by the piston's movement from the top dead center down to the bottom dead center and subsequently returning to its original position within the combustion chamber allows engineers to examine in detail the transient flow patterns inside the cylinder during the complete engine cycle. Cold flow simulations like this address many challenging problems faced by automotive engineers. Through the application of dynamic mesh adaptation, CFD simulations of internal combustion engines are being performed with greater speed and ease of use in meeting the competitive, progressive demands of the automotive industry.

1.3.6 Biomedical Science and Engineering

Nowadays, medical researchers rely on simulation tools to assist in predicting the behavior of blood flow inside the human body. Computational simulations can provide invaluable information that is extremely difficult to obtain experimentally, and they allow many variations of fluid-dynamics problems to be parametrically studied. Figures 1.8 and 1.9 illustrate just one of the many sample applications of CFD in the biomedical area, in this case where blood flows through originally stenosed and virtually stented arteries are predicted. With the breadth of physical models and advances in areas of fluid–structure interaction, particle tracking, turbulence modeling, and better meshing facilities, rigorous CFD analysis is increasingly performed to study the fluid phenomena inside the human vascular system. Medical simulations of circulatory functions offer many benefits. They can lower the chances of postoperative complications, assist in developing better surgical procedures, and deliver a good understanding of biological processes, as well as assist in design and operation of more efficient and less destructive medical equipment, such as blood pumps. For example, CFD is being increasingly employed via virtual prototyping to recommend the best

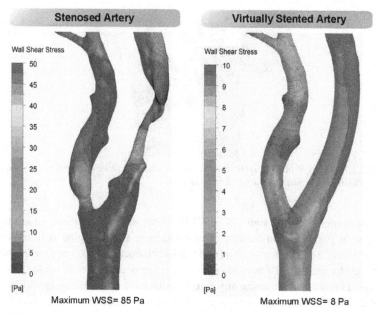

FIGURE 1.8   Example of CFD prediction of wall shear stress (WSS) for originally stenosed and virtually stented arteries.

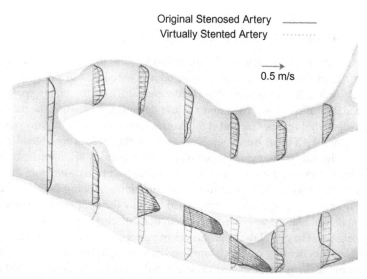

FIGURE 1.9   Example of predicted velocity profiles for originally stenosed and virtually stented arteries.

design for surgical reconstruction, such as carotid endarterectomy, and to better understand the blood flow through an aneurysm in the abdominal artery.

CFD is also receiving growing attention in the pharmaceutical industries. With ever-increasing pollution levels causing respiratory problems and frequent asthma attacks, the need has never been greater to predict and to optimize inhaled therapies. CFD can provide essential insights by simulating the entire drug-delivery process for particulate, aerosol, and gaseous drug types, from the inception of device design through airway delivery and inhalation into the lungs. Through sophisticated multi-phase models, the motion of aerosol droplets and drug particles and their transport/deposition characteristics within the airways can be predicted, and drug concentration in the lungs can be ascertained, as shown in Figure 1.10, which illustrates the modeling of particle formation/dispersion from nasal sprayers and particle transport/deposition in the nasal cavity. In order to better emulate actual fluid flow through the bronchial tree, as shown in Figure 1.11, medical images obtained from accurate CT or MRI scans are converted into a geometrical model that is subsequently used for advanced CFD flow simulations. Through these CFD simulations, deposition and uptake can thus be customized for particular drugs, delivery devices, diseases, and even individual patients. For example, Figure 1.12 illustrates the CFD prediction of pressure coefficient values for specific patient cases with and without asthma.

As demands for new and better health-care products continue to swell at a rapid pace, "health" companies will be required to perform more research and to develop promising new products. Within these diverse health-care sectors, which may include the aforementioned biomedical, pharmaceutical, and other areas associated with medical equipment and "general-health" personal products,

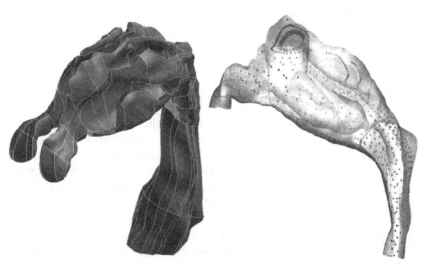

FIGURE 1.10 Example of CFD application for particle formation/dispersion from nasal sprayers and particle transport/deposition in the nasal cavity.

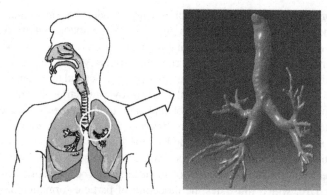

FIGURE 1.11   Example of bronchial tree geometry created from CT data for CFD simulation.

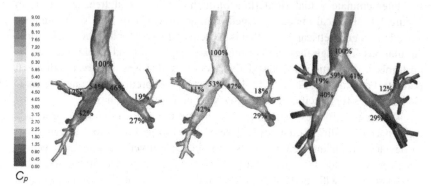

C_p

FIGURE 1.12   Example of CFD prediction of pressure coefficient (C_p) values for specific patient cases with and without asthma.

CFD is becoming an important tool in the identification and improvement of new products to meet the surging market, primarily among the aging population.

1.3.7 Chemical and Mineral Processing

World-wide, many necessities revolve around the chemical and mineral-processing industries. By applying large quantities of heat and energy to physically or chemically deform raw materials, these industries have certainly helped to mould essential products for food and health as well as vital advanced technological equipment in computing and biotechnology. In the face of increasing industrial competitiveness, these industries are confronted with major challenges in meeting the world's demands and present needs without compromising the future. This translates into making operational processes become more energy efficient, safer, and more flexible whilst better containing and reducing emissions.

For example, improving the performance of gas-sparged stirred tank reactors is considered to be of paramount importance in the chemical industry.

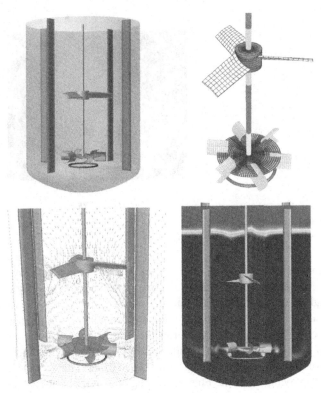

FIGURE 1.13   Example of CFD application in the simulation of a gas-sparged stirred tank reactor. *(Courtesy of ANSYS-FLUENT)*

Figure 1.13 presents the contours of different bubble-size distributions within the stirred tank, accompanied by the local flow behavior, indicated by velocity vectors, around one of the rotating blades. The detailed information about the transport of liquid and gases gained through the use of CFD and population balance approaches ensures that engineers have the best available data to work with in order to increase yield by improving fluid flows, thereby reducing operating costs and increasing system efficiency.

Also, in the world of manufacturing, no industry is more important than minerals, particularly in Australia. Mineral processing spans a wide range of activities and many of the processes involve complex fluid-flow, heat, and mass-transfer phenomena inside aggressive and hostile environments. By modeling, optimizing, and improving processes like *classification*, *separation*, and *filtration*, CFD is at the forefront of providing designs with greater efficiencies and significant production outputs. Figure 1.14 illustrates a *separation* process in mineral processing that involves the use of gas cyclones and hydro-cyclones. A gas cyclone is a commonly used apparatus that utilizes gravity and centrifugal force to separate solid particles from a gas stream. The hydro-cyclone is a similar

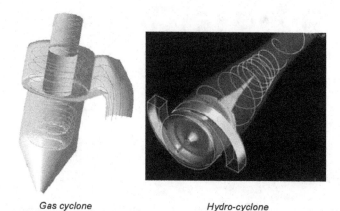

Gas cyclone   Hydro-cyclone

FIGURE 1.14   Example of CFD application in the simulation of a gas cyclone and a hydro-cyclone. *(Courtesy of ANSYS-FLUENT and ESSS and Petrobas, ANSYS-CFX)*

device; however, the operating fluid is a liquid rather than a gas. Here, the centrifugal forces generated are strong enough to cause the solids to separate from the liquid, and these separated solids fall down under gravity into the accumulator vessel situated beneath the hydro-cyclone. CFD simulations through application of advanced turbulence models are performed to ascertain the different flow physics within the cyclone geometry for the correct prediction of the fluid flows. The behavior of the particulates can then be simulated using suitable particle transport models, employing either the Eulerian or Lagrangian multi-phase approach.

The above examples represent merely the tip of the iceberg of the many CFD applications in the processing industries. Advanced CFD is also heavily involved in other important manufacturing sectors, such as extraction metallurgy as well as the oil, gas, and petrochemical industries. The recent advances in the modeling capabilities of CFD have indeed opened up many opportunities to achieve significant technological strides in the complex and demanding world of the chemical and mineral processing industries. Primarily, CFD is playing a crucial role in extrapolating a process from the laboratory and pilot-plant scale to the industrial-plant scale by combining different processes into smaller, compact, and efficient units, instead of treating them individually. This has produced an upgrade in the efficiency of a plant with the same existing constraints.

1.3.8   Civil and Environmental Engineering

Governments, research institutes, and corporations are actively seeking ways to meet environmental regulatory requirements by decreasing waste while maintaining acceptable production levels that meet increasing market demands. In many cases, CFD simulations have been at the heart of resolving many environmental issues. For instance, CFD has been used to predict the pollutant plume dispersion from a cooling tower subject to wind conditions, as shown in Figure 1.15.

In addition, CFD can assist in ensuring compliance with strict regulations during the early design stages of construction. Figure 1.16 represents pre-construction simulation for a new 22-m tank at a water treatment plant. Owing to the huge construction cost, which may exceed millions of dollars, virtual computer-aided models can

FIGURE 1.15 Example of CFD application to plume dispersion from a cooling tower. *(Courtesy of ANSYS-FLUENT)*

FIGURE 1.16 Example of CFD application to the construction of a new tank at a water treatment plant. The top right-hand corner of the figure describes the CFD simulation of the water tank that will be installed within the excavated construction site. *(Courtesy of MMI Engineering)*

be built and analyzed that greatly save time and cost in exploring all aspects of design before construction is begun. To determine the feasibility of such a construction, flow modeling is also performed (also shown in Figure 1.16), which provides insights into the flow behavior for the proposed tank that would not have been possible through physical modeling. The added understanding gained from CFD simulation provides confidence in the design proposal, thus avoiding the added costs of over-sizing and over-specification, whilst reducing risk.

Concerns about an architectural structure exposed to environmental elements have recently motivated an important study on the flow of air and water around the Itsukushima Torii (Gate), a large, 17-m-high wooden structure located in the sea near Hiroshima, Japan, as illustrated in Figure 1.17. By using the dynamic mesh model to periodically move a wall so as to act as a wave generator and using a volume-of-fluid model to track the air–water interface to replicate the motion of the sea waves, significant insights into the flow characteristics around the architectural structure were realized. The ability to capture and to better understand all the associated flow processes has certainly allowed environmental engineers the foundation for better assessment of environmental impacts on exiting structures.

1.3.9 Metallurgy

Metallurgical processes involve materials flow in different forms (solid, liquids, gases, and their mixtures) from one part of the equipment to another. Some of the typical processes include the extraction of metals from various types of iron ore to hot metal and from hot metal to steel; from copper concentrates to pure copper; and from aluminum scrap to pure aluminum or its alloys. These processes are generally complicated because of their multi-phase, high-temperature, and highly reactive characteristics. The complex phenomena of flow, heat, and mass

FIGURE 1.17 Example of CFD application to the flow of air and water around the Itsukushima Torii (Gate) located in the sea near Hiroshima, Japan. *(Courtesy of ANSYS-FLUENT)*

transport, and heterogeneous chemical reactions play very important roles in determining the overall performance of large-scale reactors that accommodate such processes.

In recent times, CFD has been very useful in studying various metallurgical processes and various aspects of a specific process. CFD has been shown to provide an insightful understanding of an existing process, of modification and optimization of the operation and design in an existing process, and of new process development. It is worthwhile mentioning that metallurgical processes are challenging for CFD modeling since many of the phenomena have not yet been properly described or incorporated into the general CFD framework. Nevertheless, many new and significant developments of multi-physics models are taking place in aptly simulating increasingly complicated industrial processes involving flow and transport of mass, momentum, energy, and chemical species in multi-phase and high-temperature reactive systems.

The use of CFD in simulating heterogeneous and slow dissolution of packed bed coke particles is illustrated in Figure 1.18. In this example, the predicted flow pattern of molten iron and carbon dissolution in the blast furnace hearth is obtained through the population balance model coupled with the flow model. By tracking the local changes in size distribution and bed porosity, many aspects related to different initial coke size distributions, the use of inert versus reactive coke types, and the use of a mix of different coke types can be investigated. Another example is the application of CFD for the redesign of off-gas systems at different ferro-silicon production plants in order to assess operating conditions and the qualitative effects of different process parameters. A typical simulation is depicted in Figure 1.19. Such prediction assists in better understanding off-gas combustion, gas collection in the hood, transport and burn-out of soot and charge particles, carbon loss, heat transfer, and particle deposition on channel walls.

1.3.10 Nuclear Safety

During the last decade, the need for more accurate computational models for relevant safety analyses of nuclear facilities has sparked an escalating interest in CFD to feasibly predict a number of important flow phenomena that

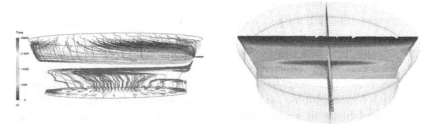

FIGURE 1.18 Example of CFD application to predict molten iron flow (*left*, timeline) and carbon dissolution (*right*, concentration) in the blast furnace hearth. (*After Yang et al., 2006*)

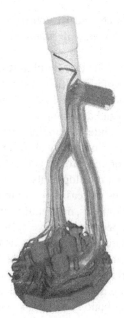

FIGURE 1.19   Example of CFD application to the flow of gas in an off-gas system at Furnace 1 at Elkem Thamshavn, Norway. *(After Johansen, 2003)*

otherwise may not have been possible through other simplified approaches. Some specific problems, such as those arising from pressurized thermal shock, coolant mixing, and thermal stripping, as well as containment issues in nuclear reactors, have certainly motivated enormous research activities for the application of CFD to analyze such problems.

CFD calculations have been performed for coolant mixing in pressurized water reactors. This problem is of significant interest to the nuclear community, particularly in attempting to understand the stationary and transient mixing of coolant in streamlined break and boron dilution scenarios, where the mixing phenomena have tremendous impact on the economical operation and structural integrity of such facilities. Figure 1.20 illustrates the case of a pump start-up due to a strong impulse-driven flow at the inlet nozzle, where the horizontal part of the flow dominates in the down comer in a pressurized water reactor. CFD predictions gave in-depth insights on the injection, which is distributed into two main jets, where the maximum of the tracer concentration at the core inlet appears at the part of the loop opposite to where the tracer is injected. In relation to small-break loss-of-coolant accidents, slug flow that can occur in the cold leg of a pressurized water reactor is potentially hazardous. Strong, oscillating pressure levels forming behind the liquid slugs can significantly affect, and subsequently weaken, the structure of the system. The calculated slug in a horizontal channel determined through the inhomogeneous multi-phase model is in good agreement with experiments, such as that in Figure 1.21. Although the

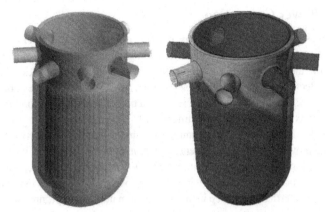

FIGURE 1.20   Example of CFD application to the prediction of turbulent mixing in the ROCOM test facility during boron dilution transients (start-up of the first coolant pump). *(After Höhne et al., 2010)*

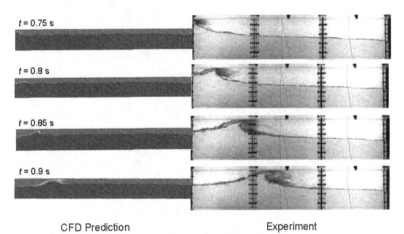

CFD Prediction   Experiment

FIGURE 1.21   Example of CFD application to predicting stratified flow in the cold leg of a pressurized water reactor. *(After Höhne et al., 2010)*

entrainment of small bubbles in front of the slug cannot be observed, the characteristic of the slug front wave rolling over and breaking is clearly identified. Based on the impending need for analysis of long, large-scale, and transient problems, CFD for performing full-containment analysis is still currently out of reach, due to insufficient available computational power and resources. Nevertheless, the use of CFD in addressing the transport of gases in multi-compartment geometry remains realistically achievable, and CFD with a sufficiently detailed mesh can give reliable answers on issues relevant to containment simulation.

1.3.11 Power Generation

In an increasingly competitive energy market, utilities and equipment manufacturers are turning to CFD to provide a technological edge through a better understanding of the equipment and processes within these industries. Although
traditional electric-power–generation sources are still widely used, renewable power
sources, such as wind energy, are emerging as a potential alternative for power generation. To maximize return on investment, CFD is being employed to optimize the
turbine blades for generating constant power under varying wind conditions, as
demonstrated by a typical three-dimensional simulation of the hydraulics in a complete Francis turbine depicted in Figure 1.22. CFD is also the only technology that
has proven to accurately model wind-farm resource distribution, especially for
highly complex terrain with steep inclines, as shown in the same figure. Signicantly,
CFD has allowed the positioning of turbines throughout an area to achieve efficient
wind capture and to minimize wake interaction. Wind resource assessments through

FIGURE 1.22   Example of CFD application to prediction of the velocity field of a wind turbine and
in the vicinity of a proposed wind farm for power. *(Courtesy of TÜV Nord e.V. and CENER)*

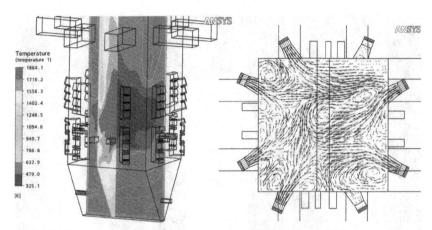

FIGURE 1.23   Example of CFD simulation on an industrial pulverized coal furnace.

CFD have allowed engineers to better study the economic viability of wind farms, where accurate results are needed in order to reduce the financial risk.

The abundance of coal in many parts of the world, such as Australia, has made this raw mineral a popular fuel over many years in the power-generation industries (see Figure 1.23). Within these industries, the burning of coal in large furnaces has largely been associated with the release of environmentally harmful and hazardous pollutants, such as CO, NOx, SOx, and mercury. In light of strict state, national, and international regulations, CFD simulations have greatly assisted engineers in identifying areas where deficiencies in design occur. More important, the causes of the ineffective operations can be established in order to reduce emissions in a cost-effective manner. With this information, design improvements or operating strategies can be implemented to maintain satisfactory levels of residual carbon-in-ash as well as to achieve better flexibility in plant operation. For example, a boiler operation is optimized through reduction of fuel consumption, pollutant emissions, slagging, and degradation of the tubes, whilst downtime is reduced through prolonged component life. For clean power technologies, gasification offers the promise of increased generation efficiency and reduced emissions. Simulations can be performed to meet the modeling needs of this important technology, as typified by the combusting fluidized coal bed in addition to other relevant unit operations, as demonstrated in Figure 1.24. By combining complete system simulations and detailed three-dimensional component analysis, engineers are at liberty to make better design decisions and, ultimately, products of improved quality at low manufacturing costs.

1.3.12   Sports

Very recently, one of the most innovative uses of CFD in the sports arena is to "design" the optimum stroke to achieve peak propulsive performance for elite swimmers, as demonstrated by the example in Figure 1.25. In aspiring to attain

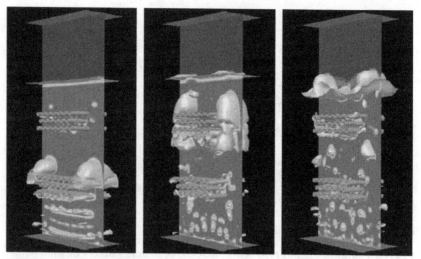

FIGURE 1.24   Example of CFD simulation of bubbles in a fluidized coal bed. *(Courtesy of ANSYS-FLUENT, Coal Gasification)*

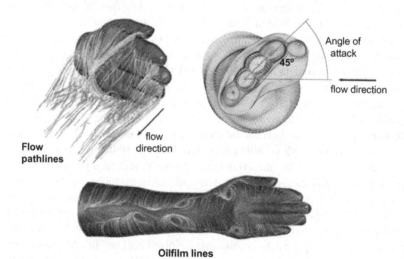

Flow pathlines

flow direction

Angle of attack

45°

flow direction

Oilfilm lines

FIGURE 1.25   Example of CFD application for designing the optimum stroke. *(Courtesy of USA Swimming, Honeywell Engines and Systems, and ANSYS-FLUENT)*

an extra edge, USA Swimming, the national governing body for competitive swimming in the United States, commissioned CFD investigations to evaluate the flow around the hand and forearm of a swimmer during the propulsion phases of the freestyle and butterfly strokes. By applying CFD, steady-state lift and drag forces for the hand and arm are determined through a sophisticated turbulence model and adaptive meshing. For this example, the force coefficients, evaluated at angles of attack ranging from $-15°$ to $195°$, and for various states of water turbulence, were found to compare very well with coefficients developed experimentally in a wind tunnel, a tow tank, and a flume. The successful comparison of the simulated results with experimental data thus validated the chosen CFD modeling technique.

In the swimming literature, the hand is often compared to an airfoil (in American English; aerofoil in British English). A polar diagram developed from CFD analyses shows otherwise, where the aerodynamic efficiency of the hand has been ascertained to be significantly less than that of an airfoil of similar aspect ratio. At velocities reached by swimmers in competitive racing, flow pathlines predicted through CFD reveal a highly three-dimensional flow with significant boundary layer separation. Large vortices form on the downstream side of the hand, and smaller tip vortices twirl off the fingertips in a manner similar to those flowing from aircraft wings or turbine blades. By undertaking additional assessments, CFD analyses can further study the effects of arm and hand acceleration and deceleration on the swimmer's ability to generate propulsive forces.

Cycling presents another sports arena where CFD has played a major role in designing the best possible bike to shave crucial milliseconds off the athletes' times. The Sports Engineering Research Group (SERG) at the University of Sheffield, working closely with British Cycling, has provided the British cycling team the means of constructing the fastest "legal" bike in the world. During the Olympic Games, the team won a total of four medals—two gold, one silver, and one bronze— a fantastic achievement for British cyclists. Their winning edge was solely attributed to SERG's aerodynamic optimization of the bike. Through CFD, SERG redesigned the Olympic bikes' forks and handlebar arrangement. CFD also helped the team to choose the most streamlined design for the aerodynamic helmet, as exemplified in Figure 1.26. By better

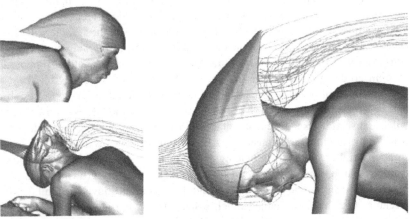

FIGURE 1.26   Example of CFD simulation in designing the ultimate aerodynamic helmet. *(Courtesy of Sports Engineering at CSES, Sheffield Hallam University, and ANSYS-FLUENT)*

understanding the flow pathlines over the aerodynamic helmet, a range of helmet designs were manufactured to accommodate different head styles, achieving the ultimate cycling efficiency. SERG's recommendations ensured the British cycling team had the competitive advantage in their quest for gold.

1.4 THE FUTURE OF COMPUTATIONAL FLUID DYNAMICS

We are witnessing a renaissance of computer simulation technology in many industrial applications. This changing landscape is partly attributable to the rapid evolution of CFD techniques and models. For example, state-of-the-art models for simulating complex fluid-mechanics problems, such as jet flames, buoyant fires, and multi-phase and/or multi-component flows, are now being progressively applied, especially through the availability of multi-purpose commercial CFD computer programs. The increasing use of these programs in industries makes clear that very demanding practical problems are now being analyzed by CFD. With decreasing hardware costs and rapid computing times, engineers are increasingly relying on the reliable yet easy-to-use CFD tools to deliver accurate results, as described in the examples in the previous sections.

Additionally, significant advances in virtual technology and electronic reporting are allowing engineers to swiftly view and interrogate the CFD predictions and to make necessary assessments and judgments on a given engineering design. In industry, CFD will eventually be so entrenched in the design process that new product development will evolve toward "zero-prototype engineering." Such a conceptual design approach is not a mere flight of the imagination, but rather a reality in the foreseeable future, especially in the automotive industry. Looking ahead, full-vehicle CFD models with underhood, climate control, and external aerodynamics will eventually be assembled into one comprehensive model to solve and analyze vehicle designs in hours instead of days. Time-dependent simulations will be routinely performed to investigate every possible design aspect. Other related "co-simulation" areas in ascertaining the structural integrity as well as the aero-acoustics of the vehicle will also be computed concurrently with the CFD models. Engineering judgment will be consistently exercised on the spot through *real-time* assessments of proposed customized design simulations in selecting the optimum vehicle.

In the area of research, the advances in computational resources are establishing large eddy simulation (LES) as the preferred methodology for many turbulence investigations of fundamental fluid-dynamics problems. Since all real-world flows are inherently unsteady, LES provides the means of obtaining such solutions and is gradually replacing traditional two-equation models in academic research. The demand for LES modeling is steadily growing. LES has made significant inroads, especially in single-phase fluid flows. In combustion research, LES has also gained much respectability, particularly in capturing complex flame characteristics because of its better accommodation of the unsteadiness of the large-scale turbulence structure affecting the combustion process. Although much effort has been focused on developing more robust CFD models to predict complicated multi-phase physics involving gas–liquid, gas–solid, liquid–solid, or gas–liquid–solid flows, LES remains in its infancy of application to these flow problems. Instead, two-equation turbulence models are still very prevalent in accounting for the turbulence within such flows. LES may be adopted as the preferred turbulence model for multi-phase flows in the future but, in the meantime, the immediate need is to further develop more sophisticated two-equation turbulence models to resolve these flows.

Based on current computational resources, numerical calculations performed through LES can be long and arduous due to the large number of grid nodal points required for computations. However, the ever-escalating trend of fast computing will permit such calculations to be performed more regularly in the foreseeable future. Also, with the model gradually moving away from the confines of academic research into the industry environment, it is not entirely surprising that LES will eventually become a common method for investigating many physical aspects of practical industrial flows.

There are many challenging prospects for the use of CFD in industry and research. Perhaps we can ponder the day when all turbulence flow problems can be resolved directly, without the consideration of any models. Direct numerical simulation (DNS) of turbulent flows in academic research and possibly in some facets of industrial applications may well become a *distinct certainty* instead of a *distinct impossibility*.

1.5 SUMMARY

The examples illustrated in this introductory chapter clearly show how computational fluid dynamics, better known by its acronym CFD, has evolved through rigorous development of numerical techniques. Despite its long-standing use within academic circles, CFD is flourishing in many industrial sectors. This unprecedented occurrence is partly due to the increasing availability and accessibility of multi-purpose commercial computer programs. The programs have certainly fed much of the swelling demand for industrial CFD applications. CFD research is also at the crossroads of progressive use. Escalating computational power has permitted the use of more sophisticated models to better resolve increasingly complex flow-transport phenomena. As multi-purpose commercial codes become more commonplace in many educational institutions, they are revolutionizing how CFD is being taught. Currently, simpler versions of these codes are being developed into dedicated educational tools for teaching and learning purposes. They are greatly assisting potential and novice users in learning CFD without having to re-invent the wheel (i.e., duplicating software efforts by writing and debugging their own computational codes).

CFD computation usually involves the generation of a set of numbers or digits that ideally provide a realistic approximation of a real-life fluid system. Nevertheless, the main outcome of any CFD exercise is that the investigator acquires an improved understanding of the flow behavior of the system in question. The main ingredients of learning CFD are to gain experience and a thorough understanding of the flow physics and the fundamentals of the numerical techniques and models. Additionally, practical guidelines for good operating practice are needed to increase competency in the use of this powerful tool. The intention of this book is therefore to specifically address all of the aforementioned issues and to equip the reader with the necessary background material for a good understanding of the internal workings of a CFD program and its successful operation.

In the next chapter, we begin by discovering and coming to grips with how a CFD code works through the various elements that constitute a complete CFD analysis. Since a greater emphasis has been placed within this book on the practicality of CFD, many practical steps and aspects exemplifying the important elements of CFD analysis are highlighted. The primary aim is to expose the reader to numerous operations that occur behind many existing commercial and shareware computer codes. With the surging demand for CFD to solve a spectrum of fluid-related problems, the ability to properly employ this methodology has never been as important as it is in the current climate.

REVIEW QUESTIONS

1.1 Which industry has CFD emerged from?

1.2 Which three disciplines is CFD derived from?

1.3 Traditionally, CFD has been used to solve aerospace and automotive engineering applications, such as drag and lift for airplanes and cars. What examples can you think of where CFD is being used within non-traditional fluid-engineering applications?

1.4 What are some of the advantages of using CFD?

1.5 What are the limitations and disadvantages of using CFD?

1.6 What CFD measurements can be obtained to assist design for safety and comfort in passenger airplanes?

1.7 How is CFD being used as a research tool, a design tool, and an educational tool in academic fields, such as thermal fluids?

1.8 How can CFD be used to improve cost-effective design procedures in the automotive industry?

1.9 The biomedical science field is turning to CFD to resolve flows within human airway and vascular systems. What advantages does CFD hold over experiments in obtaining these numerical results?

1.10 What details can CFD capture in the simulation of hydro-cyclones, a process commonly used in the minerals industry?

1.11 How is CFD being used in the civil and environmental industry?

1.12 How is CFD being employed in understanding metallurgical processes?

1.13 What kind of problems has CFD been applied to in nuclear safety analyses?

1.14 How is CFD being used in the power-generation industry? What kinds of data are collected and how are they useful in increasing the efficiency of power generation?

1.15 How can CFD influence the way swimmers improve their swimming strokes?

1.16 What competitive edge can CFD give to a cycling team?

1.17 In the future, to what extent will CFD be involved with product development in manufacturing?

1.18 What is the future of CFD?

Chapter 2

CFD Solution Procedure – A Beginning

2.1 INTRODUCTION

With the widespread availability and ease of accessibility of many commercial, shareware, and in-house computer codes, today's CFD users will most probably acquire their necessary skills and knowledge rather differently from yesterday's CFD users. CFD programs have become very prevalent, prominent, and widely used in many fields of academia, in industry, and in major research centers. Therefore, it is not surprising that potential or novice CFD users are inclined to avail themselves of available codes for learning CFD. The evolution of commercial codes toward more user-friendly environments and applications certainly reflects such demand.

Applying ready-developed CFD codes has certain advantages. Potential or novice users can initially treat these codes as black boxes (in a student setting, not in a professional setting) and operate them with the main intention of just practicing and familiarizing themselves with the many important features of the codes. Without any prior basic CFD knowledge, a first-time user can perform the necessary operations involved in setting up a fluid problem, solving the numerical problem, and managing some graphic representation of the results attained. Although the process can be regarded as very mechanical and laborious, the exposure for a first-time user to CFD is not an intimidating or a daunting experience. Interested users are generally curious and they are usually motivated to further investigate the mysterious aspects contained within the black boxes.

Fostering and cultivating the keenness and eagerness of interested users to learn CFD are important ingredients in maturing these learners as eventual expert CFD users. The acquired skills and expertise fulfill the requirements of being competent and knowledgeable in analyzing and interpreting whether the computational solution is physically realistic and numerically accurate. There have been many pitfalls in the past; over-exposure to the hardcore mathematical formulations of the fluid-flow equations and their numerical representations has certainly caused much angst among first-time users. More often than not, these users have become frustrated and disillusioned and have lamented their prospects for learning CFD. However, CFD lies at the core of comprehension of the fundamental principles of fluid-flow processes and of analyzing the computational solutions. The integration of practical experience with theoretical knowledge to learn CFD is therefore tantamount. The practical side of CFD, in arousing and sustaining the level of enthusiasm of new users, shares the same foundational importance with the theoretical component in equipping these new users with the required level of CFD knowledge for tackling fluid-flow problems.

This book aims at reconciling two diametrically opposed approaches to teaching CFD. We begin by introducing the first-time user to the salient features that are common in many commercial (and possibly in some shareware) CFD codes. The codes are usually structured around robust numerical algorithms that can tackle fluid-flow problems. In order grant easy access to their solving power, almost all current commercial (and possibly shareware) CFD packages include user-friendly graphical user interface (GUI) applications and environments for input of problem parameters and for examining the computed results. Hence, the codes provide a complete CFD analysis, consisting of three main elements:

- Pre-processor
- Solver
- Post-processor

This chapter addresses the many practical aspects of these three important elements in an attempt to uncover the numerous operations beneath many shareware and commercial CFD codes. Figure 2.1 presents a framework that illustrates the interconnectivity of the three aforementioned elements within the CFD analysis. The functions of these three elements are examined in more detail in the subsequent sections in this chapter.

2.1.1　Shareware CFD

Today's CFD users have the luxury of downloading possible shareware or freeware CFD codes from certain websites on the Internet. The website of www.cfd-online. com/Links/soft.htm/ offers the reader, under the software category, a catalog of CFD codes listed as shareware products through the **shareware, freeware** option link.

Nevertheless, first-time CFD users may wish to search the Internet to gain immediate access to an interactive CFD code. (Users may be required to register in order to freely access the interactive CFD code.) The website is http://energy. concord.org/energy2d/index.html provides simple CFD flow problems for first-time users to solve and allows colorful graphic representation of the computed results. More advanced users wishing to acquire all the necessary aspects of CFD can acquire them at www.cfd-online.com/.

2.1.2　Commercial CFD

Under the software category of the web link www.cfd-online.com/Links/soft. htm/, the reader can uncover the list of commercial codes that are currently available through the **commercial** option link. Table 2.1 presents the Internet

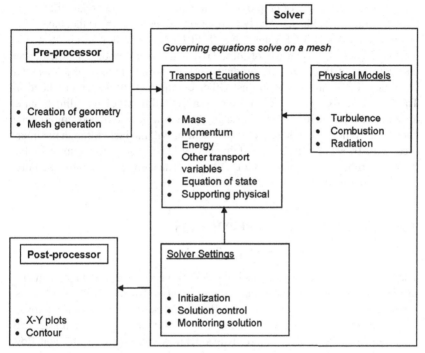

FIGURE 2.1   The interconnectivity functions of the three main elements within a CFD analysis framework.

TABLE 2.1 Internet links to some popular commercial CFD packages

Developer	Code	Distributor Web Address
ANSYS, Inc.	CFX	http://www.ansys.com/
ANSYS, Inc.	FLUENT	http://www.fluent.com/
CD-Adapco	STAR-CCM+	http://www.cd-adapco.com/
CHAM	PHOENICS	http://www.cham.co.uk/
COMSOL, Inc.	COMSOL	http://www.comsol.com/
ESI Group	CFD-ACE+	http://www.esi-group.com/
Flow Science	FLOW3D	http://www.flow3d.com/

links for some of the popular commercial CFD packages; it is by no means an exhaustive list. Commercial CFD vendors have invested much time, effort, and expense in the concerted development of user-friendly GUIs to make CFD very accessible and to facilitate its usage and application in handling very complex fluid-flow problems. We present some typical GUIs that a first-time user may encounter in the course of employing a number of the commercial CFD packages tabulated in Table 2.1. The two GUI fronts chosen are those that have been developed by ANSYS-CFX and ANSYS-FLUENT, and they are illustrated in Figures 2.2 and 2.3. (Our use of these GUI fronts is not to be construed as an endorsement of these specific products; we are simply treating them as GUI examples for illustration purposes. Other commercial packages tabulated in Table 2.1 most likely have similar user-friendly GUI fronts of appealing graphic appearances and features like those in the two aforementioned commercial packages.) While using these packages, the user will navigate through the interface fronts by accomplishing a certain number of basic steps to set up and solve the flow problem and obtain a CFD solution. The steps are described and discussed in the next section.

2.2   PROBLEM SETUP—PRE-PROCESS

2.2.1   Creation of Geometry—Step 1

The first step in any CFD analysis is the definition and creation of the geometry of the flow region, i.e., the *computational domain* for the CFD calculations. Let's consider two flow cases: a fluid flowing between two stationary parallel plates and a fluid passing through two cylinders in an open surrounding. It is important that the reader should always acknowledge the real physical flow

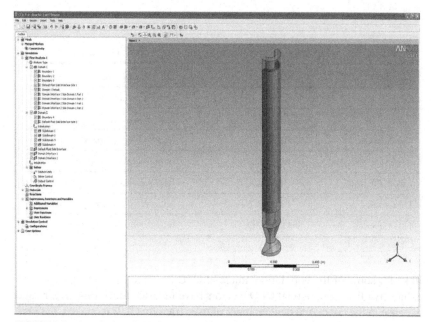

FIGURE 2.2   A typical ANSYS-CFX graphical user interface.

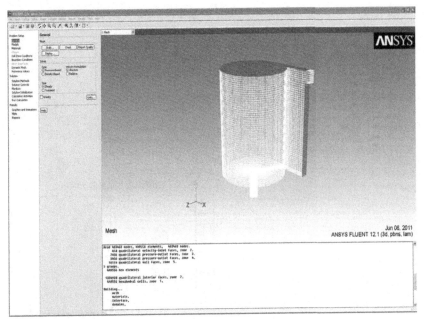

FIGURE 2.3 A typical ANSYS-FLUENT graphical user interface.

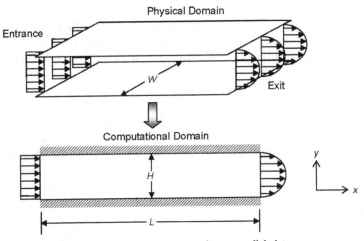

FIGURE 2.4 Case 1: Fluid flowing between two stationary parallel plates.

representation of the problem that is to be solved, as demonstrated by the respective *physical domains* in Figures 2.4 and 2.5. For the purpose of illustration, we designate the former and latter cases as Case 1 and Case 2, respectively. We shall assume that the two cylinders in Case 2 have the same length as the width W of the overall domain. We shall also assume that the width W of both

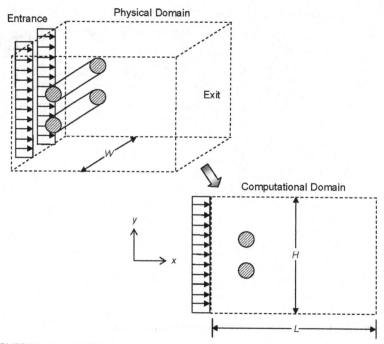

FIGURE 2.5　Case 2: Fluid passing over two cylinders in an open surrounding.

flows within the three-dimensional physical domains is sufficiently large that the flows can be taken to be invariant along this transverse direction. Hence, Case 1 and Case 2 can simply be considered as two-dimensional computational domains for CFD calculations. These two flow cases are repeatedly used as illustrative examples for demonstrating the various basic steps that are involved in the pre-process, solver, and post-process stages.

There are certain distinct dissimilarities in the nature of these two flow problems. Case 1 represents an *internal flow* problem, while Case 2 is taken typically as an *external flow* scenario. In both cases, the fluid enters at the left boundary and exits at the right boundary of the computational domains. The main difference between the two flows is accentuated by the top and bottom boundaries, which bring about the classification of internal and external flows. In Case 1, the fluid flow is bounded within a domain of rigid walls, as represented by the horizontal external walls of the two stationary parallel plates. That is not the characteristic of the fluid flow in Case 2, as the fluid can take either the inflow or the outflow boundary conditions at the top and bottom boundaries.

One important aspect that the reader should always note in the creation of the geometry for CFD calculations is to allow the flow dynamics to be sufficiently developed across the length L of the computational domains. For Case 1, we require the flow to be *fully developed* as it exits the domain. The physical interpretation and meaning of the concept of fully developed are explained in

Chapter 3. For Case 2, we need to encapsulate the occurrence of a complex *wake-making* development that persists behind the two cylinders as the flow passes over them. This phenomenon is analogous to the formation and shedding of vortices, which are commonly experienced in flow past a cylinder. In this particular case, the top and bottom boundary effects may influence the flow passing over the two cylinders; the height H of the domain needs to be prescribed at a distance that will sufficiently remove any of these boundary effects on the fluid flow surrounding the two cylinders but that is still manageable for CFD calculations. More practical guidance on this issue is addressed in Chapter 6.

2.2.2   Mesh Generation—Step 2

The second step, mesh generation, is one of the most important steps in the pre-process stage after the definition of the domain geometry. CFD requires the subdivision of the domain into a number of smaller, non-overlapping subdomains in order to solve the flow physics within the domain geometry that has been created; this results in the generation of a *mesh* (or grid) of *cells* (elements or control volumes) overlying the whole domain geometry. The essential fluid flows that are described in each of these cells are usually solved numerically, so that the discrete values of the flow properties, such as the velocity, pressure, temperature, and other transport parameters of interest, are determined. This yields the CFD solution to the flow problem that is being solved. The accuracy of a CFD solution is strongly influenced by the number of cells in the mesh within the computational domain. While, in general, increasing the number of cells will improve the accuracy of the solution, the solution is also influenced by many other factors, such as the type of mesh, the order of accuracy of the numerical method, and the adequacy of the techniques chosen to the physics of the problem. However, the accuracy of a solution is strongly dependent on the limitations imposed by the computational costs and calculation turnover times. Issues concerning numerical accuracy and computational efficiency are explored in Chapter 5.

The majority of the time spent in industry on a CFD project is usually devoted to successfully generating a mesh for the domain geometry. Most commercial CFD codes have developed their own dedicated CAD-style interface and/or facilities to import data from solid modeler packages, such as PARASO-LID, PRO/ENGINEER, SOLID EDGE, SOLID WORKS, and UNIGRAPHICS, to maximize productivity and to allow ease of geometry creation. The mesh for the created geometry can be subsequently realized through in-built powerful mesh generators that reside within the respective codes. Nevertheless, the reader should be aware that it is still up to the skills of the CFD user to design a mesh that is a suitable compromise between the desired accuracy and the solution cost. The generation of an appropriate mesh for CFD calculations is discussed further in Chapter 6.

The mesh generation step is illustrated for Case 1 and Case 2. For relatively simple geometries, such as the created geometry domain for Case 1, an overlay mesh of *structured* cells that generally comprises a regular distribution of rectangular cells can be readily realized. Figure 2.6 shows a mesh of 20 (*L*) × 20 (*H*) cells, resulting in a total of 400 cells allocated for the Case 1 geometry. For more complex geometries, meshing by triangular cells allows flexibility in mesh generation for geometries having complicated shape boundaries. Figure 2.7 illustrates a typical distribution of triangular cells within the computational domain for Case 2 geometry, with a mesh totaling 16,637 cells mapping the whole flow domain.

The specific meshes represented in Case 1 and Case 2 domain geometries should be construed only as illustrative examples for demonstrating the mesh generation step. It is not unprecedented, and is legitimate, to interchangeably use an *unstructured* mesh in place of a structured mesh for Case 1 and vice versa for Case 2. It is also not uncommon, and in some practices is a requirement, to embrace a combination of structured and unstructured meshes for more realistic simulations within flow domains that may include many inherent complex geometrical intricacies. More practical guidance on mesh generation is provided in Chapter 6.

2.2.3   Selection of Physics and Fluid Properties—Step 3

Many industrial CFD flow problems may require solutions to very complex physical flow processes, such as the accommodation of complicated chemical reactions in combusting fluid flows. The inclusion of combustion and possibly

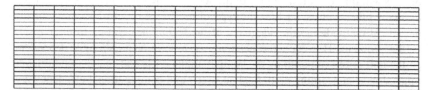

FIGURE 2.6   Structured meshing for fluid flowing between two stationary parallel plates.

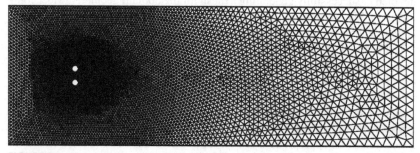

FIGURE 2.7   Unstructured meshing for fluid passing over two cylinders in an open surrounding.

radiation models in the CFD calculations is generally a prerequisite for successful modeling of these types of flows. Combustion and radiation processes have the tendency to strongly influence the local and global heat transport, which consequently affects the overall fluid dynamics within the flow domain. It is therefore imperative that the CFD user carefully identify the underlying flow physics unique to the particular fluid-flow system.

For clarity and ease of reference, a flowchart highlighting the various flow physics that may be encountered within the framework of CFD and heat transfer processes is presented in Figure 2.8. Under the main banner "Computational Fluid Dynamics & Heat Transfer," a CFD user declares initially whether simulations of the fluid-flow system are to be attained for *transient/unsteady* or *steady* solutions. He/she subsequently defines which class of fluids that the flows belong to: *inviscid* or *viscous*. *Inviscid* fluid flows are generally compressible, and the consideration of fluid compressibility in the flow physics can usually be handled through the *Panel Method*. On the other hand, *viscous* fluid flows can exist in a *laminar* or a *turbulent* state. Under these two flow conditions, prior knowledge of whether the fluids are *compressible* or *incompressible* is required. The classification of *internal* and *external* flows for viscous fluids allows the user to treat these flow problems appropriately, as discussed in section 2.2.1. Also, the transport of heat may contribute significantly to the fluid-flow process. There are three heat transfer modes: *conduction, convection,*

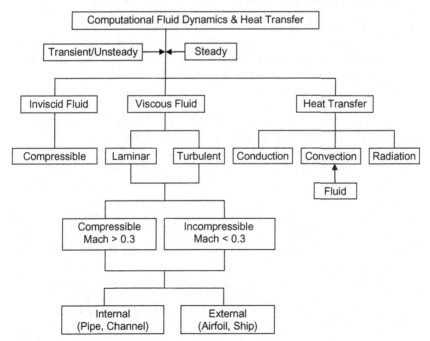

FIGURE 2.8 A flowchart encapsulating the various flow physics in CFD.

and *radiation*. For convection, the dominant mode of heat transfer will more likely be driven by convective fluid flow rather than by conduction and radiation. Nevertheless, there are circumstances where radiation and convection can co-exist and can dominate heat transfer, especially in the expansion of fires.

Let's review the basic steps that have been described thus far for the pre-process stage. In Step 1, we created two-dimensional computational domains for Case 1 and Case 2. Step 2 illustrates the generation of an overlay mesh of structured cells for Case 1 and unstructured cells of different sizes for Case 2. Step 3, the current step in the pre-process stage, consists of the identification and formulation of the flow problems in terms of the physical phenomena. A CFD user must select the appropriate flow physics, as discussed above, in order to correctly simulate the characteristics of the fluid flow. For simplicity, a steady CFD solution is considered for Case 1 and Case 2, and the fluid flows in both cases can be taken to be viscous, laminar, incompressible, and isothermal (without heat transfer). It is important that setting up the flow physics is also accompanied by ascertaining what fluid is used within the flow domain. For example, air or water has its own unique fluid and thermal properties. Therefore, the appropriate properties need to be assigned to correctly define the particular fluid in the pre-process step. Fluid properties like *density* and *viscosity* (dynamic) can usually be imposed through the GUIs in many commercial CFD codes.

2.2.4   Specification of Boundary Conditions—Step 4

The complex nature of many fluid-flow behaviors has important implications for which boundary conditions are prescribed for the flow problem. A CFD user needs to define appropriate conditions that mimic the real physical representation of the fluid flow in a solvable CFD problem.

The fourth step in the pre-process stage deals with the specification of permissible boundary conditions that are available for impending simulations. Evidently, where *inflow* and *outflow* boundaries exist within the flow domain, suitable fluid-flow boundary conditions are required to accommodate the fluid behavior upon entering and leaving the flow domain. The flow domain may also have *open* boundaries. Although the intricacies of open boundary conditions are still subject to much theoretical debate, this boundary condition remains the simplest and cheapest form to prescribe when compared with other theoretically more satisfying selections in CFD. Appropriate boundary conditions also need to be assigned for external stationary *solid wall* boundaries that border the flow geometry and the surrounding walls of possible internal obstacles within the flow domain.

We illustrate the applications of the aforementioned boundary conditions to Case 1 and Case 2 for CFD calculations. Schematic descriptions of the boundary conditions are demonstrated in Figure 2.9 for Case 1 and Figure 2.10 for Case 2.

In subsection 2.2.3, the viscous, laminar, incompressible, and isothermal fluids assumed for the purpose of illustration in both Case 1 and Case 2 require

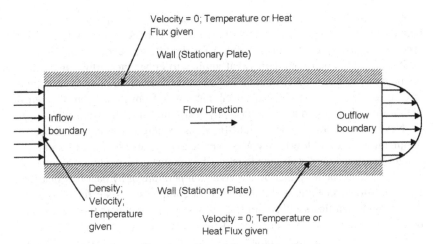

FIGURE 2.9   Boundary conditions for an internal flow problem: Case 1.

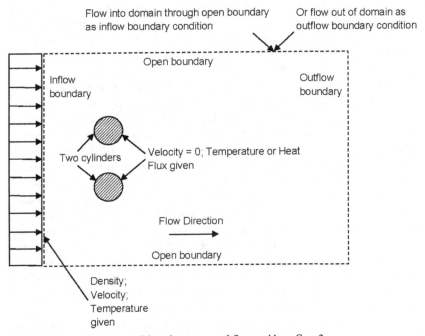

FIGURE 2.10   Boundary conditions for an external flow problem: Case 2.

the prescription of only the fluid velocity on all the bounding walls of the computational domain. By definition, the velocities are zero for the external stationary *solid walls* bounding the flow domain in Case 1 (see Figure 2.9) and the two cylindrical motionless solid walls in Case 2 (see Figure 2.10). For the inflow boundary conditions, the user is required to ascertain the inlet fluid velocity in order to stipulate the fluid entering both of these flow domains. At the outflow boundaries indicating the fluid departure, only one outlet condition, typically a specified relative pressure, is imposed. A far-field flow boundary condition can be imposed for open boundaries that applies to either the inflow or outflow boundary conditions. Care should always be exercised in handling open boundaries to ensure they are defined far enough away from the region of interest within the solution domain to obtain physically meaningful results.

Nevertheless, for a general fluid-flow case where the transport of heat is dominant within the flow domain, as may be experienced by the flows in Case 1 and Case 2, the user is obliged to prescribe the surface boundaries of the flow domains and obstacles either by the given temperature or by heat flux distributions, as shown in Figures 2.9 and 2.10. In compressible-like flows, such as buoyancy-driven flows and combusting flows (e.g., fire) where there is a strong density variation with temperature, density emerges as part of the solution everywhere within the flow domains. The density of the particular fluid flowing into these domains at the inflow boundaries needs to be specified and is usually determined directly from the imposed temperature and pressure. However, no boundary conditions are required for the density at the outflow boundaries, since the fluid density replicates the outgoing fluid characteristics through these boundaries. For open boundaries, such as in Case 2, the density can be represented by either the inflow or the outflow boundary conditions.

General-purpose CFD codes also often allow the prescription of inflow and outflow pressure or mass flow rate boundary conditions. By setting fixed pressure values, sources and sinks of mass placed at the boundaries ensure the correct mass flow into and out of the solution zone across the constant-pressure boundaries. It is also feasible to allocate directly sources and sinks of mass at the boundaries by the mass flow rates instead of pressures to retain the overall mass balance for the flow domain. To take advantage of special geometrical features that the solution region may possess, *symmetric* and *cyclic* boundary conditions can be employed to speed up the computations and enhance the computational accuracy by placing additional numbers of cells in the simplified geometry. Figure 2.11 shows the boundary geometry for which the symmetry boundary condition can be imposed for Case 1, while Figure 2.12 illustrates a generic geometry where cyclic boundary conditions may be useful. The physical meanings of the various boundary conditions that have been described herein to close the fluid-flow system are explained in Chapter 3.

Thus far, we have concentrated on the application of various boundary conditions that pertain only to *subsonic* fluid flows—flows below the speed of sound. There is another broad range of fluid flows that can possibly achieve speeds near to and above the speed of sound, and they are generally classified as *hypersonic*, *transonic*, and *supersonic* fluid flows. The prevalence of such complex flows is evident in many aerodynamic investigations. At these high speeds, the viscous regions in the flow are usually exceedingly thin and the enveloping flow in a large part of the solution domain is effectively inviscid.

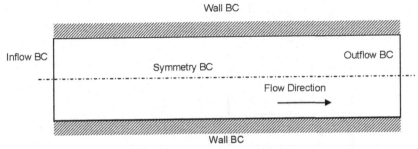

FIGURE 2.11   Definition of symmetry boundary condition for Case 1.

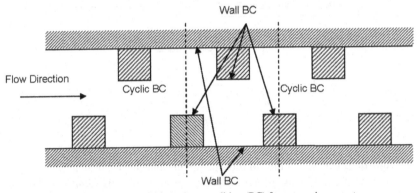

FIGURE 2.12   Definition of cyclic boundary condition (BC) for a generic geometry.

Applying boundary conditions predominantly for viscous flow to a largely inviscid flow may yield unphysical behavior of the shockwaves propagating through the flow domain. Physical description and application of boundary conditions for supersonic flows are covered in Chapters 3 and 7. For additional details, interested readers are advised to refer to Fletcher (1991) and Anderson (1995) for more discussions and prescriptions of the boundary conditions for fluid flows that are hypersonic and transonic. The majority of the commercial codes make claims about their ability to resolve all fluid-flow regimes—*subsonic*, transonic, and/or supersonic; however, they usually perform most effectively with flow below the speed of sound, i.e., subsonic, as a consequence of the boundary conditions outlined above.

2.3   NUMERICAL SOLUTION—CFD SOLVER

The appropriate use of either an in-house or a commercial CFD code requires a core understanding of the underlying numerical aspects of the *CFD solver*. This section focuses on the solver element. A CFD solver can usually be described and envisaged by the solution procedure presented in Figure 2.13. The

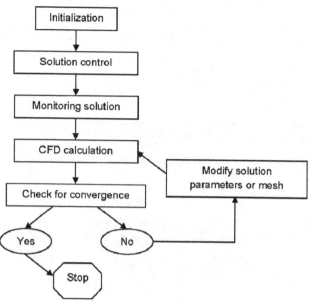

FIGURE 2.13   An overview of the solution procedure.

prerequisite processes in the solution procedure that have implications for the computational solution are *initialization, solution control, monitoring solution, CFD calculation,* and *checking for convergence.* A CFD user, whether applying in-house or commercial codes, needs to gain insights and knowledge pertaining to the workings of these prerequisite processes in order to skillfully utilize the many solver features and to better navigate the underlying "black box" operations that reside in many of these codes.

2.3.1   Initialization and Solution Control—Step 5

The fifth step of the CFD analysis encompasses two prerequisite processes within the CFD solver: *initialization* and *solution control.*

First, the underlying physical phenomena in real fluid flows, which are generally complex and non-linear within such flows, usually require treatment of the key phenomena to be resolved through an iterative solution approach. An iterative procedure generally requires all the discrete values of the flow properties, such as the velocity, pressure, temperature, and other transport parameters of interest, to be *initialized* before calculating a solution. In theory, initial conditions can be purely arbitrary. However, in practice, there are certain advantages to imposing initial conditions *intelligently.* Good initial conditions are crucial to the iterative procedure. Two reasons that a CFD user should undertake the appropriate selection of initial conditions are

- If the initial conditions are close to the final steady-state solution, the quicker the iterative procedure will converge and yield results in a shorter computational time.
- If the initial conditions are far away from reality, the computations will require longer computational efforts to reach the desired convergence. Also, improper initial conditions may lead to the iterative procedure's misbehaving and possibly "blowing up" or diverging.

Second, setting up appropriate parameters in the solution control usually entails the specification of appropriate *discretization* (*interpolation*) *schemes* and selection of suitable *iterative solvers*.

Almost all well-established and thoroughly validated general-purpose commercial codes adopt the *finite-volume method* (Chapter 4) as their standard numerical solution technique. The algebraic forms of equations governing the fluid flow within these codes are usually approximated by the application of finite-difference-type approximations to a finite-volume cell in space. At each face of the cell volume, surface fluxes of the transport variables that are required can be determined through different interpolation schemes. Some of the common interpolation schemes are *First-Order Upwind, Second-Order Upwind, Second-Order Central,* and *Quadratic Upstream Interpolation Convective Kinetics (QUICK)*. The inability of the *Central* scheme to identify the flow direction resulted in the formulation of other schemes, such as Upwind or QUICK, interpolation methods that are biased toward the upstream occurrence of the fluid flow and thus account for the flow direction. The choice of a higher-order interpolation scheme may achieve the desired level of accuracy for evaluation at the cell faces. Solution procedures like the SIMPLE, SIMPLEC, and PISO algorithms are popular in many commercial codes. The SIMPLE, SIMPLEC or PISO algorithm is geared toward guaranteeing correct linkage between the pressure and velocity, which predominantly accounts for the mass conservation within the flow domain. At present, our intention is not to dwell on the many underlying numerical properties but simply to present the interpolation schemes and pressure-velocity coupling methods that are offered as standard options in many CFD codes. It is imperative that some background knowledge of the appropriate selection of these options be acquired before any CFD calculation is performed. More discussion of, and practical guidance on, the many numerical issues pertaining to the application of interpolation schemes and pressure-velocity coupling methods are provided in Chapters 4 and 6.

Iterative solvers, so-called number-crunching engines, for numerical calculations are employed to resolve the algebraic equations. Nowadays, robust solvers, such as the Algebraic Multi Grid (AMG) algorithm and conjugate gradient methods, are standard features in many commercial codes. Other popular solvers, such as the Strongly Implicit Procedure (SIP) of Stone's method and the TDMA line-by-line procedure, are also prominently employed by many users in the CFD community. More descriptions of these solvers are provided in Chapter 4. Solver-controlling parameters that exist inside these solvers in

commercial codes tend to be optimally configured for efficient matrix calculations. The desired performance can usually be achieved through the default settings prescribed within these codes.

Step 5 hereby completes the specification of various relevant physical features pertaining to the intended fluid-flow process of a CFD problem. To exemplify the *CFD calculation* process of the solution procedure in Figure 2.13, the
Case 1 and Case 2 fluid-flow problems described in section 2.1.1 are revisited in
the next step below, to demonstrate the iterative operations that are typically
visualized through the ANSYS-CFX and ANSYS-FLUENT GUIs. We conveniently employ the default settings of the interpolation schemes and pressure-
velocity coupling methods and retain the optimal controlling parameters that
govern the performance of the iterative solvers to simply illustrate the important
features of the numerical computations.

2.3.2 Monitoring Convergence—Step 6

The sixth step of the CFD solver involves the interlinking operations of three
prerequisite processes: *monitoring solution*, *CFD calculation*, and *checking
for convergence*. Two aspects that characterize a successful CFD computational
solution are *convergence* (Chapter 5) of the iterative process and *grid independence* (Chapters 5 and 6).

Convergence can usually be assessed by progressively tracking the imbalances
that are accentuated by the advancement of the numerical calculations of the algebraic equations through each iteration step. These imbalances measure the overall
conservation of the flow properties; they are also commonly known as the *residuals*
(Chapters 4 and 5) and are generally viewed through commercial code GUIs.
Examples of GUIs by ANSYS-CFX and ANSYS-FLUENT that represent the
downward trends of the residuals for Case 1 are illustrated in Figures 2.14 and
2.15, respectively. These downward tendencies clearly point to the continual
removal, as opposed to possible accumulation, of any unwanted imbalances,

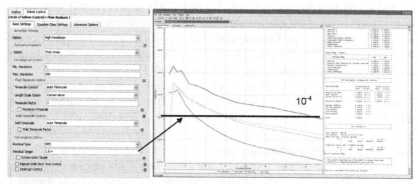

FIGURE 2.14 Typical ANSYS-CFX GUIs for monitoring convergence corresponding to the prescribed convergence criteria.

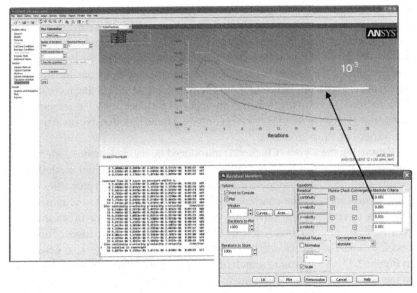

FIGURE 2.15 Typical ANSYS-FLUENT GUIs for monitoring convergence corresponding to the prescribed convergence criteria.

thereby causing the iterative process to converge rather than to diverge. A converged solution is achieved when the residuals fall below some convergence criteria or tolerance (Chapter 5) that is preset inside the solver-controlling parameters of the iterative solvers. We indicate different default settings of the tolerance values (Chapter 5) that are used by ANSYS-CFX and ANSYS-FLUENT to terminate the iterative process. They are prescribed as values of 1×10^{-4} and 1×10^{-3}, respectively, as shown in Figures 2.14 and 2.15. A converged solution is obtained quickly from either ANSYS-CFX or ANSYS-FLUENT because of the nature of the flow in Case 1 being straightforward. Besides examining the residuals, the user may use other monitoring variables, such as the lift, drag, or moment force, to ascertain the convergence of the numerical computations. Figure 2.16 shows the lift coefficient (Cl) history for Case 2 employing the commercial code ANSYS-FLUENT. It is not surprising that the fluid-flow process in Case 2, which is more complicated than that in Case 1, requires more iterations to reach convergence. No appreciable change to the lift coefficient is observed after 700 iterative steps and the computational solution is thus deemed to be converged at about 1000 iterations as the lift coefficient plateaus to a fixed value of around 0.0035. In addition to monitoring residual and variable histories, the user is well advised to also check the overall mass balance and possibly the heat balance for the fluid flow system within the computational domain. The net imbalance should be minimized to as low as possible to ensure adequate property conservation.

Progress toward a converged solution can be greatly assisted by the careful selection of various *under-relaxation factors* (Chapter 5). Most commercial codes adopt some form of under-relaxation factors to enhance the *stability* of the numerical procedure and to ensure the *convergence* of the iterative process. The incorporation of under-relaxation factors into the system of algebraic equations that govern the fluid flow is intended to significantly moderate the iteration process by limiting the change in each of the transport variables from one iterative step to the next. There are no straightforward guidelines for pertinent choices of these factors. More often than not, in-depth experience with the selection of appropriate values of these factors can only be gained by extensive investigation of a variety of flow problems.

Estimating the *errors* (Chapter 5) introduced by an inadequate mesh design for a general flow can be rather arduous. Preparing a good initial mesh design usually requires foreknowledge of, or insight into, the expected properties of the flow. In dealing with the coarseness of a mesh, the only way to eliminate the *errors* is to embrace the procedure of successive refinement of an initially coarse mesh until certain key results exhibit no appreciable changes. This process forms part of the extended arm within the solution procedure, as described in Figure 2.13. A systematic search of *grid-independent* results generally leads to the accomplishment of high-quality CFD solutions.

The reader should take note that many factors, such as *convergence, convergence criteria* or *tolerance values, residuals, stability, errors, under-relaxation factors*, and *grid independence*, underpin the many numerical considerations for the simulation of a CFD problem. These concepts may be difficult to fathom at this juncture. Supplementary discussions are provided later in other chapters to expand the knowledge of the basics already attained herein.

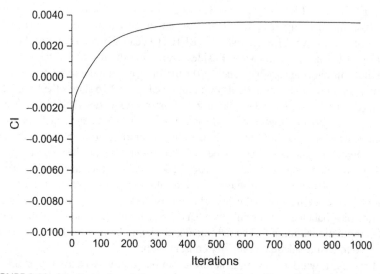

FIGURE 2.16   Monitoring the solution convergence through the lift coefficient history.

2.4 RESULT REPORT AND VISUALIZATION—POST-PROCESS

CFD has a reputation for generating vivid graphic images and, while some of the images are promotional and are usually displayed in stunning and superb colorful output, the ability to present the computational results effectively is an invaluable design tool. In this section, we concentrate on some essential computer graphic techniques frequently encountered in the presentation of CFD data. The majority of ways that the CFD results are emphasized graphically can be classified in different categories. Each of these categories, to be discussed below, assists the CFD user to better analyze and visualize the many relevant physical characteristics within the fluid-flow problem.

Commercial CFD codes like ANSYS-CFX, ANSYS-FLUENT, STAR-CD, and others often incorporate impressive visualization tools within their user-friendly GUIs to allow users to graphically view the results of a CFD calculation at the end of a computational simulation. However, there are also many excellent stand-alone applications of independent computer graphics software packages that the reader may opt to utilize for his/her CFD applications. Table 2.2 presents a list of some currently available graphics packages. The majority of these packages cater to various computer platforms although some may operate primarily in UNIX systems. Some of the commercial packages, such as FIELD-VIEW, TECPLOT, and ENSIGHT, have been specifically developed for post-processing CFD results, while others are more general-purpose visualization

TABLE 2.2 Internet links to some popular computer graphics software packages

Developer	Code	Distributor Web Address
Advanced Visual Systems	AVS, Gsharp, Toolmaster	http://www.avs.com/
Amtec Engineering	Tecplot	http://www.amtec.com/
CEI	EnSight	http://www.ceintl.com/
IBM (free, apparently)	OpenDx	http://www.opendx.org/
Intelligent Light	Fieldview	http://www.ilight.com
Numerical Algorithms Group (NAG)	Iris Explorer	http://www.nag.co.uk/
Visual Numerics	PV-Wave	http://www.roguewave.com/
Kitware	Paraview	http://www.paraview.org/
Department of Energy (DOE) Advanced Simulation and Computing Initiative (ASCI)	VisIt	https:/wcl.llnl.gov/

tools that may be applied for CFD applications. For simple graphic representations, the use of an open-source plotting package, such as GNUPLOT, is popular among many researchers in the CFD community and can be obtained for free on the Internet at www.gnuplot.info/.

For the remainder of this section, we demonstrate the different categories that can be applied to illustrate a CFD solution through a popular and versatile computer graphics software package, namely TECPLOT. These categories are mainly exemplified in the context of the solutions obtained from our previously defined flow cases: Case 1, a fluid flowing between two stationary parallel plates, and Case 2, a fluid passing through two cylinders in an open surrounding. An overview of these categories may also assist the reader in becoming accustomed to how a CFD solution can be processed and visualized within a commercial code at the end of a simulation.

The use of TECPLOT for our graphic representations of the CFD results should not be interpreted as an endorsement of a specific product. Rather, we aim simply to provide the reader with an example of a standard graphics software approach. With the rapid evolution of new graphic techniques, probing the flow behavior in a CFD problem may be better achieved through other software packages. It is entirely up to the reader to employ the appropriate graphics software package.

2.4.1   X-Y Plots

X-Y plots are mainly two-dimensional graphs that represent the variation of one dependent transport variable as compared with another, independent variable. They can usually be drawn by hand or more conveniently by many plotting packages. Such plots are the most precise and quantitative way to present the numerical data. Often, laboratory data are gathered by straight-line traverses. These graphs are therefore a popular way of directly comparing the numerical data with the experimentally measured values. Also, logarithmic scales allow the identification of important flow effects occurring especially in the vicinity of solid boundaries. These graphs are widely used for presenting line profiles of velocity and for plots of surface quantities, such as pressure and skin-friction coefficient. They are usually meant to be very easily identifiable; the reader can readily read the results without resorting to any mental or arithmetic interpolation.

An X-Y plot of a laminar velocity profile at the *fully developed* region for Case 1 is shown in Figure 2.17. The parabolic profile characterizes the flow physics typically experienced for a fluid flowing within a parallel-plate channel. Another possible way of visualizing the development of the fluid flow is through the use of successive two-dimensional graphic profiles, as shown for Case 1 in Figure 2.18. The flow distribution gradually changes from a uniform profile specified at the entrance boundary (left) to a parabolic profile as it travels downstream toward the channel exit boundary (right). (More discussion of the physical aspects of this simple flow is contained in Chapter 3.) For the more

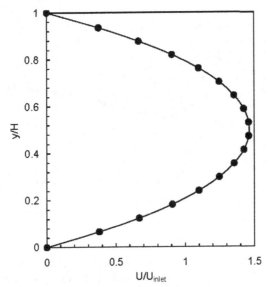

FIGURE 2.17   X-Y plot of a parabolic laminar velocity profile at the fully developed region for Case 1.

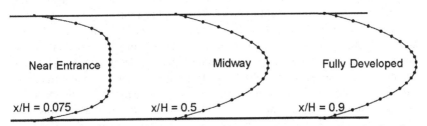

Near Entrance   Midway   Fully Developed

x/H = 0.075   x/H = 0.5   x/H = 0.9

FIGURE 2.18   Successive two-dimensional velocity profiles of a developing flow for Case 1.

complex flow structure of Case 2, the normalized horizontal velocity profile along the entire length of the computational domain, as shown in Figure 2.19, can be represented at the mid-height location to better illustrate the deceleration and acceleration characteristics of the flow behavior as the fluid passes over the two cylinders. Further downstream, the fluid flow subsequently diffuses and recovers partially toward the free-stream condition near the exit boundary (right). (The observed physical aspects of this particular flow are explored further in Chapter 3.)

2.4.2   Vector Plots

A vector plot provides the means whereby a vector quantity (usually velocity) is displayed at discrete points by an arrow, whose orientation indicates direction and whose size indicates magnitude. A vector plot generally presents a perspective

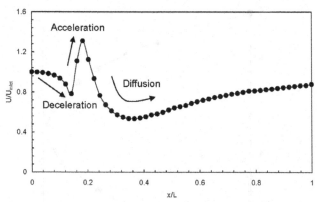

FIGURE 2.19   X-Y plot of the normalized horizontal velocity along the length L midway between two cylinders for Case 2.

view of the flow field in two dimensions. In a three-dimensional flow field, different slices of two-dimensional planes containing the vector quantities can be generated in different orientations to better scrutinize the global flow phenomena. If the mesh densities are considerably high, the CFD user can either interpolate or reduce the numbers of output locations to prevent the clustering of the arrows, "obliterating" the graphic plot.

Figure 2.20 is a typical velocity vector plot representing the fluid flowing along the parallel-plate channel. This plot gives an alternative view of the developing flow previously envisaged in Figure 2.18. The vector plot depicts the composite association of the velocity vectors with another dependent transport variable. For this particular flow case, we have arbitrarily chosen the distribution of dynamic pressure within the flow domain that impels the fluid flow. The plot also depicts the distribution between high and low pressures that are effective inside the fluid flow process. Nevertheless, the CFD user can also freely select other transport variables that better emphasize significant physical aspects of the flow phenomena. For example, where heat transfer may be important, the velocity vectors can be coupled with the temperature distribution to illustrate the transport of hot fluid within the flow domain.

FIGURE 2.20   Velocity vectors showing the flow development along the parallel-plate channel for Case 1.

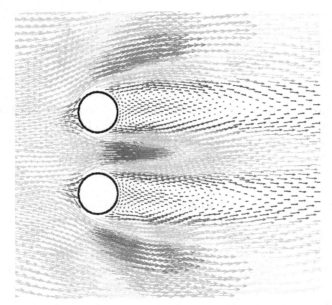

FIGURE 2.21   Velocity vectors accentuating the localized wake recirculation zones behind the two cylinders for Case 2.

For the complex wake-developing flow of Case 2, the velocity vectors, as presented in Figure 2.21, emphasize the presence of localized recirculation vortices as the fluid passes over and separates behind the proximity of the two cylinders. At the mid-height location, the varying sizes of the velocity vectors illustrating approaching flow at upstream, transitional flow in the vicinity of the two cylinders, and departing flow at downstream clearly demonstrate the three distinct flow characteristics of deceleration, acceleration, and diffusion observed in Figure 2.19. Here, the color spectrum that is indicated by the different shades of gray illustrates the spatial distribution of the dynamic pressure.

2.4.3   Contour Plots

Contour plotting is another useful and effective graphic technique that is frequently utilized in viewing CFD results. The proliferation of contour plots ever since the advent of the computer is not surprising. In CFD, contour plots are one of the most commonly found graphic representations of data. A contour line (also known as an *isoline*) can be described as a line indicative of some property that is constant in space. The equivalent representation in three dimensions is an *isosurface*. In contrast to X-Y plots, contour plots, like vector plots, provide a global description of the fluid flow encapsulated in one view. Generally, contours are plotted so that the difference between the numerical value of the dependent transport variable from one contour line to an adjacent contour line is held constant. The use of contour plots is usually not targeted at precise evaluation of the numerical values between contour lines. Although some mental and/or numerical interpolation can be performed between the contour lines in space, it is, to say the very least, an imprecise process. The actual numerical values represented by the isolines of these plots are sometimes less important than their overall disposition. In practice, the contours are usually linearly scaled. However, to better capture the hidden details in some small regions within the flow field, the reader may be required to intrepidly employ other types of scaling to reveal these isolated flow behaviors. For contour plots where the intervals are the same, the clustering of lines indicates rapid changes in the flow quantities. Such plots are particularly useful in locating propagating shocks and discontinuities.

Figures 2.22 and 2.23 are examples of *flooded* contour plots for Case 1. Here, a constant flow-field property of any transport variable is denoted by the constant intensity of the color shading. In these two plots, the so-called "rainbow-scale" color map (seen here in shades of gray) is employed to illustrate the distributions of the dimensionless resultant velocity normalized with respect to the inlet velocity and the dynamic pressure within the flow domain. The

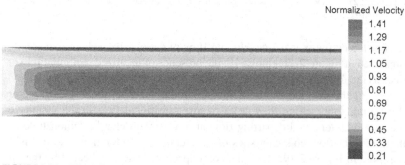

Normalized Velocity

1.41
1.29
1.17
1.05
0.93
0.81
0.69
0.57
0.45
0.33
0.21

FIGURE 2.22   Flooded contours on a rainbow-scale color map for the distribution of Case 1 normalized velocity.

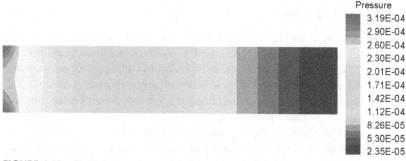

Pressure

3.19E-04
2.90E-04
2.60E-04
2.30E-04
2.01E-04
1.71E-04
1.42E-04
1.12E-04
8.26E-05
5.30E-05
2.35E-05

FIGURE 2.23   Flooded contours on a rainbow-scale color map for the distribution of Case 1 dynamic pressure.

changing flooded contours near the entrance (left boundary) in Figure 2.22 further confirm the development of the fluid flow as previously observed by the successive velocity profiles in the X-Y plot in Figure 2.18 and the velocity vector plot in Figure 2.20. In contrast, no appreciable change of velocity is observed near the exit (right boundary). The successive reduction of the pressure, as indicated by the contour plot in Figure 2.23, demonstrates the pressure gradient driving the fluid flow from the source imposed at the left boundary toward the sink located at the right boundary of the parallel-plate channel. At this point, we would like to draw the reader's attention to the color map of the pressure contour plot that is associated with the array of colors (shown in shades of gray) represented by the velocity vectors in Figure 2.20.

Figures 2.24 and 2.25 further exemplify another type of contour plot for the flow-field situation of Case 2. The former is the line contour representation of the pressure coefficient, while the latter demonstrates flooded contours on a gray-scale color map for the dynamic pressure distribution in complex flow around two cylinders (Case 2). In both figures, contours that are tightly clustered around the two cylinders clearly indicate the presence of vigorous flow activity as the fluid passes over the top and bottom cylinders. The manifestation of positive and negative pressure coefficients accompanied by the high and low dynamic pressures in the vicinity of the two cylindrical surfaces causes the flow stream velocities to change significantly across the curved surfaces. If the

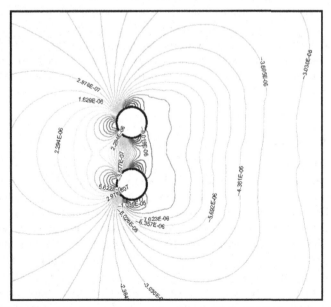

FIGURE 2.24   Line contours on a rainbow-scale color map for the distribution of Case 2 pressure coefficient.

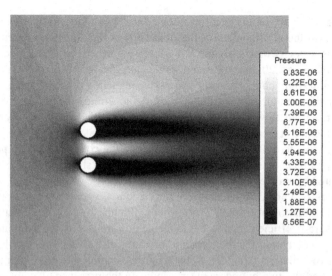

Pressure
9.83E-06
9.22E-06
8.61E-06
8.00E-06
7.39E-06
6.77E-06
6.16E-06
5.55E-06
4.94E-06
4.33E-06
3.72E-06
3.10E-06
2.49E-06
1.88E-06
1.27E-06
6.56E-07

FIGURE 2.25   Flooded contours on a gray-scale color map for the distribution of Case 2 dynamic pressure.

pressure *decreases* in the downstream direction, then the boundary layer thickness reduces; this case is termed a *favorable pressure gradient*. If, however, the pressure *increases* in the downstream direction, then the boundary layer thickens rapidly; this case is termed an *adverse pressure gradient*. The adverse pressure gradient, together with the shear forces acting at a sufficient length, cause the boundary layer to come to rest. The flow separates from the surface, leading to the formation of reversed-flow eddies, as represented by the apparent recirculation vortices (wakes) seen in the velocity vector plot of Figure 2.21. (Physical explanations of these flow behaviors are discussed further in Chapter 3.)

2.4.4   Other Plots

The application of *streamlines* in the pre-processing CFD stage, as in all aspects of fluid dynamics, is another exceptional tool for examining the nature of a flow in either two or three dimensions. By definition, streamlines are parallel to the mean velocity vector, where they trace the flow pattern using *massless* particles. (Synonyms for streamlines that are widely used in many graphics software packages include *streamtraces*, *streaklines*, and *pathlines*.) For example, they can generally be obtained by integrating the three spatial velocity components expressed in a three-dimensional Cartesian frame: $dx/dt = u$, $dy/dt = v$, and $dz/dt = w$. In more complex flow problems, such as multi-phase flows that involve the transport of solid particles, the *particle tracks* associated with discrete particles of a certain diameter and mass being injected inside the parent

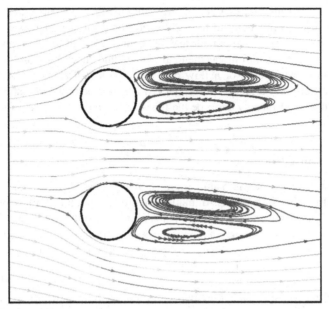

FIGURE 2.26 Streamline plot emphasizing the definitive localized wake recirculation zones behind the two cylinders for Case 2.

fluid fall in this same category. Here, important information on the particle residence time, particle velocity magnitude, and other properties can be duly extracted. The example of a streamline plot shown in Figure 2.21 defines the basic flow topology of localized recirculation zones behind the two cylinders, as previously identified from the velocity vector plot in Figure 2.21. This tool can often reveal important features that can be obscured in some isolated flow regimes, which is clearly demonstrated by the more definitive representation of the observed wake-developing vortices through the streamline plot. As in the velocity vector plot, the color spectrum (shown in shades of gray) corresponds to the spatial distribution of the dynamic pressure.

2.4.5 Data Report and Output

It is generally impractical to view the *raw data* of a CFD simulation, especially on a mesh that may entail thousands or millions of grid points, except possibly on a reasonably small "mapped" grid. However, alphanumeric reporting approaches need to be adopted because reports may be helpful for qualitatively checking the attained numerical solution and/or for extracting the quantitative results for post-analysis purposes. Important variables, such as surface fluxes, forces, and integrals, can be evaluated at each respective boundary encompassing the computational domain. Some relevant data of significant interest are the evaluation of the mass flux in/out of each boundary, which provides a clear indication of whether mass is conserved within the flow domain, and the determination of the components of forces/moments on a surface, such as wall shear stresses that may reveal important features connected with the flow physics of separation/reattachment/impingement. In some fluid mechanics problems, the transient/unsteady nature of a flow may be important. Data for a number of pertinent transport variables can be tracked and dumped into formatted/unformatted files to adequately describe the time-development response of the fluid-flow process.

2.4.6   Animation

CFD data fit very neatly into animation—*moving pictures* of the data produced through the CFD simulation. Animation, like other graphic display tools, not only represents a technical record of quantitative results, but also is a work of art. Some examples where animation has assisted in enhancing the physical representation of the fluid-flow processes are the movement of particles with the fluid in multi-phase flows, moving geometries, such as mixing tanks, and shock propagation in high-speed flows. Nowadays, a collection of short-length animated videos or movies bristling with multiple frames of brightly colored representations of CFD simulations can be found on many websites. Animation is undoubtedly an effective visualization tool for education and marketing purposes. The ever-increasing availability of these moving pictures is indeed a tribute to the tireless efforts of computer graphics developers to *bring fluid flow to life*. An example of animated CFD results for a developing free-standing fire within a flickering period can be found in Chapter 7; the results illustrate the puffing behavior and other physical aspects of fire dynamics in a way that helps the reader appreciate the pivotal role of animation in CFD.

2.5   SUMMARY

Chapter 2 aims to expose potential or novice users to a powerful tool or technique for tackling problems associated with fluid flows. We have intended to capture the quintessence of how the CFD discipline has evolved—the prevalence of many commercial, shareware, and in-house computer codes is unquestionably a testimonial to a dynamically evolving discipline. The widespread availability of these codes has certainly provided a favorable environment for students or new users learning CFD. Our intention throughout this chapter has been to provide proper guidance and possible supplementary knowledge emphasizing the many practical aspects within the framework of CFD analysis.

Although reference has been made to particular commercial codes, this chapter is not intended to replace the user manuals of the CFD codes. The descriptions and explanations of basic practical steps are geared toward assisting the reader to achieve the goal of attaining the eventual computational solution. At the completion of this chapter, we strongly encourage readers to begin applying any available CFD codes even without any *prior* knowledge of the fundamentals in order to familiarize themselves with the many facets of CFD.

A complete CFD analysis consists of *pre-processor*, *solver*, and *post-processor* stages. It simply encompasses the procedures of appropriately setting up the flow problem, solving and monitoring the solution, and analyzing the CFD results at the end of the simulation. Nevertheless, in the midst of these rather mechanically driven "black-box" operations, there are many fundamental principles underlying each of the three elements. The reader may wish to carefully review the many basic practical steps that have been carried out in handling Case 1 and Case 2 flow problems within this chapter. While performing the review, the reader should refer to the list of questions presented below, which may assist him/her in contemplating and perhaps becoming more aware of the numerous theoretical and numerical issues that are profoundly embedded inside the CFD analysis. The questions are

- What are the physical flow processes of the CFD problem? (Chapter 3)
- How is flow physics described in mathematical equations? (Chapter 3)
- What are the equations governing fluid flow and heat transfer? (Chapter 3)
- Why are boundary conditions important, and how are they applied? (Chapter 3)
- What are the physical meanings of the boundary conditions? (Chapter 3)
- How are the mathematical equations solved? (Chapter 4)
- Why does a flow domain need to be subdivided into many smaller, non-overlapping subdomains or a computational mesh/grid? (Chapter 4)
- How are computational methods/techniques employed? (Chapter 4)
- What is the meaning of monitoring curves? (Chapter 5)
- How is the numerical procedure terminated? (Chapter 5)
- What are solution errors? (Chapter 5)
- How is a computational solution judged to be correct, numerically accurate, and physically meaningful? (Chapter 5)
- When dealing with more complex flow problems, are there any other available methods/techniques, practical experiences, or general guidelines that can assist in overcoming convergence difficulties? (Chapter 6)
- Are there any additional illustrative examples using CFD and showing how the solution can be better analyzed? (Chapter 7)
- What are the future advancements in CFD? (Chapter 8)

The various chapters indicated in parentheses at the end of each question are primarily intended to indicate to the reader where the theoretical and numerical considerations of CFD are established in other chapters of this book. The chapters aim to comprehensively answer these questions while acknowledging the many practical aspects that are covered in this chapter. Needless to say, applying CFD in practice goes hand in hand with understanding the basic equations governing the fluid-flow process. These equations are based on conservation laws and theories. Their physical significance and implications, as well as respective formulations, are addressed in the next chapter.

REVIEW QUESTIONS

2.1 How are commercial codes allowing CFD analyses to be carried out with ease for the novice user?

2.2 What are the main elements involved in a complete CFD analysis?

2.3 Why is it important to correctly define the computational domain for the fluid-flow problem? Give an example.

2.4 What is the consequence of using a very fine mesh (i.e., a very large number of cells) compared with using a coarse mesh (i.e., a small number of cells)?

2.5 What is the main difference between a structured and unstructured mesh and when is each type applied to physical domains?

2.6 What types of boundary conditions can be imposed on the computational domain?

2.7 What type of boundary can be used for a computational boundary that represents an open physical boundary?

2.8 What advantages can a *symmetry* boundary condition and a *cyclic* boundary condition provide and when can they be applied?

2.9 What is the main purpose of a *CFD solver*?

2.10 What are the advantages of providing intelligent values for the initial solution?

2.11 What is an iteration process, and how is it performed?

2.12 What does the convergence criterion control?

2.13 What is the main purpose of the *post-processing* stage?

2.14 What are the advantages of using X-Y plots? Give examples of what CFD results X-Y plots can capture.

2.15 Why are contour plots best used to display the distribution of a variable?

2.16 What is the meaning of a streamline? What advantages do streamlines have over other plot types?

Chapter 3

Governing Equations for CFD—Fundamentals

3.1 INTRODUCTION

CFD is fundamentally based on the *governing equations* of fluid dynamics. The governing equations represent mathematical statements of the *conservation laws of physics*. The purpose of this chapter is to introduce the derivation and discussion of these equations, where the following physical laws are adopted:

- Mass is conserved for the fluid.
- Newton's second law: The rate of change of momentum equals the sum of forces acting on the fluid.
- First law of thermodynamics: The rate of change of energy equals the sum of the rate of heat addition to the fluid and the rate of work done on the fluid.

It is important that anyone concerned with CFD have some understanding of the physical phenomena of fluid motion, as it is these phenomena that CFD analyzes and predicts. All of CFD is based on the governing equations; we must therefore begin with the most basic description of the fluid-flow processes and the meaning and significance of each of the *terms* within them. After the governing equations are obtained, forms particularly suited for use in formulating CFD solutions will be delineated. The physical aspects of the boundary conditions and their appropriate mathematical statements will also be developed, since the appropriate *numerical* form of the physical boundary condition is strongly dependent on the particular mathematical form of the governing equations and numerical algorithm used. The goals of this chapter are to remove some of the mysteries surrounding the prediction of a fluid in motion through computer-based tools and to give the reader a solid understanding of the equations governing fluid transport.

3.2 THE CONTINUITY EQUATION
3.2.1 Mass Conservation

One conservation law that is pertinent to fluid flow is *matter may be neither created nor destroyed*. Consider an arbitrary control volume V fixed in space and time as shown in Figure 3.1. The fluid moves through the fixed control

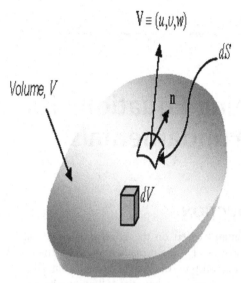

$$V \equiv (u, \upsilon, w)$$

Volume, V

dS

n

dV

FIGURE 3.1   Finite-control volume fixed in space.

volume, flowing across the control surface. *Mass conservation* requires that the rate of change of mass within the control volume is equivalent to the mass flux crossing the surface S of volume V (we define V as velocity, which is different from italic *V*, which represents a volume). In **integral form**,

$$\frac{d}{dt} \int_V \rho \, dV = - \int_S \rho V \cdot n \, dS \tag{3.1}$$

where **n** is the unit normal vector. We can apply Gauss's divergence theorem, which equates the volume integral of a divergence of a vector into an area integral over the surface that defines the volume. This is stated as

$$\int_V div \rho \, V \, dV = - \int_S \rho V \cdot n \, dS \tag{3.2}$$

Using the above theorem, the surface integral in Eq. (3.1) may be replaced by a volume V integral; hence the equation becomes

$$\int_V \left[\frac{\partial \rho}{\partial t} + \nabla \cdot (\rho V) \right] dV = 0 \tag{3.3}$$

where $\nabla \cdot (\rho V) \equiv div \, \rho V$. Since Eq. (3.3) is valid for any size of volume V, the implication is that

$$\frac{\partial \rho}{\partial t} + \nabla \cdot (\rho V) = 0 \tag{3.4}$$

Equation (3.4) is the mass conservation. In the Cartesian coordinate system, it can be expressed as

$$\frac{\partial \rho}{\partial t} + \frac{\partial(\rho u)}{\partial x} + \frac{\partial(\rho v)}{\partial y} + \frac{\partial(\rho w)}{\partial z} = 0 \qquad (3.5)$$

where the fluid velocity **V** at any point in the flow field is described by the local velocity components u, v, and w, which are, in general, functions of location (x, y, z) and time (t).

Alternatively, consider the scenario of a fluid flowing between two stationary parallel plates, as illustrated in Figure 3.2. An infinitesimally small control volume $\Delta x \Delta y \Delta z$ fixed in space (enlarged to the right of the figure) is analyzed where the mass conservation statement applies to the (u, v, w) flow field. Transport due to such motion is often referred to as *advection*. The conservation law requires that, for unsteady flow, *the rate of increase of mass within the fluid element equals the net rate at which mass enters the control volume* (inflow − outflow); in other words,

$$\frac{dm}{dt} = \sum_{in} \dot{m} - \sum_{out} \dot{m} \qquad (3.6)$$

The rate at which mass enters the control volume through the surface perpendicular to x may be expressed as $(\rho u)\,\Delta y\,\Delta z$, where ρ is the local density of the fluid, and similarly through the surfaces perpendicular to y and z as $(\rho v)\,\Delta x \Delta z$ and $(\rho w)\,\Delta x \Delta y$, respectively. The rate at which the mass leaves the surface at $x + \Delta x$ may be expressed through Taylor expansion as

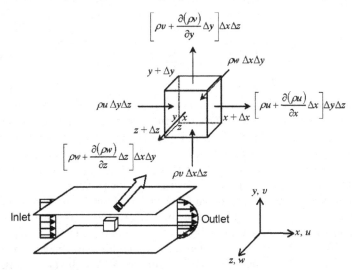

FIGURE 3.2   The conservation of mass in an infinitesimal control volume of a fluid flow between two stationary parallel plates.

$$\left[(\rho u) + \frac{\partial(\rho u)}{\partial x}\Delta x\right]\Delta y\Delta z + O(\Delta x, \Delta V) \quad \text{where} \quad \Delta V = \Delta x\Delta y\Delta z \qquad (3.7)$$

Similarly, the rate at which mass leaves the surfaces at $y + \Delta y$ and $z + \Delta z$ may also be expressed as

$$\left[(\rho v) + \frac{\partial(\rho v)}{\partial y}\Delta y\right]\Delta x\Delta z + O(\Delta y, \Delta V)$$

and

$$\left[(\rho w) + \frac{\partial(\rho w)}{\partial z}\Delta z\right]\Delta x\Delta y + O(\Delta z, \Delta V) \qquad (3.8)$$

Since the mass of the fluid element m is given by $\rho\,\Delta x\Delta y\Delta z$, Eq. (3.6) becomes

$$\frac{\partial(\rho\Delta x\Delta y\Delta z)}{\partial t} = (\rho u)\Delta y\Delta z + (\rho v)\Delta x\Delta z + (\rho w)\Delta x\Delta y$$

$$-\left[(\rho u) + \frac{\partial(\rho u)}{\partial x}\Delta x\right]\Delta y\Delta z - \left[(\rho v) + \frac{\partial(\rho v)}{\partial y}\Delta y\right]\Delta x\Delta z$$

$$-\left[(\rho w) + \frac{\partial(\rho w)}{\partial z}\Delta z\right]\Delta x\Delta y + \Delta V\ O(\Delta x, \Delta y, \Delta z) \qquad (3.9)$$

In the limit, canceling terms and dividing by the constant-size $\Delta x\Delta y\Delta z$, we obtain

$$\frac{\partial\rho}{\partial t} + \frac{\partial(\rho u)}{\partial x} + \frac{\partial(\rho v)}{\partial y} + \frac{\partial(\rho w)}{\partial z} = 0 \qquad (3.10)$$

Equation (3.10) is exactly the same form as derived in Eq. (3.5). This equation is precisely the **partial differential form** of the *continuity equation*. We have shown that the integral form in Eq. (3.1) can, after some manipulation, yield the partial differential form. This specific differential form is usually called the *conservation form*. Both Eq. (3.1) and Eq. (3.10) are in conservation form; the manipulation performed does not alter the situation.

In Chapter 2, a two-dimensional CFD analysis is performed for a channel flow that is described by the fluid flow between two stationary parallel plates (see Figure 3.2). This is made possible by the assumption that the dimension in the z coordinate direction is sufficiently large that the flow remains invariant along this coordinate direction. Since the fluid is taken to be incompressible, the density ρ is constant, i.e., the spatial and temporal variations in density are neglected relative to those velocity components of u, v, and w. We can obtain the continuity equation in two dimensions for an incompressible flow as

$$\cancel{\frac{\partial \rho}{\partial t}}^{=0} + \rho \left(\frac{\partial u}{\partial x} + \frac{\partial v}{\partial y} + \cancel{\frac{\partial w}{\partial z}}^{=0} \right) = 0$$

$$\underbrace{}_{constant\ density}$$

Flow invariant in
the z direction

or

$$\boxed{\frac{\partial u}{\partial x} + \frac{\partial v}{\partial y} = 0} \qquad\qquad (3.11)$$

3.2.2   Physical Interpretation

Let us examine the physical meaning of the continuity equation, Eq. (3.11), as applied to an infinitesimally small control volume for the two-dimensional case of the fluid flow between two parallel plates, to illustrate the fundamental physical principle. Two situations are considered.

Consider the first situation, if $\partial u/\partial x > 0$, then the velocity at the surface at $x + \Delta x$ is greater than the velocity at the surface x, i.e., $u(x + \Delta x) > u(x)$. Since more fluid is physically *leaving* the control volume than *entering* along the x direction, there should be more fluid entering than leaving along the y direction. Here, $\partial v/\partial y < 0$ and the velocity at the surface $y + \Delta y$ is less than the velocity at the surface y, i.e., $v(y + \Delta y) < v(y)$.

Alternatively, for the second situation, if $\partial u/\partial x < 0$, then the velocity at the surface at $x + \Delta x$ is less than the velocity at the surface x, i.e., $u(x + \Delta x) < u(x)$. Since more fluid is physically entering the control volume than *leaving* along the x direction, there should be more fluid leaving than entering along the y direction. Here, $\partial v/\partial y > 0$ and the velocity at the surface $y + \Delta y$ is greater than the velocity at the surface y, i.e., $v(y + \Delta y) > v(y)$.

Both situations satisfy the continuity equation: $\partial u/\partial x + \partial v/\partial y = 0$ (*mass conservation*).

Example 3.1

Consider a laminar boundary layer that can be approximated as having a velocity profile $u(x) = U_\infty y/\delta$ where $\delta = cx^{1/2}$, c is a constant, U_∞ is the free-stream velocity, and δ is the boundary-layer thickness. With reference to the two-dimensional fluid flow over a flat plate, as shown in Figure 3.1.1, determine the velocity v (vertical component) inside the boundary layer.

Solution

As the boundary layer grows downstream, the horizontal velocity u is gradually slowed down due to viscous effect and the no-slip condition at the surface of the flat plate. In order to satisfy the continuity equation, the vertical velocity v should be positive and acting to remove the fluid from the boundary layer.

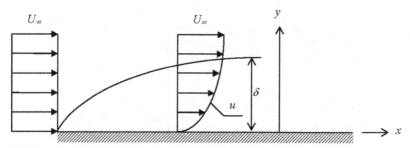

FIGURE 3.1.1   Two-dimensional flow over a flat plate.

We begin the analysis by substituting the velocity profile $u(x)$ into Eq. (3.13), yielding

$$\frac{\partial}{\partial x}\left(\frac{U_\infty y}{cx^{1/2}}\right) + \frac{\partial v}{\partial y} = 0 \quad \Rightarrow \quad -\frac{U_\infty y}{2cx^{3/2}} = -\frac{\partial v}{\partial y}$$

Integrating the vertical velocity v with respect to y, the equation becomes

$$v = \frac{U_\infty y^2}{4cx^{3/2}} = \left(\frac{U_\infty y}{cx^{1/2}}\right)\left(\frac{y}{4x}\right) = \frac{uy}{4x}$$

Discussion

This physically means that the velocity ratio $u = y/4x$ increases away from the surface at a fixed x location, i.e., it decreases further downstream at a fixed y location. At the edge of the boundary layer, $y = \delta = cx^{1/2}$; the velocity ratio v/u equals $c/4x^{1/2}$. If the constant c is assumed unity, the boundary-layer thickness δ and the velocity ratio v/u as a function of the horizontal distance x from the leading edge of the flat plate can be described, and they are illustrated in Figures 3.1.2 and 3.1.3. The latter figure further illustrates the decrease of the velocity ratio v/u further downstream of

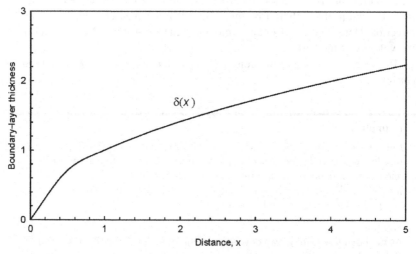

FIGURE 3.1.2   Boundary-layer thickness δ as a function of the horizontal distance x.

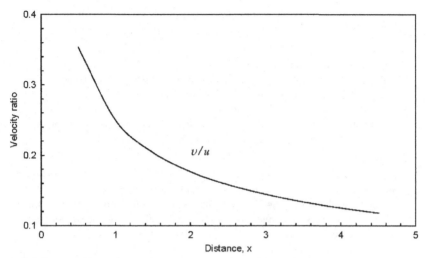

FIGURE 3.1.3 Velocity ratio v/u as a function of the horizontal distance x.

the fluid flow over the flat plate. At some downstream distance x, the change in the horizontal u velocity is appreciably small. Here, $\partial u/\partial x \to 0$, which also leads to $\partial v/\partial y \to 0$, hence satisfying the continuity equation, Eq. (3.11).

Example 3.2

Consider the CFD case in Chapter 2 for the steady, two-dimensional, incompressible, laminar flow between two stationary parallel plates with the following dimensions: height $H = 0.1$ m and length $L = 0.5$ m (see Figure 3.2.1). Using CFD, plot the velocity vector along the channel length. Discuss the physical meaning of the continuity equation by plotting the velocity components u and v close to the bottom wall surface along the channel length with the working fluid

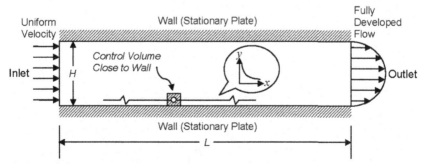

FIGURE 3.2.1 Two-dimensional laminar flow between two stationary parallel plates.

taken as air and a uniformly distributed velocity profile of 0.01 m/s applied at the channel inlet (u_{in}).

Solution

The problem is described as follows:

From the CFD simulation, the flow field along the channel length is illustrated in Figure 3.2.2. It is observed that the flow gradually changes from a uniform profile at the inlet surface to a parabolic profile as it travels downstream along the channel. The resultant velocity profiles of u and v close to the bottom wall surface along the channel length are given in Figures 3.2.3 and 3.2.4.

Discussion

It is observed that within the *hydrodynamic entrance region*, i.e., $x < 3H$, the horizontal velocity u decreases along the channel length. This means that $\partial u/\partial x < 0$ close to the wall surface. The air flow is slowed along the x direction due to the no-slip boundary condition imposed near the wall as it flows over the wall surface. More fluid is therefore physically entering than leaving the flow domain along the x direction. At the same time, there should be more fluid leaving than entering along the y direction; the vertical velocity v increases, which implies that $\partial v/\partial y > 0$ in order to conserve the mass. On the other hand, when the flow is in the *fully developed region*, i.e., $x \geq 3H$, it is observed that the horizontal velocity u does not

FIGURE 3.2.2   Velocity profile distribution along the channel length.

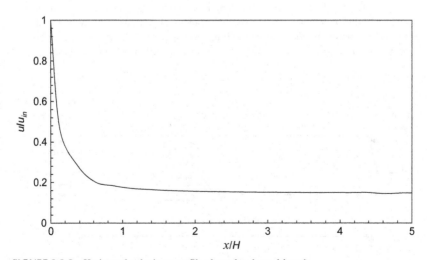

FIGURE 3.2.3   Horizontal velocity u profile along the channel length.

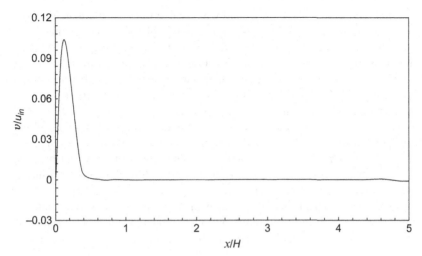

FIGURE 3.2.4 Vertical velocity v profile along the channel length.

appreciably change along the x direction, i.e., $\partial u/\partial x = 0$. In accordance with the continuity equation, $\partial v/\partial y$ must also be zero within this flow region. This is clearly reflected by the constant vertical velocity v, as shown in Figure 3.2.3, in the *fully developed region*. The physical meanings of hydrodynamic entrance region and fully developed region are further discussed in Example 3.5.

3.2.3 Comments

In most fluid mechanics textbooks, the principle of mass conservation is often explained by a fluid flowing in a pipe (see Figure 3.3). The mass entering a pipe, denoted by the mass flow rate $\dot{m}_1$, is equivalent to the product of the density, inlet velocity, and cross-sectional area, i.e., $\rho u_1 A_1$. This mass must equal the mass flow rate leaving the pipe, which is denoted by $\dot{m}_2$ given as $\rho u_2 A_2$. If we take, for instance, that the cross-sectional areas are the same for both the inlet and outlet surfaces of the pipe, i.e., $A_1 = A_2$, and based on mass conservation, the outlet velocity u_2 must equal the inlet velocity u_1. If the cross-sectional areas are different at both ends of the pipe, take for example $A_1 = 2A_2$, then $u_2 = 2u_1$, which means that the flow is *accelerated*. On the other hand, if $A_1 = \frac{1}{2}A_2$, then $u_2 = \frac{1}{2}u_1$, which means that the flow is *decelerated*.

FIGURE 3.3 A pipe flow scenario.

This principle of mass conservation as applied to the whole domain in one dimension is also applicable to any small control volume that is used in CFD to numerically solve the partial differential equations. In two dimensions, the mass flow may not be conserved in one direction, but, overall, it will be conserved throughout the control volume by either removing or adding the mass in the other direction. It is important to persevere with the physical meaning of the mathematical equations in mind when studying and analyzing CFD results—a philosophy that we urge the reader to embrace. Indeed, this philosophy is extrapolated to all mathematical equations and operations encompassing any physical problems. We have explored at some length the physical meaning of the continuity equation, Eq. (3.11), and in the following sections we further explore the effects of various terms, such as *advection*, *diffusion*, and so forth, in the other mathematical equations. Whatever the terms are, we strongly encourage the reader at all times to continuously adopt the philosophy of understanding the physical meaning of the terms in the equations that are being dealt with.

3.3   THE MOMENTUM EQUATION

3.3.1   Force Balance

In deriving this physical law, we begin by considering the fluid element as described in Figure 3.2 for mass conservation. *Newton's second law of motion* states that the sum of forces acting on the fluid element, as illustrated in Figure 3.4, equals the product of its mass and the acceleration of the element. There are essentially three scalar relations along the x, y, and z directions of the Cartesian frame for which the fundamental law can be invoked. We begin by considering the x component of Newton's second law,

$$\sum F_x = ma_x \tag{3.12}$$

where F_x and a_x are the force and acceleration along the x direction. The acceleration a_x on the right-hand side of Eq. (3.12) is simply the time rate change of u, which is given by the substantial derivative (see Appendix A for a more detailed description). Thus,

$$a_x = \frac{Du}{Dt} \tag{3.13}$$

Recalling that the mass of the fluid element m is $\rho \, \Delta x \Delta y \Delta z$, the rate of increase of x-momentum is

$$\rho \frac{Du}{Dt} \Delta x \Delta y \Delta z \tag{3.14}$$

On the left-hand side of Eq. (3.12), there are two sources of the force that the moving fluid element experiences. They are *body forces* and *surface forces*. The body forces that may influence the rate of change of the fluid momentum are

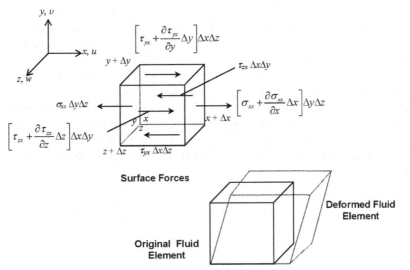

FIGURE 3.4 Surface forces acting on the infinitesimal control volume for the velocity component u. Deformed fluid element due to the action of the surface forces.

gravity, centrifugal, Coriolis, and electromagnetic forces. These effects are usually incorporated by introducing them into the momentum equations as additional source terms to the contribution of the surface forces. The surface forces for the velocity component u, as seen in Figure 3.4, that deform the fluid element are due to the normal stress σ_{xx} and tangential stresses τ_{yx} and τ_{zx} acting on the surfaces of the fluid element. Combining the sum of these surface forces on the fluid element (see Appendix A for a more detailed derivation) and the time rate change of u from Eq. (3.14) into Eq. (3.12), the x-momentum equation becomes

$$\rho \frac{Du}{Dt} = \frac{\partial \sigma_{xx}}{\partial x} + \frac{\partial \tau_{yx}}{\partial y} + \frac{\partial \tau_{zx}}{\partial z} + \sum F_x^{body\ forces} \qquad (3.15)$$

In a similar fashion, the y-momentum and z-momentum equations can be obtained as

$$\rho \frac{Dv}{Dt} = \frac{\partial \tau_{xy}}{\partial x} + \frac{\partial \sigma_{yy}}{\partial y} + \frac{\partial \tau_{zy}}{\partial z} + \sum F_y^{body\ forces} \qquad (3.16)$$

and

$$\rho \frac{Dw}{Dt} = \frac{\partial \tau_{xz}}{\partial x} + \frac{\partial \tau_{yz}}{\partial y} + \frac{\partial \sigma_{zz}}{\partial z} + \sum F_z^{body\ forces} \qquad (3.17)$$

The normal stresses σ_{xx}, σ_{yy}, and σ_{zz} in Eqs. (3.15)–(3.17) are due to the combination of pressure p and normal viscous stress components τ_{xx}, τ_{yy}, and τ_{zz} acting perpendicular to the control volume. The remaining terms contain the tangential viscous stress components. In many fluid flows, a suitable model for the viscous stresses is introduced. They are usually a function of the local deformation rate (or strain rate) that is expressed in terms of the velocity gradients. The formulation of the appropriate stress–strain relationships for a Newtonian fluid is found in Appendix A.

For the two-dimensional case of the fluid flow between parallel plates (the flow being invariant along the z direction), the case investigated is a *constant-property* fluid flow. This implies that the density is constant and body forces, particularly due to gravity (for example, density variation due to buoyancy), need not be considered in the equations. By invoking the continuity equation, the momentum equations with the inclusion of the stress–strain relationships can be reduced to

$$
\underbrace{\frac{Du}{Dt}}_{acceleration} = -\underbrace{\frac{1}{\rho}\frac{\partial p}{\partial x}}_{pressure\ gradient} + \underbrace{v\frac{\partial^2 u}{\partial x^2} + v\frac{\partial^2 u}{\partial y^2}}_{diffusion} + \frac{\partial}{\partial x}\left[\lambda\left(\underbrace{\frac{\partial u}{\partial x} + \frac{\partial v}{\partial y}}_{continuity\ equation}\right)^{=0}\right]
$$

$$
+ v\frac{\partial}{\partial x}\left[\underbrace{\frac{\partial u}{\partial x} + \frac{\partial v}{\partial y}}_{continuity\ equation}\right]^{=0} + \underbrace{\frac{\partial v}{\partial x}\frac{\partial u}{\partial x} + \frac{\partial v}{\partial y}\frac{\partial v}{\partial x}}_{constant\ property}^{=0} + \underbrace{\sum F_x^{body\ forces}}_{body\ forces}
$$

$$
\underbrace{\frac{Dv}{Dt}}_{acceleration} = -\underbrace{\frac{1}{\rho}\frac{\partial p}{\partial y}}_{pressure\ gradient} + \underbrace{v\frac{\partial^2 v}{\partial x^2} + v\frac{\partial^2 v}{\partial y^2}}_{diffusion} + \frac{\partial}{\partial y}\left[\lambda\left(\underbrace{\frac{\partial u}{\partial x} + \frac{\partial v}{\partial y}}_{continuity\ equation}\right)^{=0}\right]
$$

$$
+ v\frac{\partial}{\partial y}\left[\underbrace{\frac{\partial u}{\partial x} + \frac{\partial v}{\partial y}}_{continuity\ equation}\right]^{=0} + \underbrace{\frac{\partial v}{\partial x}\frac{\partial u}{\partial x} + \frac{\partial v}{\partial y}\frac{\partial v}{\partial y}}_{constant\ property}^{=0} + \underbrace{\sum F_y^{body\ forces}}_{body\ forces}
$$

or

$$
\boxed{\underbrace{\frac{\partial u}{\partial t}}_{local\ acceleration} + \underbrace{u\frac{\partial u}{\partial x} + v\frac{\partial u}{\partial y}}_{advection} = -\underbrace{\frac{1}{\rho}\frac{\partial p}{\partial x}}_{pressure\ gradient} + \underbrace{v\frac{\partial^2 u}{\partial x^2} + v\frac{\partial^2 u}{\partial y^2}}_{diffusion}} \tag{3.18}
$$

$$
\boxed{\underbrace{\frac{\partial v}{\partial t}}_{local\ acceleration} + \underbrace{u\frac{\partial v}{\partial x} + v\frac{\partial v}{\partial y}}_{advection} = -\underbrace{\frac{1}{\rho}\frac{\partial p}{\partial y}}_{pressure\ gradient} + \underbrace{v\frac{\partial^2 v}{\partial x^2} + v\frac{\partial^2 v}{\partial y^2}}_{diffusion}} \tag{3.19}
$$

Equations (3.18) and (3.19), derived from *Newton's second law*, where v is the kinematic viscosity ($v = \mu/\rho$), describe the conservation of momentum in the fluid flow and are also known as the Navier–Stokes equations. The physical significance of each of the terms in Eqs. (3.18) and (3.19) is examined and explained in the next section.

3.3.2   Physical Interpretation

Consider the motion of the fluid of a piston compressing the air within an enclosed cylinder. From the viewpoint of point A in Figure 3.5, the fluid is locally accelerating due to the increase of the air velocity within the shrinking volume changing in time. In relation to the x-momentum, Eq. (3.18), this is represented by the acceleration term $\partial u/\partial t$ for the horizontal velocity component u. Similarly, if the motion of fluid of the piston is inverted vertically, the local rate of increase of the air velocity is represented by the term $\partial v/\partial t$, which denotes the local acceleration of the vertical component v of Eq. (3.19).

The above example describes the motion of the fluid changing locally with time. We further explore the physical meaning of the fluid's accelerating locally in space. Consider the motion of fluid through a Venturi, as indicated in Figure 3.6. As the fluid travels between the locations B and C along the x direction, and assuming that the velocity in itself is not fluctuating with time, the horizontal velocity component u has a local acceleration in space where the velocity is accelerating between the incremental distance of locations B and C, i.e., the velocity gradient in the term $u\partial u/\partial x$ of Eq. (3.18) is increasing. Similarly, if the Venturi is vertically oriented, then the vertical velocity component v has a local acceleration gradient in space where the velocity gradient in the term $v\partial v/\partial y$ is increasing between the incremental distance of locations B and C. We usually refer to the fluid sweeping past point B and on its way to point C in the flow field as the *advection* term of the momentum equations.

We further investigate the physical interpretations of the pressure gradient and diffusion terms in the momentum equations, Eqs. (3.18) and (3.19), in the following examples.

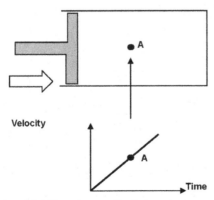

FIGURE 3.5   The motion of fluid in a piston mechanism.

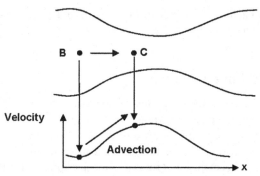

FIGURE 3.6 The motion of fluid through a Venturi.

Example 3.3

Consider an incompressible, inviscid, laminar flow past a circular cylinder of diameter d in Figure 3.3.1. The flow variation along the approaching stagnation streamline (A – B) can be expressed as

$$u(x) = U_\infty \left(1 - \frac{R^2}{x^2} \right)$$

With reference to Eqs. (3.24) and (3.25), determine the total acceleration experienced by the fluid as it flows along the stagnation streamline. Also determine the pressure distribution along the streamline by deriving Bernoulli's equation and the stagnation pressure at the stagnation point.

Solution

For an inviscid flow, the shear stresses are zero, i.e., $\tau = 0$ for all shear stresses. The diffusion terms are therefore zero. Along the stagnation streamline, the vertical velocity component v is zero. Also, since the continuity equation applies, and considering x-momentum Eq. (3.24), the following terms drop off from the equation:

$$\frac{\partial u}{\partial t} + u\frac{\partial u}{\partial x} + v\underbrace{\frac{\partial u}{\partial y}}_{=0} = -\frac{1}{\rho}\frac{\partial p}{\partial x} + v\underbrace{\frac{\partial^2 u}{\partial x^2}}_{=0} + v\underbrace{\frac{\partial^2 u}{\partial y^2}}_{=0}$$

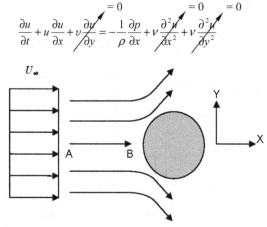

FIGURE 3.3.1 Fluid motion over a circular cylinder.

The preceding equation reduces to

$$\frac{\partial u}{\partial t} + u\frac{\partial u}{\partial x} = -\frac{1}{\rho}\frac{\partial p}{\partial x}$$

The total acceleration of the fluid comprises the sum of the local acceleration and advection terms and is driven by the pressure gradient in the x direction. Assuming that the upstream velocity $U_\infty = 1$ m/s and a radius $R = 1$ m, the total acceleration is given by

$$a_x = \frac{\partial u}{\partial t} + u\frac{\partial u}{\partial x} = \left(1 - \frac{1}{x^2}\right)\left(\frac{2}{x^3}\right)$$

The equation derived represents another form of the momentum equation, which is Euler's equation for fluid flow. For a steady flow, Bernoulli's equation can be obtained by integrating the equation along a streamline. In other words,

$$u\frac{\partial u}{\partial x} = -\frac{1}{\rho}\frac{\partial p}{\partial x} \quad\Rightarrow\quad \frac{p(x)}{\rho} + \frac{u^2(x)}{2} = \frac{p_\infty}{\rho} + \frac{U_\infty^2}{2}$$

The above equation can be re-arranged, and taking the upstream pressure p_∞ to be atmospheric, it becomes

$$p(x) - p_{atm} = \frac{\rho}{2}\left(U_\infty^2 - u^2(x)\right) = \frac{\rho}{2}\left[1 - \left(1 - \frac{1}{x^2}\right)^2\right]$$

Discussion
Along the streamline, the velocity profile $u(x)$ given above can be represented by the profile described in Figure 3.3.2. The velocity drops very rapidly as the fluid

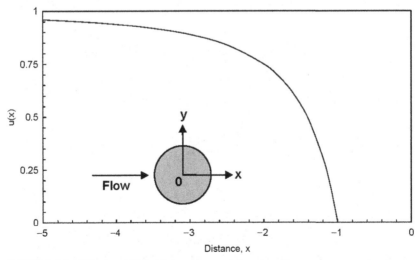

FIGURE 3.3.2 Velocity profile $u(x)$ along the stagnation streamline.

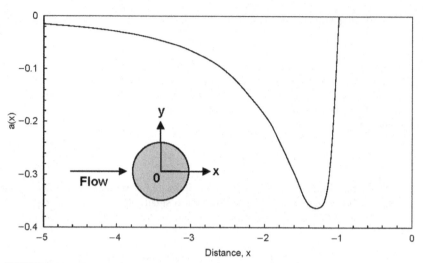

FIGURE 3.3.3 Total acceleration profile $a(x)$ along the stagnation streamline.

approaches the cylinder. At the surface of the cylinder, the velocity is zero (stagnation point) and the surface pressure is a maximum.

The total acceleration profile depicted in Figure 3.3.3 also shows the strong deceleration of the fluid as it approaches the cylinder. The maximum deceleration occurs at $x = -1.29$ m with a magnitude of -0.372 m/s^2.

The pressure difference $p(x) - p_{atm}$ derived above is illustrated in Figure 3.3.4, which demonstrates that the pressure increases as the fluid approaches the stagnation point. With the density ρ taken to be unity, it reaches the maximum value of 0.5, i.e., $p_{stag} - p_{atm} = (1/2)\rho U_\infty^2$ as $u(x) \to 0$ near the stagnation point.

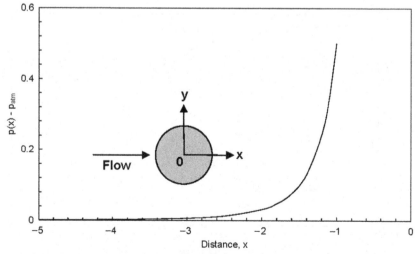

FIGURE 3.3.4 Pressure difference $p(x) - p_{atm}$ along the stagnation streamline.

Example 3.4

Consider a steady, incompressible, laminar flow through the parallel-plate channel, as investigated in Example 3.2. For a constant-property fluid with a fully developed flow, determine the velocity profile subject to the boundary condition where the vertical component v is zero everywhere.

Solution

The horizontal velocity component u depends only on x. The axial dependence on the horizontal velocity may be obtained by solving the appropriate form of the x-momentum equation, Eq. (3.24). Since the vertical component v is zero everywhere, the continuity equation, Eq. (3.11), reduces to

$$\frac{\partial u}{\partial x} + \underbrace{\frac{\partial v}{\partial y}}_{=0} = 0 \qquad \Rightarrow \qquad \frac{\partial u}{\partial x} = 0$$

This indicates that the velocity u is only a function of y. The x-momentum Eq. (3.24) becomes

$$\underbrace{\frac{\partial u}{\partial t}}_{=0} + u\underbrace{\frac{\partial u}{\partial x}}_{=0} + v\underbrace{\frac{\partial u}{\partial y}}_{=0} = -\frac{1}{\rho}\frac{\partial p}{\partial x} + v\underbrace{\frac{\partial^2 u}{\partial x^2}}_{=0} + v\frac{\partial^2 u}{\partial y^2}$$

Since $v = \mu/\rho$, the above equation reduces to

$$\frac{\partial^2 u}{\partial y^2} = \frac{1}{\mu}\frac{\partial p}{\partial x}$$

Hence, the momentum conservation requirement is just a simple balance between the shear and pressure forces in this case. Integrating once yields the velocity gradient $\partial u/\partial y$ with respect to y:

$$\frac{\partial u}{\partial y} = \frac{1}{\mu}\frac{\partial p}{\partial x}y + C_1$$

Integrating again yields the horizontal velocity u with respect to y:

$$u(y) = \frac{1}{2\mu}\left(\frac{\partial p}{\partial x}\right)y^2 + C_1 y + C_2$$

It is noted that the pressure gradient $\partial p/\partial x$ is treated as a constant as far as the integration is concerned since it is not a function of y (we refer to the y-momentum equation later in this example).

The two boundary conditions required to determine the constants C_1 and C_2 are

$$u = 0 \qquad \text{at} \qquad y = \frac{H}{2} \quad \text{(no-slip)}$$

$$\frac{\partial u}{\partial y} = 0 \qquad \text{at} \qquad y = 0 \quad \text{(symmetry)}$$

From the symmetry condition, based on the velocity gradient $\partial u/\partial y$ with respect to y, the constant C_1 is zero.

$$0 = \frac{1}{\mu}\left(\frac{\partial p}{\partial x}\right) \cdot 0 + C_1 \quad \Rightarrow \quad C_1 = 0$$

The constant C_2 can be determined by applying the no-slip condition above into the velocity profile $u(y)$ equation. Therefore,

$$0 = \frac{1}{2\mu}\left(\frac{\partial p}{\partial x}\right)\left(\frac{H}{2}\right)^2 + C_2 \quad \Rightarrow \quad C_2 = -\frac{1}{2\mu}\left(\frac{\partial p}{\partial x}\right)\left(\frac{H}{2}\right)^2$$

The velocity profile $u(y)$ becomes

$$u(y) = \frac{1}{2\mu}\left(\frac{\partial p}{\partial x}\right)y^2 - \frac{1}{2\mu}\left(\frac{\partial p}{\partial x}\right)\left(\frac{H}{2}\right)^2 \quad \Rightarrow$$

$$u(y) = \frac{3}{2}\frac{H^2}{12\mu}\left(-\frac{\partial p}{\partial x}\right)\left[1 - \frac{y^2}{(H/2)^2}\right] \quad \Rightarrow$$

$$\boxed{u(y) = \frac{3}{2}U_m\left[1 - \frac{y^2}{(H/2)^2}\right]} \qquad (3.4\text{-A})$$

In this equation, the average velocity U_m is given as

$$U_m = \frac{H^2}{12\mu}\left(-\frac{\partial p}{\partial x}\right) \qquad (3.4\text{-B})$$

The volume flow rate q passing between the plates can be obtained from the relationship

$$q = U_m H = -\frac{H^3}{12\mu}\left(\frac{\partial p}{\partial x}\right) \qquad (3.4\text{-C})$$

The pressure gradient $\partial p/\partial x$ is negative, as the pressure decreases in the direction of the flow. If we let Δp represent the pressure drop between the inlet and outlet of the channel at a distance l apart, then

$$\frac{\Delta p}{l} = -\frac{\partial p}{\partial x}$$

and the volume flow rate can be expressed as

$$q = \frac{H^3 \Delta p}{12\mu l}$$

The flow is proportional to the pressure gradient, inversely proportional to the viscosity, and strongly dependent on the gap width ($\sim H^3$). In terms of the average velocity, it becomes

$$U_m = \frac{H^2}{12\mu}\frac{\Delta p}{l}$$

The above equations provide convenient relationships for relating the pressure drop along the channel between the parallel plates and the rate of flow or average velocity. The maximum velocity U_{max} occurs midway ($y = 0$) between the two plates so that

$$U_{max} = \frac{3}{2}U_m$$

Based on the boundary condition where the vertical component v is zero everywhere, and if the body force due to gravity is considered in the y-momentum equation, Eq. (3.19), the equation becomes

$$\cancel{\frac{\partial v}{\partial t}}^{=0} + u\cancel{\frac{\partial v}{\partial x}}^{=0} + v\cancel{\frac{\partial v}{\partial y}}^{=0} = -\frac{1}{\rho}\frac{\partial p}{\partial y} + v\cancel{\frac{\partial^2 v}{\partial x^2}}^{=0} + v\cancel{\frac{\partial^2 v}{\partial y^2}}^{=0} - g \quad \Rightarrow$$

Gravity

$$\frac{\partial p}{\partial y} = -\rho g$$

The gravitational force acts downward, which results in the negative body force being represented in the above equation. This equation can be integrated to yield

$$p = -\rho g y + \{C_3 = f(x)\}$$

which shows that the pressure varies hydrostatically in the y direction. The constant C_3 in the above equation can be expressed by a function $f(x)$, where it can be related to the varying pressure gradient $\partial p/\partial x$ along the x direction and some reference pressure p_o.

$$f(x) = \left(\frac{\partial p}{\partial x}\right)x + p_o$$

where p_o is at a location $x = y = 0$ and the pressure variation throughout the fluid can be obtained from

$$p = -\rho g y + \left(\frac{\partial p}{\partial x}\right)x + p_o \qquad (3.4\text{-D})$$

Discussion

Details of a steady, laminar flow between infinite parallel plates are completely predicted by the solution formulated above from the Navier–Stokes equations. For instance, if the pressure gradient, viscosity, and plate spacing are specified, then the velocity profile can be determined, as well as the average velocity and flow rate—Eqs. (3.4-A), (3.4-B), and (3.4-C). Also, for a given fluid and reference pressure p_o, the pressure at any point can be predicted—Eq. (3.4-D). The physical significance of the velocity distribution as seen in Eq. (3.4-A) is that the profile is *parabolic*. The CFD simulation of Example 3.2 also predicts this particular velocity profile. The parabolic profile has been confirmed by many experiments: Nakayama (1988) experimentally displayed a line of bubbles resembling a parabolic shape occurring at the downstream end of a channel. This fully developed flow distribution is known as the *Hagen-Poiseuille flow*, after the first two investigators who reported this flow behavior (Hagen, 1939, Poiseuille, 1940). This relatively simple example of an exact solution illustrates the detailed information about the flow field that can be obtained. The resultant velocity profile above being parabolic has physical implications, as it is a typical distribution experienced for laminar flows in parallel channels.

Example 3.5

Consider the two-dimensional CFD case of an incompressible laminar flow between two stationary parallel plates to illustrate the physical meaning of the momentum equations. Here, the dimensions of the channel flow are given as height $H = 0.1$ m and length $L = 1$ m. Using CFD, discuss the development of the velocity profiles between the inlet and the *hydrodynamic entry length* (L_E) with air (density $\rho = 1.2$ kg/m^3) as the working fluid, for the following conditions:

(a) A fixed inlet velocity $u_{in} = 0.01$ m/s and dynamic viscosities $\mu_1 = 4 \times 10^{-5}$ kg/m $\cdot$ s?> and $\mu_2 = 10^{-5}$ kg/m $\cdot$ s.

(b) A fixed dynamic viscosity $\mu = 4 \times 10^{-5}$ kg/m $\cdot$ s and inlet velocities $u_{in1} = 0.01$ m/s and $u_{in2} = 0.04$ m/s.

Solution

The problem is described as follows (see Figure 3.5.1):

(a) By definition, the region from the inlet surface to the point at which the boundary layer merges at the centerline is called the *hydrodynamic entrance region* and the length of this region is called the *hydrodynamic entry length*. This entrance region is also referred as the *hydrodynamically developing flow*, since the velocity profile is still developing. Beyond this region, the velocity profile is fully developed and remains unchanged; it is called the *fully developed region*. The fluid flow at this instance is considered fully developed (see Figure 3.5.1).

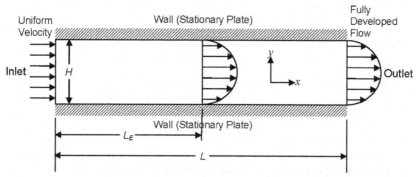

FIGURE 3.5.1 Hydrodynamic entry length (L_E) location from the inlet surface in a two-dimensional laminar flow between two stationary parallel plates.

Based on the CFD simulation, the velocity profiles at different downstream locations x/H from the inlet surface for varying dynamic viscosities are given in Figures 3.5.2 and 3.5.3. The former refers to the case where the flow has a higher dynamic viscosity than the latter.

(b) The velocity profiles at different downstream locations x/H from the inlet surface for varying inlet velocities are given in Figures 3.5.4 and 3.5.5. The former refers

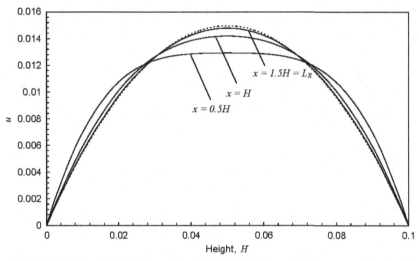

FIGURE 3.5.2 Case A: Velocity profiles at different locations of x/H for $u_{in} = 0.01$ m/s and dynamic viscosity $\mu_1 = 4 \times 10^{-5}$ kg/m · s.

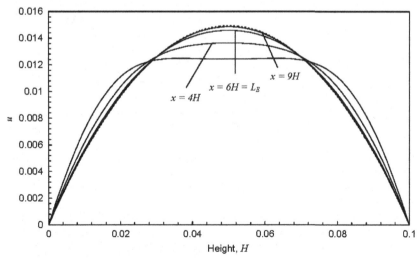

FIGURE 3.5.3 Case B: Velocity profiles at different locations of x/H for $u_{in} = 0.01$ m/s and dynamic viscosity $\mu_2 = 10^{-5}$ kg/m · s.

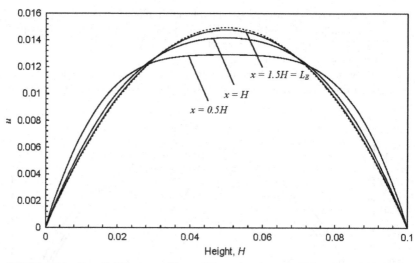

FIGURE 3.5.4   Case C: Velocity profiles at different locations of x/H for $u_{in1} = 0.01$ m/s and dynamic viscosity $\mu = 4 \times 10^{-5}$ kg/m · s.

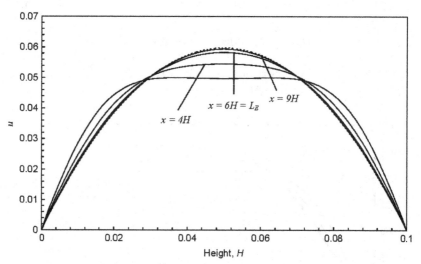

FIGURE 3.5.5   Case D: Velocity profiles at different locations of x/H for $u_{in2} = 0.04$ m/s and dynamic viscosity $\mu = 4 \times 10^{-5}$ kg/m · s.

to the case where the flow has a lower velocity than the latter. The case presented in Figure 3.5.4 is obviously exactly the same as the case presented in Figure 3.5.2. It is repeated here for comparison against the higher-velocity case.

Discussion

We rewrite the steady two-dimensional x-momentum equation, Eq. (3.24), as follows:

$$\underbrace{\rho u \frac{\partial u}{\partial x} + \rho v \frac{\partial u}{\partial y}}_{inertia} = \underbrace{-\frac{\partial p}{\partial x}}_{pressure} + \underbrace{\mu \left(\frac{\partial^2 u}{\partial x^2} + \frac{\partial^2 u}{\partial y^2} \right)}_{friction}$$

By comparing the results in Figure 3.5.2 (Case A) with Figure 3.5.3 (Case B), where the same *inertia* force is applied, it can be seen that the higher dynamic viscosity apparent in Case A produces a higher *friction* force, which inhibits the fluid momentum. This leads to a quicker transition of the flow to a fully developed stage at a shorter distance. Case B has less frictional resistance, thus allowing a slower development of the hydrodynamic boundary layer and reaches fully developed flow at a longer distance. The pressure gradients are the same for both cases. However, when the fluid has the same friction forces, as demonstrated by the results in Figure 3.5.4 (Case C) and Figure 3.5.5 (Case D), the results for the higher velocity, i.e., higher inertia force, are surprisingly exactly the same as those of the case of higher friction force in Figure 3.5.3. The higher *inertia* force in Case C yields the effect of reaching the same hydrodynamic entry length as imposing a higher *friction* due to the higher wall shear stress resisting the fluid flow, as in Case A. All the cases investigated demonstrate the relative physical contributions of the inertia and friction forces competing with each other to conserve the momentum transfer.

3.3.3 Comments

The principle of conservation of momentum has been explored with investigations of the various contributions from the advection and diffusion terms in the momentum equations that affect the fluid flow. The reader should be aware of the importance of keeping in mind the physical meaning of the mathematical equations and the roles they play in CFD analyses. In CFD, the concept of *dynamic similarity* is frequently adopted. This involves normalizing the mathematical equations to yield the non-dimensional governing equations. For instance, we observe in Example 3.5 that, through the various combinations of different inlet velocities and dynamic viscosities, the same fluid flow effect is obtained in reference to the development of the flow in yielding the same hydrodynamic entrance region. We can now even change the air density from $\rho_1 = 1.2$ kg/m^3 to $\rho_2 = 4.8$ kg/m^3 while fixing the inlet velocity and dynamic viscosity at 0.01 m/s and $\mu = 4 \times 10^{-5}$ kg/m · s, respectively; the same results are obtained. This is because the increase of density contributes to the increase of the inertia force, which has the same effect as increasing the inlet velocity. There appears to be some homogeneity of the fluid-flow behavior that results in the combination of these physical variables. One important non-dimensional parameter that describes the flow characteristics is the Reynolds number (Re), which is defined as the ratio of the inertia force to the friction force, i.e.,

$$Re = \frac{Inertia\ force}{Friction\ force} = \frac{\rho u_{in} H}{\mu} \qquad (3.20)$$

It is observed that Eq. (3.20) encapsulates the three variables, density, dynamic viscosity, and inlet velocity. Through different combinations of these variables, the same hydrodynamic entrance region will be achieved if the resultant Reynolds number is the same. Another important use of the Reynolds number is to indicate whether the flow is laminar or turbulent. For a channel flow, the flow will remain laminar if the critical Reynolds number is below 1400. Further physical interpretation of this non-dimensional parameter is demonstrated in Example 3.9.

3.4   THE ENERGY EQUATION

3.4.1   Energy Conservation

The equation for the conservation of energy is derived from the consideration of the *first law of thermodynamics*:

$$\begin{matrix} \text{Time rate of} \\ \text{change of energy} \end{matrix} = \begin{matrix} \text{Net rate of} \\ \text{heat added} \end{matrix} \left(\sum \dot{Q} \right) + \begin{matrix} \text{Net rate of} \\ \text{work done} \end{matrix} \left(\sum \dot{W} \right) \quad (3.21)$$

As discussed in Appendix A, the time rate of change of any arbitrary variable property ϕ is defined as *the product of the density and the substantial derivative of ϕ*. In keeping with our derivation of the Navier–Stokes (momentum) equations, we refer again to the elemental volume in the Cartesian frame, as described in Figure 3.2. The time rate of change of energy for the moving fluid element is just simply

$$\rho \frac{DE}{Dt} \Delta x \Delta y \Delta z \qquad (3.22)$$

The two terms represented by $\Sigma \dot{Q}$ and $\Sigma \dot{W}$ describe the net rate of heat addition to the fluid within the control volume and the net rate of work done by surface forces on the fluid. We first consider the effects in the x direction, as illustrated in Figure 3.7. The rate of work done and heat added in the y and z directions automatically follow from the analysis of the x direction.

The rate of work done on the control volume in the x direction is equivalent to the product of the surface forces (caused by the normal viscous stress σ_{xx} and tangential viscous stresses τ_{yx} and τ_{zx}) and the velocity component u. The formulae for the net rate of work done in this direction as well as for the other coordinate directions due to the contributions of the normal and tangential surface forces are detailed in Appendix A. When we combine all the

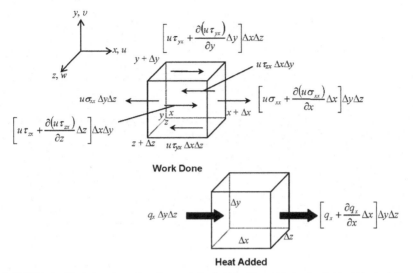

FIGURE 3.7   Work done by surface forces on the fluid and heat added to the fluid within the infinitesimal control volume. Only the fluxes in the x direction are shown.

contributions of the surface forces in the x, y, and z directions, and substitute these expressions along with the time rate of change of energy E, from Eq. (3.22) into Eq. (3.21), the equation for the conservation of energy is given as

$$\rho \frac{DE}{Dt} = \frac{\partial(u\sigma_{xx})}{\partial x} + \frac{\partial(v\sigma_{yy})}{\partial y} + \frac{\partial(w\sigma_{zz})}{\partial z}$$

$$+ \frac{\partial(u\tau_{yx})}{\partial y} + \frac{\partial(u\tau_{zx})}{\partial z} + \frac{\partial(v\tau_{xy})}{\partial x} + \frac{\partial(v\tau_{zy})}{\partial z} + \frac{\partial(w\tau_{xz})}{\partial x} + \frac{\partial(w\tau_{yz})}{\partial y} \qquad (3.23)$$

$$- \frac{\partial q_x}{\partial x} - \frac{\partial q_y}{\partial y} - \frac{\partial q_z}{\partial z}$$

The energy fluxes q_x, q_y, and q_z in Eq. (3.23) can be formulated by applying *Fourier's law of heat conduction*, which relates the heat flux to the local temperature gradient:

$$q_x = -\lambda \frac{\partial T}{\partial x} \quad q_y = -\lambda \frac{\partial T}{\partial y} \quad q_z = -\lambda \frac{\partial T}{\partial z} \qquad (3.24)$$

where λ is the thermal conductivity. By substituting Eq. (3.24) into Eq. (3.23), and applying the normal stresses described in Appendix A, the energy equation becomes

$$\rho \frac{DE}{Dt} = \frac{\partial}{\partial x}\left[\lambda \frac{\partial T}{\partial x}\right] + \frac{\partial}{\partial y}\left[\lambda \frac{\partial T}{\partial y}\right] + \frac{\partial}{\partial z}\left[\lambda \frac{\partial T}{\partial z}\right]$$
$$- \frac{\partial(up)}{\partial x} - \frac{\partial(vp)}{\partial y} - \frac{\partial(wp)}{\partial z} + \Phi$$

$$(3.25)$$

The effects due to the viscous stresses in the energy equation are described by the dissipation function Φ, which can be shown to be

$$\Phi = \frac{\partial(u\tau_{xx})}{\partial x} + \frac{\partial(u\tau_{yx})}{\partial y} + \frac{\partial(u\tau_{zx})}{\partial z}$$

$$+ \frac{\partial(v\tau_{xy})}{\partial x} + \frac{\partial(v\tau_{yy})}{\partial y} + \frac{\partial(v\tau_{zy})}{\partial z} + \frac{\partial(w\tau_{xz})}{\partial x} + \frac{\partial(w\tau_{yz})}{\partial y} + \frac{\partial(w\tau_{zz})}{\partial z}$$

The dissipation function represents a source of energy due to deformation work done on the fluid. This work is extracted from the mechanical energy that causes fluid movement, which is converted into heat.

Thus far, we have not defined the specific energy E of a fluid. Often, the energy of a fluid is defined as the sum of the *internal energy*, *kinetic energy*, and *gravitational potential energy*. We can regard the gravitational force as a body force and we can include the effects of potential energy changes as a source term. For compressible flows, the energy equation is often re-arranged to give an equation for the *enthalpy*. More detailed discussion of the relationship between the specific energy E and enthalpy h can be found in Appendix A as well as in the textbook by Cengel (2003).

Let us consider the special case where the fluid is incompressible and the continuity equation applies. *Neglecting* the kinetic energy, the enthalpy h can be reduced to C_pT, where C_p is the specific heat and is assumed to be constant. Equation (3.25) can be expressed as

$$\rho C_p \frac{DT}{Dt} = \frac{\partial}{\partial x}\left[\lambda \frac{\partial T}{\partial x}\right] + \frac{\partial}{\partial y}\left[\lambda \frac{\partial T}{\partial y}\right] + \frac{\partial}{\partial z}\left[\lambda \frac{\partial T}{\partial z}\right] + \frac{\partial p}{\partial t} + \Phi \qquad (3.26)$$

In most practical fluid-engineering problems, the local time derivative of pressure $\partial p/\partial t$ and the dissipation function Φ can be neglected, and Eq. (3.26) reduces to

$$\rho C_p \underbrace{\frac{DT}{Dt}}_{acceleration} = \underbrace{\frac{\partial}{\partial x}\left[\lambda \frac{\partial T}{\partial x}\right] + \frac{\partial}{\partial y}\left[\lambda \frac{\partial T}{\partial y}\right] + \frac{\partial}{\partial z}\left[\lambda \frac{\partial T}{\partial z}\right]}_{diffusion} \qquad (3.27)$$

For ease of understanding the equation, a two-dimensional form of Eq. (3.27) will be derived. Assuming that the temperature is invariant along the z direction and the thermal conductivity k is constant, the equation for the conservation of energy in two dimensions can be expressed as

$$\underbrace{\frac{\partial T}{\partial t}}_{local\ acceleration} + \underbrace{u\frac{\partial T}{\partial x} + v\frac{\partial T}{\partial y}}_{advection} = \underbrace{\frac{\lambda}{\rho C_p}\frac{\partial^2 T}{\partial x^2} + \frac{\lambda}{\rho C_p}\frac{\partial^2 T}{\partial y^2}}_{diffusion} \qquad (3.28)$$

3.4.2   Physical Interpretation

Physically, Eq. (3.28) defines the temperature of a differential fluid control volume as it travels past a point, taking into consideration the local acceleration derivative (where the temperature in itself may be fluctuating with time at a given point) and also the advection derivative (where the temperature changes spatially from one point to another). To reinforce the physical meaning of these derivatives, imagine you are sitting on a high seat, close to the ceiling in a sauna room, where the buoyant heat flow causes the air to be hottest. You decide to move to a lower seat in the sauna and make your way down to the floor. As you descend, the air is a little cooler and you experience a temperature decrease— this is analogous to the advection derivative in Eq. (3.28). Additionally, as you sit on the lower seat, the sauna door opens as someone enters and you feel a sudden rush of cold air. The temperature around you immediately drops for that moment—this is analogous to the local *acceleration* derivative in Eq. (3.28). The net temperature change you experience is therefore a combination of both the act of descending from the highest seat to the floor and also the rush of cool air as the sauna door opens.

The remaining term in Eq. (3.28) represents the temperature flow due to heat conduction (the diffusion derivative), in which the thermophysical property k is the thermal conductivity of the fluid. We examine the physical interaction of this term with the previous two derivatives. To further reinforce the physical meaning, imagine the problem of the fluid flowing between parallel plates in Figure 3.2 with the plates now heated. If the surrounding fluid velocity is very low, the surrounding fluid temperature within the channel increases due to the heat flowing from the flat plate into the bulk fluid—this is analogous to the heat conduction through the heat diffusion derivative dominating over the local acceleration and advection derivatives of the fluid. However, if the surrounding fluid velocity is high, the heat of the fluid is carried away by the relatively cooler fluid. The high temperatures are found only near the hot surface of the flat plates—this is analogous to the local *acceleration* and *advection* derivatives dominating over the heat diffusion derivative.

Example 3.6

To illustrate the application of the energy equation (3.28), consider the steady heat conduction across an infinite long solid slab with a finite thickness as illustrated in Figure 3.6.1. Determine the analytical expressions based on the boundary conditions:

(a) $x = 0$, $T = T_0$ and $x = L$, $T = T_L$

(b) $x = 0$, $T = T_0$ and $x = L$, $q_L = -k \dfrac{\partial T(L)}{\partial x}$

Solution

Since the heat transfer is steady and it is a solid problem, the acceleration and advection terms in Eq. (3.34) vanish.

$$\underset{=0}{\cancel{\frac{\partial T}{\partial t}}} + \underset{=0}{\cancel{u\frac{\partial T}{\partial x}}} + \underset{=0}{\cancel{v\frac{\partial T}{\partial y}}} = \frac{\lambda}{\rho C_p}\frac{\partial^2 T}{\partial x^2} + \frac{\lambda}{\rho C_p}\frac{\partial^2 T}{\partial y^2}$$

The above equation reduces to

$$\frac{\partial^2 T}{\partial x^2} + \frac{\partial^2 T}{\partial y^2} = 0$$

For an infinitely long slab, we can restrict ourselves to a one-dimensional analysis, and the equation can be further simplified to

$$\frac{\partial^2 T}{\partial x^2} + \underset{=0}{\cancel{\frac{\partial^2 T}{\partial y^2}}} = 0 \qquad \Rightarrow \qquad \frac{\partial^2 T}{\partial x^2} = 0$$

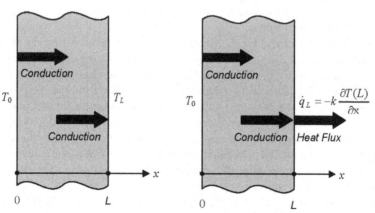

FIGURE 3.6.1   Heat conduction across the solid slab subjected to various temperature boundary conditions.

Integrating the differential equation once with respect to x yields

$$\frac{\partial T}{\partial x} = C_1$$

where C_1 is an arbitrary constant. Subsequent integration of the above equation yields

$$T(x) = C_1 x + C_2$$

which is the analytical expression of the differential equation. The general solution contains two unknown constants, C_1 and C_2. We need two equations, which can be determined through the boundary conditions imposed on the left and right surfaces of the solid slab.

Analysis

(a)   For this case, the left-side boundary condition is applied to the general solution by replacing the x's by zero and $T(x)$ by T_0. In other words,

$$T_0 = C_1 \times 0 + C_2 \quad \Rightarrow \quad C_2 = T_0$$

The right-side boundary condition is applied by replacing the x's by L and $T(x)$ by T_L, which gives

$$T_L = C_1 L + C_2 \quad \Rightarrow \quad T_L C_1 L + T_0 \quad \Rightarrow \quad C_1 = \frac{T_L - T_0}{L}$$

Substituting the C_1 and C_2 expressions into the general solution, we obtain

$$T(x) = \frac{T_L - T_0}{L} x + T_0$$

(b)   For this case, the left-side boundary condition is the same as in Case (a). Thus,

$$C_2 = T_0$$

On the right-side boundary condition, a heat flux $\dot{q}_L$ is specified rather than a specified temperature T_L. Noting that

$$\frac{\partial T}{\partial x} = C_1$$

the application of the boundary condition yields

$$-\lambda \frac{\partial T(L)}{\partial x} = \dot{q}_L \quad \Rightarrow \quad -\lambda C_1 = \dot{q}_L \quad \Rightarrow \quad C_1 = -\frac{\dot{q}_L}{\lambda}$$

Substituting the above equations into the general solution, we obtain

$$T(x) = -\frac{\dot{q}_L}{\lambda} x + T_0$$

Discussion

The above equations are the analytical solutions for the temperature distribution across the finite thickness L in the solid slab. They satisfy not only the one-dimensional partial differential form of the energy equation but also the two specified boundary conditions. Both general solutions exhibit a linear distribution whose slopes are $T_L - T_0/L$ and $-\dot{q}_L/k$, respectively. During the integration process, in establishing the general one-dimensional partial differential form of the energy equation, the thermal diffusivity $\alpha = \lambda/\rho\, C_p$ disappears. This may imply to the reader that the heat conduction across the slab is not influenced by any thermophysical properties, such as the thermal conductivity k, density ρ, and specific heat C_p of the solid material. Nevertheless, in Case (b), the final expression of the general solution demonstrates the influence of the material thermal conductivity that is obtained from the physical boundary condition based on Fourier's law. A better understanding of the types of physical boundary conditions for the temperature is addressed at the end of this chapter.

Example 3.7

Consider a two-dimensional CFD case of the incompressible laminar flow between two stationary parallel plates to illustrate the physical meaning of the energy equation. Using CFD and the dimensions of height $H = 0.1$ m and length $L = 1$ m, with air as the working fluid (density $\rho = 1.2$ kg/m^3), a specified uniform temperature of 330 K at the inlet, and a wall temperature maintained at 300 K, discuss the development of the temperature profiles between the inlet and the *thermal entry length* (L_T) for the following cases:

(a) A fixed inlet velocity $u_{in} = 0.01$ m/s and thermal conductivities $\lambda_1 = 0.04$ W/m · K and $\lambda_2 = 0.01$ W/m · K.

(b) A fixed thermal conductivity $\lambda = 0.04$ W/m · K and inlet velocities $u_{in1} = 0.01$ m/s and $u_{in2} = 0.1$ m/s.

Solution

The problem is described as follows (see Figures 3.7.1):

(a) By definition, the region from the inlet surface to the point at which the thermal boundary layer merges at the centerline is called the *thermal entrance region* and the length of this region is called the *thermal entry length*. Flow in this entrance region is also called the *thermally developing flow,* since the dimensionless temperature profile $(T_s - T)/(T_s - T_m)$, where T_m denotes the mean temperature, is still developing. Beyond this region, the dimensionless temperature profile is fully developed and remains unchanged. Similar to the fluid flow, the region in which the flow is thermally developed, and thus both the velocity and dimensionless temperature profiles remain unchanged, is called *fully developed flow* (see Figure 3.7.1).

Based on the CFD simulation, the dimensionless temperature profiles at different downstream locations x/H from the inlet surface for varying thermal conductivities are given in Figures 3.7.2 and 3.7.3. The former refers to the case where the flow has a higher thermal conductivity than the latter.

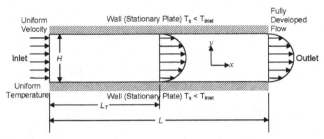

FIGURE 3.7.1 Thermal entry length (L_T) location from the inlet surface in a two-dimensional laminar flow between two stationary parallel plates.

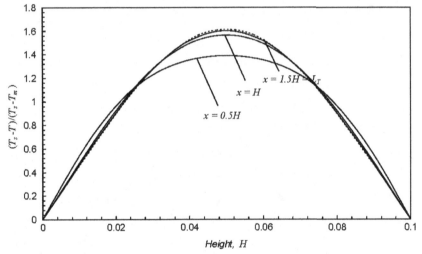

FIGURE 3.7.2 Dimensionless temperature profiles at different locations of x/H for $u_{in} = 0.01$ m/s and thermal conductivity $\lambda_1 = 0.04$ W/m · K.

(b) The temperature profiles at different downstream locations x/H from the inlet surface for varying inlet velocities are given in Figures 3.7.4 and 3.7.5. The former refers to the case where the flow has a lower velocity than the latter. For comparison purposes, the case presented in Figure 3.7.4 is of course exactly the same as the case presented in Figure 3.7.2 and is repeated here against the higher-velocity case.

Discussion

As in Example 3.5 for the momentum equations, we can rewrite the steady two-dimensional energy equation, Eq. (3.34), as follows:

$$\underbrace{u\frac{\partial T}{\partial x} + v\frac{\partial T}{\partial y}}_{advection} = \underbrace{\frac{\lambda}{\rho C_p}\left(\frac{\partial^2 T}{\partial x^2} + \frac{\partial^2 T}{\partial y^2}\right)}_{diffusion}$$

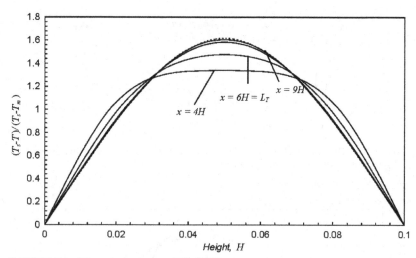

FIGURE 3.7.3   Dimensionless temperature profiles at different locations of x/H for $u_{in} = 0.01$ m/s and thermal conductivity $\lambda_2 = 0.01$ W/m · K.

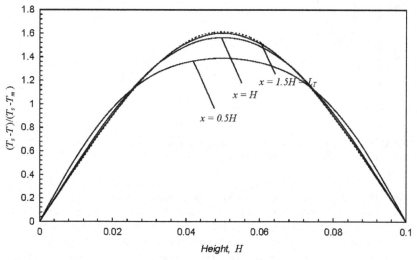

FIGURE 3.7.4   Dimensionless temperature profiles at different locations of x/H for $u_{in1} = 0.01$ m/s and thermal conductivity $\lambda = 0.04$ W/m · K.

The competition between the advection and diffusion in the energy equation is highlighted from the above results through the variation of the thermal conductivities and inlet velocities. For the case of a higher thermal conductivity, the diffusion term dominates. This leads to a shorter thermal entry length (i.e., the fully developed stage is reached at a shorter distance) due to a quicker development of the

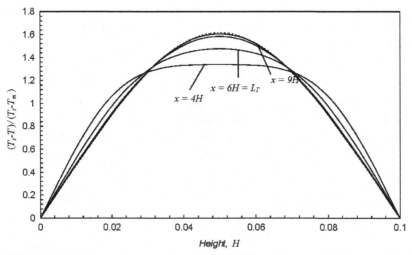

FIGURE 3.7.5 Dimensionless temperature profiles at different locations of x/H for $u_{in2} = 0.1$ m/s and thermal conductivity $k = 0.04$ W/m $\cdot$ K.

thermal boundary layer. This appears to have the same effect as that for the case of a higher friction force that inhibits the fluid momentum. However, a lower thermal conductivity induces less resistance to the heat flow, thus allowing a slower development of the thermal boundary layer, and the flow reaches the fully developed stage at a longer distance. A higher inlet velocity culminates in a higher advection overcoming the diffusion effect of the heat flow, resulting in a longer development of the thermal entry length. It is noted that the thermal entry lengths for all the cases in the current example correspond exactly to the same hydrodynamic entry lengths obtained in Example 3.5. More explanation of the relationship between these hydrodynamic and thermal behaviors is given in the next section. All the cases investigated demonstrate the relative physical contributions of the advection and diffusion effects competing with each other to conserve the energy transfer.

3.4.3 Comments

Here, the principle of conservation of energy, as with the conservation of momentum, has been explored with investigations of the various contributions of the advection and diffusion terms in the energy equation. In Example 3.7, there also appears to be some homogeneity for the thermal behavior that results in the combination of the various flow variables. The dimensionless parameter previously introduced by Eq. (3.20) in Section 3.3.3., the Reynolds number (Re), depicts the contribution between the inertia force and the friction force, which describes the flow characteristics. Here, we introduce another important dimensionless parameter, the Prandtl number (*Pr*), which denotes the ratio between the *molecular diffusivity of momentum* and the *molecular diffusivity of heat*, i.e.,

$$Pr = \frac{Molecular \; diffusivity \; of \; momentum}{Molecular \; diffusivity \; of \; heat} = \frac{\nu}{\alpha} = \frac{\mu C_p}{\lambda} \qquad (3.29)$$

During laminar flow, the magnitude of the dimensionless Prandtl number is a measure of the relative growth of the velocity and thermal boundary layers. For fluids with $Pr \approx 1$, such as gases, the two boundary layers essentially coincide with each other. From Example 3.7, if the specific heat of air is taken to be 1000 J/kg, various combinations of the dynamic viscosity and thermal conductivity of air can enable the Prandtl number to be unity. Here, the two boundary layers coincide with each other and the same hydrodynamic and thermal entry lengths are obtained. For fluids with $Pr \gg 1$, such as water or oils, the velocity boundary layer outgrows the thermal boundary layer, while the opposite is true for fluids with $Pr \ll 1$, such as liquid metals. In thermal flow problems, we can therefore infer that the heat-transfer characteristics are controlled by the combination of the Prandtl number and Reynolds number (representing the advection term). Further interpretation of the Prandtl number is demonstrated in Example 3.9.

3.5   THE ADDITIONAL EQUATIONS FOR TURBULENT FLOW

3.5.1   What Is Turbulence?

Many, if not most, flows of engineering significance are turbulent. The turbulent flow regime is, therefore, not just of theoretical interest among academics but a problematic source for engineers who need to capture the effects of turbulence in solving everyday problems.

Flows in the laminar regime are completely described by the continuity and momentum equations as aforementioned. In simple cases, they can be solved analytically (Examples 3.1, 3.3, and 3.4). More complex flows may have to be tackled numerically with CFD techniques. It is well known that small disturbances associated with disturbances in the fluid streamlines of a laminar flow can eventually lead to a chaotic and random state of motion—a turbulent condition. These disturbances may originate from the free stream of the fluid motion, or they may be induced by the surface roughness, where they may be amplified in the direction of flow, in which case turbulence will occur. The onset of turbulence depends on the ratio of inertia force to viscous force, which is indicated by the Reynolds number, Eq. (3.20). At a low Reynolds number, inertia forces are smaller than viscous forces. The naturally occurring disturbances are dissipated away and the flow remains laminar. At a high Reynolds number, the inertia forces are sufficiently large to amplify the disturbances, and a transition to turbulence occurs. Here, the motion becomes intrinsically unstable even with constant imposed boundary conditions. The velocity and all other flow properties are varying in a random and chaotic way.

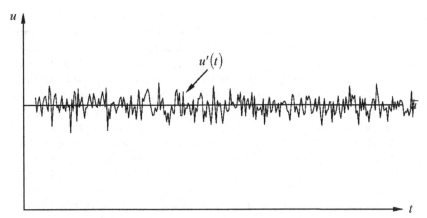

FIGURE 3.8   Velocity fluctuating with time at some point in a turbulent flow.

Turbulence is associated with the existence of *random fluctuations* in the fluid (see Figure 3.6). This behavior can be exemplified by a typical point velocity measurement as a function of time at some location in the turbulent flow, as shown in Figure 3.8. The random nature of flow precludes computations based on the equations that describe the fluid motion. Although the conservation equations remain applicable, the dependent variable, such as the transient velocity distribution in Figure 3.8, must be interpreted as an instantaneous velocity—a phenomenon that is impossible to predict, as the fluctuating velocity occurs randomly with time. Instead, the velocity can be decomposed into a steady mean value $\bar{u}$ with a fluctuating component $u'(t)$ superimposed on it: $u(t) = \bar{u} + u'(t)$. In general, it is most attractive to characterize a turbulent flow by the mean values of flow properties ($\bar{u}$, $\bar{v}$, $\bar{w}$, $\bar{p}$, etc.) with its corresponding statistical fluctuating property (u', v', w', p', etc.).

Turbulent fluctuations always have a *three-dimensional* spatial character. Visualizations of turbulent flows have revealed rotational flow structures, so-called turbulent eddies, with a wide range of length and velocity scales called turbulent scales. The largest eddies have a characteristic velocity and a characteristic length of the same order as the velocity and length scale of the mean flow. This suggests that, for turbulent flows where $ul \gg v$ (i.e., inertia effects dominate viscous effects), the largest eddies whose scales are comparable with the mean flow are dominated by inertia effects rather than viscous effects. The large eddies are, therefore, effectively inviscid. Transport of eddies is attained by the extraction of energy from the mean flow by a process called vortex stretching. The presence of velocity gradients in the mean flow causes deformation of the fluid, such as shear and linear strain and rotation, which "stretch" eddies that are appropriately aligned by forcing one end of the eddies to move faster than the other. During vortex stretching, the angular momentum is conserved and the stretching work done by the mean flow on the large eddies provides the energy that maintains the turbulence. These larger eddies then breed new instabilities, creating smaller eddies that are transported mainly by vortex stretching from the larger eddy rather than from the mean flow. Thus the energy is handed down from the larger eddy to the smaller eddy. This process continues until the eddies become so small that viscous effects become important (the eddy length scales $ul \ll v$). Work is performed against the action of the viscous stresses, so that the energy associated with the eddy motions is dissipated and converted into thermal internal energy. The continual transfer of energy from the larger eddy to smaller and smaller eddies is termed *energy cascade*. Larger eddies are flow dependent as they are generated from mean flow characteristics; thus their turbulent scales are large compared with viscosity, causing the structure of the eddy to be highly *anisotropic* (i.e., varying in all directions). Small eddies have much smaller turbulent scales (with scales up to the order of 10^{-4}) compared with viscosity causing the flow to be *isotropic* since the diffusive effect of viscosity dominates and smears out the directionality of the flow structure.

3.5.2   *k-ε* Two-Equation Turbulence Model

There is a crucial difference when modeling the physical phenomena between laminar and turbulent flow. For the latter, the appearance of turbulence eddies occurs over a wide range of length scales. A typical flow domain having a cross-sectional area of 0.1 m by 0.1 m with a high Reynolds number turbulent flow might contain eddies down to 10 or 100 μm in size. In order to describe the flow processes at all length scales, we would require computing meshes of 10^9 to 10^{12} grid points. Also, the fastest events can take place with a frequency of the order 10 kHz, which we would need to discretize time into steps of about 100 μs.

With present-day computing power, the computing requirements for a direct numerical solution (DNS) of the time-dependent Navier–Stokes equations of fully turbulent flows at high Reynolds numbers are still truly phenomenal. Meanwhile, engineers require computational procedures that can supply adequate information about the turbulent processes, but wish to avoid the need to predict all the effects associated with each and every eddy in the flow. This category of CFD users is almost always satisfied with information about the time-averaged properties of the flow (mean velocities, mean pressures, mean stresses, etc.). Since engineers are content to focus their attention on mean quantities, by adopting a suitable time-averaging operation on the momentum equations, we are able to discard all details concerning the state of the flow contained in the instantaneous fluctuations. This process of obtaining mean quantities is applied to the incompressible, two-dimensional equations of continuity and the *conservative* form of momentum and energy that produces the time-averaged governing equations, more popularly known as the Reynolds-averaged Navier–Stokes (RANS) equations, yields

$$\frac{\partial \bar{u}}{\partial x} + \frac{\partial \bar{v}}{\partial y} = 0 \qquad (3.30)$$

$$\frac{\partial \bar{u}}{\partial t} + \frac{\partial (\bar{u}\bar{u})}{\partial x} + \frac{\partial (\bar{v}\bar{u})}{\partial y} = -\frac{1}{\rho}\frac{\partial \bar{p}}{\partial x} + \frac{\partial}{\partial x}\left(v\frac{\partial \bar{u}}{\partial x}\right) + \frac{\partial}{\partial y}\left(v\frac{\partial \bar{u}}{\partial y}\right)$$

$$+\frac{\partial}{\partial x}\left[v\frac{\partial \bar{u}}{\partial x}\right] + \frac{\partial}{\partial y}\left[v\frac{\partial \bar{v}}{\partial x}\right] - \left[\frac{\partial (\overline{u'u'})}{\partial x} + \frac{\partial (\overline{u'v'})}{\partial y}\right]$$

$$(3.31)$$

$$\frac{\partial v}{\partial t} + \frac{\partial (\bar{u}\bar{v})}{\partial x} + \frac{\partial (\bar{v}\bar{v})}{\partial y} = -\frac{1}{\rho}\frac{\partial \bar{p}}{\partial y} + \frac{\partial}{\partial x}\left(v\frac{\partial \bar{v}}{\partial x}\right) + \frac{\partial}{\partial y}\left(v\frac{\partial \bar{v}}{\partial y}\right)$$

$$+\frac{\partial}{\partial x}\left[v\frac{\partial \bar{u}}{\partial y}\right] + \frac{\partial}{\partial y}\left[v\frac{\partial \bar{v}}{\partial y}\right] - \left[\frac{\partial (\overline{u'v'})}{\partial x} + \frac{\partial (\overline{v'v'})}{\partial y}\right]$$

$$(3.32)$$

$$\frac{\partial \bar{T}}{\partial t} + \frac{\partial (\bar{u}\bar{T})}{\partial x} + \frac{\partial (\bar{v}\bar{T})}{\partial y} = \frac{\partial}{\partial x}\left(\frac{k}{\rho C_p}\frac{\partial \bar{T}}{\partial x}\right) + \frac{\partial}{\partial y}\left(\frac{k}{\rho C_p}\frac{\partial \bar{T}}{\partial y}\right) - \left[\frac{\partial \overline{u'T'}}{\partial x} + \frac{\partial \overline{v'T'}}{\partial y}\right]$$

$$(3.33)$$

where $\bar{u}$, $\bar{v}$, $\bar{p}$, and $\bar{T}$ are mean values and u', v', p', and T' are turbulent fluctuations. The term $k/\rho C_p$ in Eq. (3.33) is the thermal diffusivity α of the fluid. The equations above are similar to those formulated for laminar flows, except for the presence of additional terms of the form $\overline{a'b'}$. As a result, we have three additional unknowns (in three dimensions, we have nine additional unknowns), known as the Reynolds stresses, in the time-averaged momentum equations. Similarly, the time-averaged temperature equation shows extra terms $\overline{u'T'}$ and $\overline{v'T'}$ (in three dimensions, we have an extra term $\overline{w'T'}$).

The time-averaged equations can be solved if the Reynolds stresses and extra temperature transport terms can be related to the mean flow and heat quantities. It was proposed by Boussinesq (1868) that the Reynolds stresses could be linked to the mean rates of deformation. We obtain

$$-\rho\overline{u'u'} = 2\mu_T\frac{\partial \bar{u}}{\partial x} - \frac{2}{3}\rho k \qquad -\rho\overline{v'v'} = 2\mu_T\frac{\partial \bar{v}}{\partial y} - \frac{2}{3}\rho k$$

$$-\rho\overline{u'v'} = \mu_T\left(\frac{\partial \bar{v}}{\partial x} + \frac{\partial \bar{u}}{\partial y}\right)$$

$$(3.34)$$

The right-hand side is analogous to *Newton's law of viscosity*, except for the appearance of turbulent or eddy viscosity μ_T and turbulent kinetic energy k.

In Eq. (3.40), the turbulent momentum transport is assumed to be proportional to the mean gradients of velocity. Similarly, the turbulent transport of temperature is taken to be proportional to the gradient of the mean value of the transported quantity. In other words,

$$-\rho\overline{u'T'} = \Gamma_T \frac{\partial \bar{T}}{\partial x} \quad -\rho\overline{v'T'} = \Gamma_T \frac{\partial \bar{T}}{\partial y} \tag{3.35}$$

where Γ_T is the turbulent diffusivity. Since the turbulent transport of momentum and heat is due to the same mechanisms—eddy mixing—the value of the turbulent viscosity can be taken to be close to that of the turbulent viscosity μ_T. Based on the definition of the turbulent Prandtl number Pr_T, we obtain

$$Pr_T = \frac{\mu_T}{\Gamma_T}$$

Experiments have established that this ratio is often nearly constant. Most CFD procedures assume this to be the case and use values of Pr_T around unity.

Since the complexity of turbulence in most engineering flow problems precludes the use of any simple formulae, it is possible to develop similar transport equations to accommodate the turbulent quantity k and other turbulent quantities, one of which is the rate of dissipation of turbulent energy ε. Here we indicate the form of a typical two-equation turbulence model that is commonly used in handling many turbulent fluid-engineering problems, the *standard k-ε model* by Launder and Spalding (1974).

Some preliminary definitions are required first. The turbulent kinetic energy k and rate of dissipation of turbulent energy ε can be defined and expressed in Cartesian tensor notation as

$$k = \frac{1}{2}u_i'u_i' \text{ and } \varepsilon = v_T \overline{\left(\frac{\partial u_i'}{\partial x_j}\right)\left(\frac{\partial u_i'}{\partial x_j}\right)} \text{ where } i,j = 1,2,3$$

From the local values of k and ε, a local turbulent viscosity μ_T can be evaluated as

$$\mu_T = \frac{C_\mu \rho k^2}{\varepsilon} \tag{3.36}$$

and the kinematic turbulent or eddy viscosity is denoted by $v_T = \mu_T/\rho$.

By substituting the Reynolds stress expressions in Eq. (3.20) and the extra temperature transport terms in Eq. (3.35) into the governing Eqs. (3.30), (3.31), (3.32), and (3.33), and by removing the overbar that by default indicates average quantities, we obtain

$$\frac{\partial u}{\partial x} + \frac{\partial v}{\partial y} = 0 \tag{3.37}$$

$$\frac{\partial u}{\partial t} + \frac{\partial(uu)}{\partial x} + \frac{\partial(vu)}{\partial y} = -\frac{1}{\rho}\frac{\partial p}{\partial x} + \frac{\partial}{\partial x}\left[(v+v_T)\frac{\partial u}{\partial x}\right] + \frac{\partial}{\partial y}\left[(v+v_T)\frac{\partial u}{\partial y}\right]$$
$$+ \frac{\partial}{\partial x}\left[(v+v_T)\frac{\partial u}{\partial x}\right] + \frac{\partial}{\partial y}\left[(v+v_T)\frac{\partial v}{\partial x}\right] \tag{3.38}$$

$$\frac{\partial v}{\partial t} + \frac{\partial(uv)}{\partial x} + \frac{\partial(vv)}{\partial y} = -\frac{1}{\rho}\frac{\partial p}{\partial y} + \frac{\partial}{\partial x}\left[(v+v_T)\frac{\partial v}{\partial x}\right] + \frac{\partial}{\partial y}\left[(v+v_T)\frac{\partial v}{\partial y}\right]$$

$$+ \frac{\partial}{\partial x}\left[(v+v_T)\frac{\partial u}{\partial y}\right] + \frac{\partial}{\partial y}\left[(v+v_T)\frac{\partial v}{\partial y}\right] \tag{3.39}$$

$$\frac{\partial T}{\partial t} + \frac{\partial(uT)}{\partial x} + \frac{\partial(vT)}{\partial y} = \frac{\partial}{\partial x}\left[\left(\frac{v}{Pr}+\frac{v_T}{Pr_T}\right)\frac{\partial T}{\partial x}\right] + \frac{\partial}{\partial y}\left[\left(\frac{v}{Pr}+\frac{v_T}{Pr_T}\right)\frac{\partial T}{\partial y}\right]$$

$$\tag{3.40}$$

The term v/Pr appearing in the temperature equation, Eq. (3.40), is obtained from the definition of the laminar Prandtl number, which is already defined in Eq. (3.29) as $Pr = v/\alpha$ where $\alpha = k/\rho C_p$. Interestingly, the time-averaged equations above have the same form as those developed for the laminar equations except for the additional turbulent viscosity found in the diffusion and non-pressure-gradient terms for the momentum equations and also found in the diffusion term for the energy equation. Hence, the solution to turbulent flow in engineering problems entails greater diffusion than is imposed by the turbulent nature of the fluid flow.

The *non-conservative* governing equations can also be similarly derived, resulting in

$$\frac{\partial u}{\partial x} + \frac{\partial v}{\partial y} = 0 \tag{3.41}$$

$$\frac{\partial u}{\partial t} + u\frac{\partial u}{\partial x} + v\frac{\partial u}{\partial y} = -\frac{1}{\rho}\frac{\partial p}{\partial x} + \frac{\partial}{\partial x}\left[(v+v_T)\frac{\partial u}{\partial x}\right] + \frac{\partial}{\partial y}\left[(v+v_T)\frac{\partial u}{\partial y}\right]$$

$$+ \frac{\partial}{\partial x}\left[(v+v_T)\frac{\partial u}{\partial x}\right] + \frac{\partial}{\partial y}\left[(v+v_T)\frac{\partial v}{\partial x}\right] \tag{3.42}$$

$$\frac{\partial v}{\partial t} + u\frac{\partial v}{\partial x} + v\frac{\partial v}{\partial y} = -\frac{1}{\rho}\frac{\partial p}{\partial y} + \frac{\partial}{\partial x}\left[(v+v_T)\frac{\partial v}{\partial x}\right] + \frac{\partial}{\partial y}\left[(v+v_T)\frac{\partial v}{\partial y}\right]$$

$$+ \frac{\partial}{\partial x}\left[(v+v_T)\frac{\partial u}{\partial y}\right] + \frac{\partial}{\partial y}\left[(v+v_T)\frac{\partial v}{\partial y}\right] \tag{3.43}$$

$$\frac{\partial T}{\partial t} + u\frac{\partial T}{\partial x} + v\frac{\partial T}{\partial y} = \frac{\partial}{\partial x}\left[\left(\frac{v}{Pr}+\frac{v_T}{Pr_T}\right)\frac{\partial T}{\partial x}\right] + \frac{\partial}{\partial y}\left[\left(\frac{v}{Pr}+\frac{v_T}{Pr_T}\right)\frac{\partial T}{\partial y}\right] \tag{3.44}$$

Additional differential transport equations that are required for the standard k-ε model, for the case of a constant fluid property and expressed in non-conservation form, are

$$\frac{\partial k}{\partial t} + u\frac{\partial k}{\partial x} + v\frac{\partial k}{\partial y} = \frac{\partial}{\partial x}\left(\frac{v_T}{\sigma_k}\frac{\partial k}{\partial x}\right) + \frac{\partial}{\partial y}\left(\frac{v_T}{\sigma_k}\frac{\partial k}{\partial y}\right) + P - D \qquad (3.45)$$

$$\frac{\partial \varepsilon}{\partial t} + u\frac{\partial \varepsilon}{\partial x} + v\frac{\partial \varepsilon}{\partial y} = \frac{\partial}{\partial x}\left(\frac{v_T}{\sigma_\varepsilon}\frac{\partial \varepsilon}{\partial x}\right) + \frac{\partial}{\partial y}\left(\frac{v_T}{\sigma_\varepsilon}\frac{\partial \varepsilon}{\partial y}\right) + \frac{\varepsilon}{k}(C_{\varepsilon 1}P - C_{\varepsilon 2}D) \qquad (3.46)$$

where the production term P is formulated as

$$P = 2v_T\left[\left(\frac{\partial u}{\partial x}\right)^2 + \left(\frac{\partial v}{\partial y}\right)^2\right] + v_T\left(\frac{\partial u}{\partial y} + \frac{\partial v}{\partial x}\right)^2$$

and the destruction term D is given by ε. The physical significance of the above equations is that the rate of change and the advection transport of k or ε equals the diffusion transport combined with the rate of production and destruction of k or ε. The equations contain five adjustable constants, C_μ, σ_k, σ_ε, $C_{\varepsilon 1}$, and $C_{\varepsilon 2}$. These constants have been arrived at by comprehensive data fitting for a wide range of turbulent flows (Launder and Spalding, 1974):

$$C_\mu = 0.09, \quad \sigma_k = 1.0, \quad \sigma_\varepsilon = 1.3, \quad C_{\varepsilon 1} = 1.44, \quad C_{\varepsilon 2} = 1.92.$$

The production and destruction of turbulent kinetic energy are always closely linked in the k-equation, Eq. (3.45). The dissipation rate ε is large where the production of k is large. The model equation, Eq. (3.45), assumes that the production and destruction terms are proportional to the production and destruction terms of the k-equation. Adoption of such terms ensures that ε increases rapidly if k increases rapidly and that it decreases sufficiently fast to avoid non-physical (negative) values of turbulent kinetic energy if k decreases. The factor ε/k in the production and destruction terms makes these terms dimensionally correct in the ε-equation, Eq. (3.46).

Example 3.8

Consider the two-dimensional CFD case of the flow between two stationary parallel plates to demonstrate the laminar and turbulent nature of the fluid flow (see Figure 3.8.1). Using CFD with the dimensions of height $H = 0.1$ m and length $L = 10$ m of the channel, observe the velocity and viscosity profiles in the fully developed region with the working fluid taken as air (density $\rho = 1.2$ kg/m^3 and dynamic viscosity $\mu = 2 \times 10^{-5}$ kg/m · s) for inlet velocities of $u_{in1} = 0.02$ m/s and $u_{in2} = 1$ m/s.

Solution

The problem is described as follows (see Figure 3.8.1):

For an inlet velocity of $u_{in1} = 0.02$ m/s, the Reynolds number determined from Eq. (3.26) is given as 120. The flow is laminar. However, for an inlet velocity

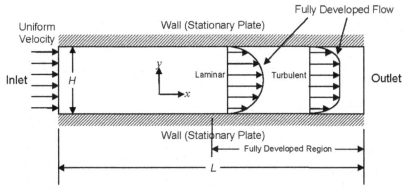

FIGURE 3.8.1   Illustration of a laminar and turbulent flow in a two-dimensional fluid flow between two stationary parallel plates.

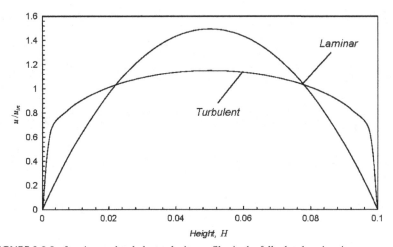

FIGURE 3.8.2   Laminar and turbulent velocity profiles in the fully developed region.

of $u_{in2} = 1$ m/s, the Reynolds number is calculated to be 6000. This Reynolds number is well above the critical Reynolds number of 1400; hence the flow is turbulent.

Based on the CFD simulation, the laminar and turbulent velocity profiles in the fully developed region are given in Figure 3.8.2. The laminar (dynamic) and turbulent viscosities corresponding to the same location of the velocity profiles are shown in Figure 3.8.3.

Discussion

It is observed that there is a significant difference in the flow structure for the laminar and turbulent flow regimes. In the fully developed region, the velocity profile for laminar flow is parabolic, but the velocity profile for turbulent flow is rather blunt, primarily at the center of the channel, with a steep velocity gradient near the wall surface. The turbulent velocity profile being blunt or flat at the center is due to the effect of the high turbulent viscosity diffusing the flow in the momentum

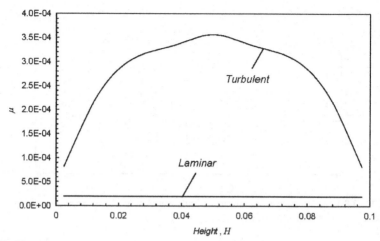

FIGURE 3.8.3   Laminar and turbulent viscosities in the fully developed region.

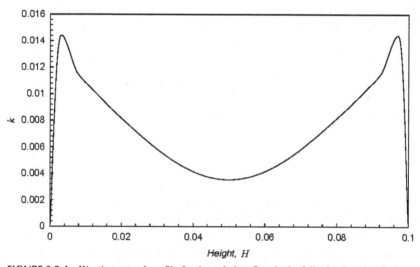

FIGURE 3.8.4   Kinetic energy k profile for the turbulent flow in the fully developed region.

equations (this is exemplified in Figure 3.8.3, where the turbulent viscosity is significantly higher by orders of magnitude than the laminar [dynamic] viscosity at the center of the channel). To further illustrate the physical characteristics of the turbulent viscosity that is derived in Eq. (3.36), profiles of the turbulent kinetic energy and dissipation along the height of the channel at the same location of the velocity profiles are plotted, which are respectively depicted in Figures 3.8.4 and 3.8.5. The results show that the turbulent kinetic energy k and dissipation rate ε peak near the walls. Away from the wall and toward the center of the channel, the kinetic energy decreases, but at a slower rate than the dissipation rate. Even though the kinetic energy is high near the wall, the dissipation rate is much higher, thereby resulting

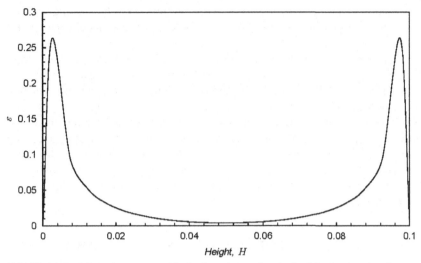

FIGURE 3.8.5   Dissipation rate ε profile for the turbulent flow in the fully developed region.

in a lower turbulent viscosity. At the center, the kinetic energy is higher than the dissipation rate, which indicates a greater generation of turbulence, thereby resulting in a higher turbulent viscosity.

3.5.3   Comments

The two-equation k-ε model is the most widely used and validated turbulence model. The model's performance has been assessed against a number of practical flows. It has achieved notable successes in predicting thin shear layers, boundary layers, and duct flows without the need for case-by-case adjustment of the model constants. It has also been shown to perform extremely well where the Reynolds shear stresses are important in confined flows. This accommodates a wide range of flows with industrial engineering applications, which explains the model's popularity among CFD users. Extension of the model to incorporate the buoyancy effects has led to application of the model to study environmental flows, such as pollutant dispersion in the atmosphere and in lakes and modeling of enclosure fires.

Despite the many successful applications in handling industrial problems, the standard k-ε model demonstrates only moderate agreement when predicting unconfined flows. The weakness of the model is particularly amplified for weak shear layers—far-wake and mixing-layer unconfined separated flows. Also, for the case of the axisymmetric jets in stagnant surroundings, the spreading rate is severely over-predicted where, in major parts of these flows, the rate of production of the turbulent kinetic energy is much less than the rate of dissipation. The difficulties can, however, be overcome by making ad hoc adjustments to the model constants, thereby reducing the model's generality and robustness.

Numerous problems are also experienced with the model in predicting swirling flows and flows with large, rapid, extra strains (for example, highly curved boundary layers and diverging passages), since the model is unable to fully describe the subtle effects of the streamline curvature on turbulence. One major weakness of the standard k-ε model is the assumption of an *isotropic* eddy viscosity. Owing to the deficiencies of the treatment of the normal stresses, secondary flows that exist in long non-circular ducts, which are driven by anisotropic normal Reynolds stresses, cannot be predicted. Finally, the model is oblivious to body forces due to rotation of the frame of reference.

3.6   GENERIC FORM OF THE GOVERNING EQUATIONS FOR CFD

From the governing equations derived above for either the laminar or turbulent conditions, there are significant commonalities between the various equations. Here we present the three-dimensional form of the governing equations for the conservation of mass, momentum, energy, and the turbulent quantities. If we introduce a general variable ϕ and express all the fluid-flow equations in the conservative *incompressible* form, the equation can be written as

$$\frac{\partial \phi}{\partial t} + \frac{\partial (u\phi)}{\partial x} + \frac{\partial (v\phi)}{\partial y} + \frac{\partial (w\phi)}{\partial z} = \frac{\partial}{\partial x}\left[\Gamma\frac{\partial \phi}{\partial x}\right] + \frac{\partial}{\partial y}\left[\Gamma\frac{\partial \phi}{\partial y}\right] + \frac{\partial}{\partial z}\left[\Gamma\frac{\partial \phi}{\partial z}\right] + S_\phi$$

(3.47)

while in the conservative *compressible* form, the equation is given by

$$\frac{\partial (\rho\phi)}{\partial t} + \frac{\partial (\rho u\phi)}{\partial x} + \frac{\partial (\rho v\phi)}{\partial y} + \frac{\partial (\rho w\phi)}{\partial z} = \frac{\partial}{\partial x}\left[\Gamma\frac{\partial \phi}{\partial x}\right] + \frac{\partial}{\partial y}\left[\Gamma\frac{\partial \phi}{\partial y}\right]$$
$$+ \frac{\partial}{\partial z}\left[\Gamma\frac{\partial \phi}{\partial z}\right] + S_\phi \qquad (3.48)$$

Equations (3.47) and (3.48) are the so-called transport equations for the property ϕ. Each of them illustrates the various physical transport processes occurring in the fluid flow: The local acceleration and advection terms on the left-hand side are respectively equivalent to the diffusion term (Γ = diffusion coefficient) and the *source* term (S_ϕ) on the right-hand side. Tables 3.1 and 3.2 present the governing equations for incompressible and compressible flows in the Cartesian framework. In order to bring forth the common features, we have, of course, combined the terms that are not shared between the equations inside the source terms. It is noted that the additional source terms in the

TABLE 3.1 The governing equations for incompressible flow in Cartesian coordinates

Mass Conservation Equation

(m) $\dfrac{\partial u}{\partial x} + \dfrac{\partial v}{\partial y} + \dfrac{\partial w}{\partial z} = 0$

Momentum Equations

(M_x) $\dfrac{\partial u}{\partial t} + \dfrac{\partial (uu)}{\partial x} + \dfrac{\partial (vu)}{\partial y} + \dfrac{\partial (wu)}{\partial z} = \dfrac{\partial}{\partial x}\left[(v+v_T)\dfrac{\partial u}{\partial x}\right] + \dfrac{\partial}{\partial y}\left[(v+v_T)\dfrac{\partial u}{\partial y}\right] + \dfrac{\partial}{\partial z}\left[(v+v_T)\dfrac{\partial u}{\partial z}\right] + \left(S_u = -\dfrac{1}{\rho}\dfrac{\partial p}{\partial x} + S'_u\right)$

(M_y) $\dfrac{\partial v}{\partial t} + \dfrac{\partial (uv)}{\partial x} + \dfrac{\partial (vv)}{\partial y} + \dfrac{\partial (wv)}{\partial z} = \dfrac{\partial}{\partial x}\left[(v+v_T)\dfrac{\partial v}{\partial x}\right] + \dfrac{\partial}{\partial y}\left[(v+v_T)\dfrac{\partial v}{\partial y}\right] + \dfrac{\partial}{\partial z}\left[(v+v_T)\dfrac{\partial v}{\partial z}\right] + \left(S_v = -\dfrac{1}{\rho}\dfrac{\partial p}{\partial y} + S'_v\right)$

(M_z) $\dfrac{\partial w}{\partial t} + \dfrac{\partial (uw)}{\partial x} + \dfrac{\partial (vw)}{\partial y} + \dfrac{\partial (ww)}{\partial z} = \dfrac{\partial}{\partial x}\left[(v+v_T)\dfrac{\partial w}{\partial x}\right] + \dfrac{\partial}{\partial y}\left[(v+v_T)\dfrac{\partial w}{\partial y}\right] + \dfrac{\partial}{\partial z}\left[(v+v_T)\dfrac{\partial w}{\partial z}\right] + \left(S_w = -\dfrac{1}{\rho}\dfrac{\partial p}{\partial z} + S'_w\right)$

Energy Equation

(E) $\dfrac{\partial T}{\partial t} + \dfrac{\partial (uT)}{\partial x} + \dfrac{\partial (vT)}{\partial y} + \dfrac{\partial (wT)}{\partial z} = \dfrac{\partial}{\partial x}\left[\left(\dfrac{v}{Pr} + \dfrac{v_T}{Pr_T}\right)\dfrac{\partial T}{\partial x}\right] + \dfrac{\partial}{\partial y}\left[\left(\dfrac{v}{Pr} + \dfrac{v_T}{Pr_T}\right)\dfrac{\partial T}{\partial y}\right] + \dfrac{\partial}{\partial z}\left[\left(\dfrac{v}{Pr} + \dfrac{v_T}{Pr_T}\right)\dfrac{\partial T}{\partial z}\right] + S_T$

Turbulence Equations

(k) $\dfrac{\partial k}{\partial t} + \dfrac{\partial (uk)}{\partial x} + \dfrac{\partial (vk)}{\partial y} + \dfrac{\partial (wk)}{\partial z} = \dfrac{\partial}{\partial x}\left[\dfrac{v_T}{\sigma_k}\dfrac{\partial k}{\partial x}\right] + \dfrac{\partial}{\partial y}\left[\dfrac{v_T}{\sigma_k}\dfrac{\partial k}{\partial y}\right] + \dfrac{\partial}{\partial z}\left[\dfrac{v_T}{\sigma_k}\dfrac{\partial k}{\partial z}\right] + (S_k = P - D)$

(ε) $\dfrac{\partial \varepsilon}{\partial t} + \dfrac{\partial (u\varepsilon)}{\partial x} + \dfrac{\partial (v\varepsilon)}{\partial y} + \dfrac{\partial (w\varepsilon)}{\partial z} = \dfrac{\partial}{\partial x}\left[\dfrac{v_T}{\sigma_\varepsilon}\dfrac{\partial \varepsilon}{\partial x}\right] + \dfrac{\partial}{\partial y}\left[\dfrac{v_T}{\sigma_\varepsilon}\dfrac{\partial \varepsilon}{\partial y}\right] + \dfrac{\partial}{\partial z}\left[\dfrac{v_T}{\sigma_\varepsilon}\dfrac{\partial \varepsilon}{\partial z}\right] + \left(S_\varepsilon = \dfrac{\varepsilon}{k}(C_{\varepsilon 1}P - C_{\varepsilon 2}D)\right)$

where $P = 2v_T\left[\left(\dfrac{\partial u}{\partial x}\right)^2 + \left(\dfrac{\partial v}{\partial y}\right)^2 + \left(\dfrac{\partial w}{\partial z}\right)^2\right] + v_T\left[\left(\dfrac{\partial u}{\partial y} + \dfrac{\partial v}{\partial x}\right)^2 + \left(\dfrac{\partial v}{\partial z} + \dfrac{\partial w}{\partial y}\right)^2 + \left(\dfrac{\partial w}{\partial x} + \dfrac{\partial u}{\partial z}\right)^2\right]$ and $D = \varepsilon$

TABLE 3.2 The governing equations for compressible flow in Cartesian coordinates

Mass Conservation Equation

(m) $\dfrac{\partial \rho}{\partial t} + \dfrac{\partial(\rho u)}{\partial x} + \dfrac{\partial(\rho v)}{\partial y} + \dfrac{\partial(\rho w)}{\partial z} = 0$

Momentum Equations

$(M_x)\ \dfrac{\partial(\rho u)}{\partial t} + \dfrac{\partial(\rho u u)}{\partial x} + \dfrac{\partial(\rho v u)}{\partial y} + \dfrac{\partial(\rho w u)}{\partial z} = \dfrac{\partial}{\partial x}\left[(\mu+\mu_T)\dfrac{\partial u}{\partial x}\right] + \dfrac{\partial}{\partial y}\left[(\mu+\mu_T)\dfrac{\partial u}{\partial y}\right] + \dfrac{\partial}{\partial z}\left[(\mu+\mu_T)\dfrac{\partial u}{\partial z}\right] + \left(S_u = -\dfrac{\partial p}{\partial x} + S_u'\right)$

$(M_y)\ \dfrac{\partial(\rho v)}{\partial t} + \dfrac{\partial(\rho u v)}{\partial x} + \dfrac{\partial(\rho v v)}{\partial y} + \dfrac{\partial(\rho w v)}{\partial z} = \dfrac{\partial}{\partial x}\left[(\mu+\mu_T)\dfrac{\partial v}{\partial x}\right] + \dfrac{\partial}{\partial y}\left[(\mu+\mu_T)\dfrac{\partial v}{\partial y}\right] + \dfrac{\partial}{\partial z}\left[(\mu+\mu_T)\dfrac{\partial v}{\partial z}\right] + \left(S_v = -\dfrac{\partial p}{\partial y} + S_v'\right)$

$(M_z)\ \dfrac{\partial(\rho w)}{\partial t} + \dfrac{\partial(\rho u w)}{\partial x} + \dfrac{\partial(\rho v w)}{\partial y} + \dfrac{\partial(\rho w w)}{\partial z} = \dfrac{\partial}{\partial x}\left[(\mu+\mu_T)\dfrac{\partial w}{\partial x}\right] + \dfrac{\partial}{\partial y}\left[(\mu+\mu_T)\dfrac{\partial w}{\partial y}\right] + \dfrac{\partial}{\partial z}\left[(\mu+\mu_T)\dfrac{\partial w}{\partial z}\right] + \left(S_w = -\dfrac{\partial p}{\partial z} + S_w'\right)$

Energy Equation

(E)

$\dfrac{\partial(\rho h)}{\partial t} + \dfrac{\partial(\rho u h)}{\partial x} + \dfrac{\partial(\rho v h)}{\partial y} + \dfrac{\partial(\rho w h)}{\partial z} = \dfrac{\partial}{\partial x}\left[\lambda\dfrac{\partial T}{\partial x}\right] + \dfrac{\partial}{\partial y}\left[\lambda\dfrac{\partial T}{\partial y}\right] + \dfrac{\partial}{\partial z}\left[\lambda\dfrac{\partial T}{\partial z}\right] + \dfrac{\partial}{\partial x}\left[\dfrac{\mu_T}{Pr_T}\dfrac{\partial h}{\partial x}\right] + \dfrac{\partial}{\partial y}\left[\dfrac{\mu_T}{Pr_T}\dfrac{\partial h}{\partial y}\right] + \dfrac{\partial}{\partial z}\left[\dfrac{\mu_T}{Pr_T}\dfrac{\partial h}{\partial z}\right] + \dfrac{\partial p}{\partial t} + \Phi + S_T$

Turbulence Equations

$(k)\ \dfrac{\partial(\rho k)}{\partial t} + \dfrac{\partial(\rho u k)}{\partial x} + \dfrac{\partial(\rho v k)}{\partial y} + \dfrac{\partial(\rho w k)}{\partial z} = \dfrac{\partial}{\partial x}\left[\dfrac{\mu_T}{\sigma_k}\dfrac{\partial k}{\partial x}\right] + \dfrac{\partial}{\partial y}\left[\dfrac{\mu_T}{\sigma_k}\dfrac{\partial k}{\partial y}\right] + \dfrac{\partial}{\partial z}\left[\dfrac{\mu_T}{\sigma_k}\dfrac{\partial k}{\partial z}\right] + (S_k = P - D)$

$(\varepsilon)\ \dfrac{\partial(\rho \varepsilon)}{\partial t} + \dfrac{\partial(\rho u \varepsilon)}{\partial x} + \dfrac{\partial(\rho v \varepsilon)}{\partial y} + \dfrac{\partial(\rho w \varepsilon)}{\partial z} = \dfrac{\partial}{\partial x}\left[\dfrac{\mu_T}{\sigma_\varepsilon}\dfrac{\partial \varepsilon}{\partial x}\right] + \dfrac{\partial}{\partial y}\left[\dfrac{\mu_T}{\sigma_\varepsilon}\dfrac{\partial \varepsilon}{\partial y}\right] + \dfrac{\partial}{\partial z}\left[\dfrac{\mu_T}{\sigma_\varepsilon}\dfrac{\partial \varepsilon}{\partial z}\right] + \left(S_\varepsilon = \dfrac{\varepsilon}{k}(C_{\varepsilon 1}P - C_{\varepsilon 2}D)\right)$

where $P = 2\mu_T\left[\left(\dfrac{\partial u}{\partial x}\right)^2 + \left(\dfrac{\partial v}{\partial y}\right)^2 + \left(\dfrac{\partial w}{\partial z}\right)^2\right] + \mu_T\left[\left(\dfrac{\partial u}{\partial y}+\dfrac{\partial v}{\partial x}\right)^2 + \left(\dfrac{\partial v}{\partial z}+\dfrac{\partial w}{\partial y}\right)^2 + \left(\dfrac{\partial w}{\partial x}+\dfrac{\partial u}{\partial z}\right)^2\right]$

$-\dfrac{2}{3}\mu_T\left(\dfrac{\partial u}{\partial x}+\dfrac{\partial v}{\partial y}+\dfrac{\partial w}{\partial z}\right)^2 - \dfrac{2}{3}\rho\mu_T k\left(\dfrac{\partial u}{\partial x}+\dfrac{\partial v}{\partial y}+\dfrac{\partial w}{\partial z}\right)$ and $D = \rho\varepsilon$

momentum equations S'_u, S'_v, and S'_w comprise the pressure and non-pressure gradient terms and other possible sources, such as gravity, that influence the fluid motion, while the additional source term S_T in the energy equation may contain heat sources or sinks within the flow domain. It is also noted that the production P in the incompressible form of the turbulence equations can be obtained from its compressible counterpart by invoking the incompressible form of the continuity equation and division by constant density.

For compressible flow, the density and temperature can be evaluated through the equations of state, which provide the linkage between the energy equation and the mass and momentum equations. For a perfect gas, the equations of state are $p = \rho RT$, where R is the gas constant, and $e = C_v T$, where C_v is the specific heat of constant volume. For the dynamic viscosity and thermal conductivity, the variables can usually be determined via a linear or polynomial dependence on temperature.

The state equation is usually used as the starting point for computational procedures in either the finite-difference or finite-volume methods. Algebraic expressions of this equation for the various transport properties are formulated and hereafter solved. For incompressible flow, by setting the transport property ϕ equal to $1, u, v, w, T, k$, and ε and selecting appropriate values for the diffusion coefficient Γ and source terms S_ϕ, we obtain the special forms presented in Table 3.3 for each of the partial differential equations for the conservation of mass, momentum, energy, and the turbulent quantities. In Table 3.4, by setting

TABLE 3.3 General form of governing equations for incompressible flow in Cartesian coordinates

Φ	Γ_Φ	S_Φ
1	0	0
u	$v + v_T$	$-\dfrac{1}{\rho}\dfrac{\partial p}{\partial x} + S'_u$
v	$v + v_T$	$-\dfrac{1}{\rho}\dfrac{\partial p}{\partial y} + S'_v$
w	$v + v_T$	$-\dfrac{1}{\rho}\dfrac{\partial p}{\partial z} + S'_w$
T	$\dfrac{v}{Pr} + \dfrac{v_T}{Pr_T}$	S_T
k	$\dfrac{v_T}{\sigma_k}$	$P - D$
ε	$\dfrac{v_T}{\sigma_\varepsilon}$	$\dfrac{\varepsilon}{k}(C_{\varepsilon 1}P - C_{\varepsilon 2}D)$

TABLE 3.4 General form of governing equations for compressible flow in Cartesian coordinates

Φ	Γ_Φ	S_Φ
1	0	0
u	$\mu + \mu_T$	$-\dfrac{\partial p}{\partial x} + S'_u$
v	$\mu + \mu_T$	$-\dfrac{\partial p}{\partial y} + S'_v$
w	$\mu + \mu_T$	$-\dfrac{\partial p}{\partial z} + S'_w$
h	$\dfrac{\mu_T}{Pr_T}$	$\dfrac{\partial}{\partial x}\left[\lambda\dfrac{\partial T}{\partial x}\right] + \dfrac{\partial}{\partial y}\left[\lambda\dfrac{\partial T}{\partial y}\right] + \dfrac{\partial}{\partial z}\left[\lambda\dfrac{\partial T}{\partial z}\right] + \dfrac{\partial p}{\partial t} + \Phi + S_T$
k	$\dfrac{\mu_T}{\sigma_k}$	$P - D$
ε	$\dfrac{\mu_T}{\sigma_\varepsilon}$	$\dfrac{\varepsilon}{k}(C_{\varepsilon 1}P - C_{\varepsilon 2}D)$

the transport property ϕ equal to 1, u, v, w, h, k, and ε and selecting appropriate values for Γ and S_ϕ, we nonetheless obtain the special forms for the compressible form of the partial differential equations for the conservation of mass, momentum, energy, and the turbulent quantities.

Although we have systematically walked through the derivation of the complete set of governing equations in detail from basic conservation principles, the final general form pertaining to the fluid motion, heat transfer, etc., conforms simply to the generic form of Eq. (3.47) for incompressible flow and Eq. (3.48) for compressible flow. These equations are important generic transport equations because they can accommodate increasing complexity within the CFD model for solving more complicated problems generally found in engineering applications. Let us focus on some typical complex engineering flow problems that are of significant interest, such as multi-step combustion processes of swirling turbulent reactive flows in combustors and multi-phase flows involving interactions between gas bubbles and liquids in bubble columns. Solutions to these processes can easily be obtained by modeling them through additional transport equations expressed in the simple generic form of Eq. (3.47) for incompressible flow and Eq. (3.48) for compressible flow. For reactive flows, the transport of the various *chemical species* can be handled by the additional scalar quantities representing each of the reactive species and appropriately formulating the reaction rates in the source terms to account for the chemical reaction processes that are occurring. For bubbly flows, additional transport equations of the *number density* can be formulated and solved for the various gas bubble sizes that migrate along with the liquid in the bubble columns.

Understanding CFD is not meant to be an arduous process. On the contrary, Eqs. (3.47) and (3.48), originally formulated from first principles, reinforce the inherent *simplicity* that is embraced for any transport property that may be required to be solved within the CFD framework.

In Sections 3.3.3 and 3.4.3, we introduced the dimensionless parameters, such as the Reynolds number (Re) and Prandtl number (Pr), that may be useful for describing some similar physical phenomena of the flow and heat-transfer processes. In the next example it is demonstrated that the governing equations of mass, momentum, and energy can be non-dimensionalized to reduce the number of parameters that appear in the equations. Other similar characteristics of the fluid flow are also discussed.

Example 3.9

Consider the *dynamic similarity* of the partial differential equations that govern a two-dimensional CFD case for a steady, incompressible, laminar flow between two stationary parallel plates. The channel dimensions are height $H = 0.1$ m and length $L = 1$ m. This is the same model geometry as previously investigated in Examples 3.7 and 3.8.

(a) Non-dimensionalize the continuity, momentum, and energy equations given by Eqs. (3.13), (3.24), (3.25), and (3.34).
(b) Using CFD, determine the flow field for both air and water with the same Reynolds number while adjusting the inlet velocity. Discuss the velocity profiles in the fully developed region.
(c) Also determine the temperature field for both air and water with the same Reynolds number as in Case (b). Discuss the non-dimensional temperature profiles in the fully developed region with a prescribed uniform temperature of 330 K at the inlet and a wall temperature specified at 300 K.

Solution
(a) The non-dimensional form of the governing equations (and boundary conditions) can be achieved by dividing all the dependent and independent flow variables by relevant and meaningful constant quantities. For lengths, the variable can be divided by a characteristic length H (which is the width of the channel), all velocities by a reference velocity u_{in} (which is the inlet velocity), pressure by ρu_{in}^2 (which is twice the dynamic pressure for the channel), and temperature by a suitable temperature difference (which is $T_\infty - T_s$ for the channel). We therefore obtain

$$x^* = \frac{x}{H}, \qquad y^* = \frac{y}{H}, \qquad u^* = \frac{u}{u_{in}}, \qquad v^* = \frac{v}{u_{in}},$$

$$p^* = \frac{p}{\rho u_{in}^2}, \qquad \text{and} \qquad T^* = \frac{T - T_s}{T_\infty - T_s}$$

where the asterisks denote the non-dimensional variables. Introducing these variables into the governing equations of mass, momentum, and energy produces

$$\frac{\partial u^*}{\partial x^*} + \frac{\partial v^*}{\partial y^*} = 0 \quad — \quad \text{continuity}$$

$$u^* \frac{\partial u^*}{\partial x^*} + v^* \frac{\partial u^*}{\partial y^*} = \frac{1}{Re}\left(\frac{\partial^2 u^*}{\partial x^{*2}} + \frac{\partial^2 u^*}{\partial y^{*2}}\right) - \frac{\partial p^*}{\partial x^*} \quad — \quad \text{x-momentum}$$

$$u^* \frac{\partial v^*}{\partial x^*} + v^* \frac{\partial v^*}{\partial y^*} = \frac{1}{Re}\left(\frac{\partial^2 v^*}{\partial x^{*2}} + \frac{\partial^2 v^*}{\partial y^{*2}}\right) - \frac{\partial p^*}{\partial y^*} \quad — \quad \text{y-momentum}$$

$$u^* \frac{\partial T^*}{\partial x^*} + v^* \frac{\partial T^*}{\partial y^*} = \frac{1}{RePr}\left(\frac{\partial^2 T^*}{\partial x^{*2}} + \frac{\partial^2 T^*}{\partial y^{*2}}\right) \quad — \quad \text{energy}$$

Discussion

A major advantage of non-dimensionalizing the governing equations is the significant reduction of parameters to be considered. By grouping the dimensional parameters, originally eight ($H, u_{in}, T_\infty, T_s, k, \rho, \mu$, and C_p), the non-dimensionalized problem now involves only two parameters (Re and Pr). Hence, for a given geometry, problems having the same values for the similarity parameters will have identical solutions. Another advantage of the use of similarity parameters is that results from a large number of experiments can be grouped and reported conveniently in terms of such parameters.

(b) The physical significance of the Reynolds number (Re) is investigated herein. As previously defined in Eq. (3.20), this dimensionless number requires the values of density (ρ) and dynamic viscosity (μ). For air, the values are $\rho = 1.2 \text{ kg/m}^3$ and $\mu = 2 \times 10^{-5} \text{ kg/m} \cdot \text{s}$, while for water, they are $\rho = 1000 \text{ kg/m}^3$ and $\mu = 10^{-3} \text{ kg/m} \cdot \text{s}$. If a laminar flow is assumed for both fluids, e.g., $Re = 120$, the inlet velocities for air and water are 0.02 m/s and 0.0012 m/s.

Based on CFD simulations, the axial dimensional and non-dimensional velocity profiles for air and water in the fully developed region are given in Figures 3.9.1 and 3.9.2.

Discussion

Although the two flows have different fluid properties, one being air and the other being water, we obtain the same flow behavior. This is because they have the same Reynolds numbers, and the non-dimensional governing equations of the x-momentum and y-momentum are identical, which leads to the same numerical results. If we consider another fluid-mechanics example that is the flow of air or water over a flat plate (Figure 3.9.3) having different lengths but with the same inlet velocities at the same Reynolds numbers, these two geometrically similar bodies have the same physical phenomena, since they have the same friction coefficients (C_f).

(c) The physical significance of the Prandtl number (Pr) is investigated here. As previously defined in Eq. (3.35), it requires the values of the kinematic viscosity (ν) and thermal diffusivity (α). For air, the values are $\nu = 1.667 \times 10^{-5} \text{ m}^2/\text{s}$ and $\alpha = 1.667 \times 10^{-5} \text{ m}^2/\text{s}$, while for water they are

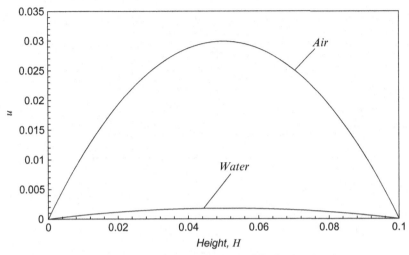

FIGURE 3.9.1 Dimensional axial velocity profiles in the fully developed region.

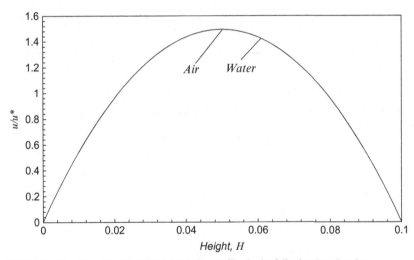

FIGURE 3.9.2 Non-dimensional axial velocity profiles in the fully developed region.

$v = 1 \times 10^{-6} \, \text{m}^2/\text{s}$ and $\alpha = 1.435 \times 10^{-7} \, \text{m}^2/\text{s}$. This yields $Pr = 1$ for air and $Pr \approx 7$ for water.

Based on the CFD simulations, the non-dimensional temperature profiles for air and water in the fully developed region are given in Figure 3.9.4.

Discussion
With reference to the energy diffusion term, $1/Re \, Pr$ in the non-dimensional energy equation, the Reynolds numbers of air and water are the same; however, the

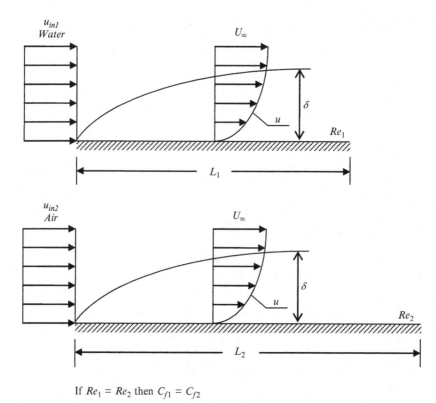

If $Re_1 = Re_2$ then $C_{f1} = C_{f2}$

FIGURE 3.9.3   Two geometrically similar bodies having the same friction coefficients at the same Reynolds numbers.

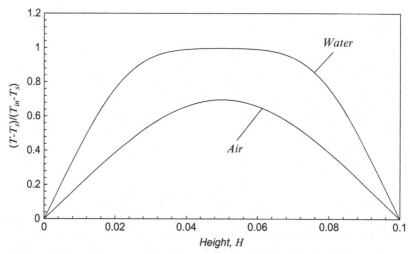

FIGURE 3.9.4   Non-dimensional temperature profiles in the fully developed region.

Prandtl number of air is found to be much less than that for water, leading to heat flow being diffused more in air than in water. Therefore, with less diffusion, more advection is encouraged in water than in air, which leads to a higher temperature in the fully developed region. Nevertheless, we have the same momentum diffusion for the velocity fields, since the Reynolds numbers are the same for both fluids in the x-momentum and y-momentum equations; both fluids therefore have the same velocity profiles.

3.7   PHYSICAL BOUNDARY CONDITIONS OF THE GOVERNING EQUATIONS

In Chapter 2, we introduced various types of boundary conditions that are required to perform the numerical calculations of fluid flow in a channel. Here, we explore the physical meanings of the boundary conditions and other important boundary conditions that may be required to close the fluid-flow system.

The continuity, momentum, energy, and turbulent quantities of the previous equations govern the flow and heat transfer of a fluid. They are the same equations whether the flow is over high-rise buildings or bridges, through subsonic wind tunnels, or within various narrow channels of intricate electronic components in a computer. However, the flow fields are quite *different* for each of these cases, although the governing equations are the *same*. The reason for the difference lies in the specification of the *boundary conditions*.

The boundary conditions, and sometimes the *initial conditions,* dictate the particular solutions to be obtained from the governing equations. For example, when all the geometrical shapes of the high-rise buildings are treated, when certain physical boundary conditions are applied on the surfaces, and when appropriate boundary conditions are imposed on the far free stream, then the resulting solution of the governing partial differential equations listed in Table 3.1 will yield the complex flow field over the high-rise buildings. In contrast, the solution of the flow field within heated channels in a computer casing demands other physical boundary conditions—for example, the specific geometrical shape and location of air entrainment and exhaust ports within the computer box.

We have formulated the governing equations that have been developed and described in the previous sections. However, the real driver for any particular solution is the boundary conditions. This has particular significance in CFD because any numerical solution of the governing equations must result in astrong and compelling numerical representation of the proper boundary conditions.

Let us now review the physical boundary conditions for a subsonic viscous flow. We focus initially on the so-called *no-slip* condition. Here, the boundary condition on a solid surface assumes zero relative velocity between the surface and the fluid immediately at the surface. If the surface is stationary, with the flow moving past it, then all the velocity components can be taken to be zero. In other words,

$$u = v = w = 0 \text{ at the surface} \tag{3.49}$$

In Chapter 2 and Examples 3.2, 3.5, 3.7, 3.8, and 3.9, we were introduced to the inflow boundary condition for a channel flow. The solution of the governing equations for the transport property ϕ for most flows requires at least one velocity component to be given at the inflow boundary. For the channel flow, this is provided by the *Dirichlet* boundary condition on the velocity in the x direction:

$$u = f \text{ and } v = w = 0 \text{ at the inflow boundary} \tag{3.50}$$

where f can either be specified as a constant value or a velocity profile at the surface. Computationally, Dirichlet conditions can be applied accurately as long as f is continuous. We were also introduced in Chapter 2, as well as in Examples 3.2, 3.5, 3.7, 3.8, and 3.9, to the use of an outflow boundary condition. Commonly, outflow boundaries are positioned at locations where the flow is approximately unidirectional and where surface stresses take known values. In a fully developed flow exiting from the channel, there is no change in the velocity component in the direction across the boundary. To satisfy stress continuity, the shear forces along the surface are taken to be zero; this gives the outflow condition

$$\frac{\partial u}{\partial n} = \frac{\partial v}{\partial n} = \frac{\partial w}{\partial n} = 0 \text{ at the outflow boundary} \tag{3.51}$$

where n is the direction normal to the surface, which for the channel flow problem is the x direction. This condition is usually known as the *Neumann* boundary condition. Physically, in reference to the continuity equation, Eq. (3.10), it is clear that the appropriate boundary conditions (3.49), (3.50), and (3.51) that are imposed at any location on the surface of the channel walls close the system mathematically and satisfy local and overall mass conservation.

There is an analogous no-slip condition associated with the temperature at the surface. If the material temperature of the surface is denoted by T_w, then the temperature fluid layer immediately in contact with the surface is also T_w. In a given problem where the wall temperature is known, the Dirichlet boundary condition applies and the fluid temperature is

$$T = T_w \text{ at the wall} \tag{3.52}$$

The application of this boundary condition is illustrated by Example 3.6, part (a), and Examples 3.8 and 3.9. However, if the wall temperature is not known (e.g., if the temperature is changing as a function of time due to the heat transfer to or from the surface), then Fourier's law of heat condition can be applied to provide the necessary boundary condition at the surface. If we denote the instantaneous wall heat flux as q_w, then according to Fourier's law

$$q_w = -\left(k\frac{\partial T}{\partial n}\right)_w \quad \text{at the wall} \tag{3.53}$$

The application of this boundary condition was also illustrated by Example 3.6, in Case (b). Here, the changing surface temperature T_w is responding to the thermal response of the wall material through the heat transfer to the wall q_w. This type of boundary condition, as far as the flow is concerned, is a boundary condition of the temperature *gradient* at the wall. For the case where there is no heat transfer to the surface, the wall temperature is by definition called an adiabatic wall temperature T_{adia}. The proper boundary condition comes from Eq. (3.53) with $q_w = 0$; hence

$$\left(\frac{\partial T}{\partial n}\right)_w = 0 \text{ at the wall} \tag{3.54}$$

This condition falls in line with the Neumann boundary condition for the velocity at the outflow boundaries. On the inflow and outflow boundaries of the flow domain, it is common to have the temperature specified at the inflow boundary and the adiabatic condition adopted at the outflow boundary.

Other commonly used boundary conditions include the open boundary condition. If we refer back to the description of the boundary conditions for the flow over high-rise buildings at the beginning of this section, the far free-stream boundary requires the application of an open boundary, which simply states that the normal gradient of any of the transport property ϕ is zero, i.e., $\partial\phi/\partial n = 0$. Gresho and Sani (1990) reviewed the intricacies of open boundary conditions in an incompressible flow and stated that there are some *theoretical concerns* regarding this boundary condition. However, its success in CFD practice left them to recommend it as the simplest and cheapest method when compared with theoretically more satisfying selections. Furthermore, symmetric and cyclic boundary conditions can be employed to take advantage of special geometrical features of the solution region. For the symmetry boundary condition, the normal velocity at the surface is zero while the normal gradients of the other velocity components are zero, where the latter condition also applies for any scalar quantity. For the cyclic boundary condition, the transport property of the one surface ϕ_1 is equivalent to the transport property of the second surface ϕ_2, i.e., $\phi_1 = \phi_2$, depending on which two surfaces of the flow domain experience periodicity.

For the turbulent quantities, Dirichlet and Neumann boundary conditions for the turbulent kinetic energy k and its dissipation ε are usually applied at the inflow and outflow boundaries. However, special treatment of the boundary conditions for k and ε is required to properly account for the laminar sublayer that exists near the wall surface in turbulent flow. We discuss the wall treatment for the turbulent quantities in detail in Chapter 6.

For a supersonic flow, Dirichlet and Neumann boundary conditions can be applied at the inflow boundary but not at the outflow boundary. All variables and properties at the outflow boundary are determined from the upstream flow. If the flow is assumed to be inviscid, the flow velocity is usually taken to be a finite, non-zero value at the solid surface. This means that the wall shear stress is zero. Also, there cannot be any mass flow into and out of the wall. The component velocity perpendicular to the wall is thus zero and the flow at the surface is tangent to the solid surface.

3.8   SUMMARY

In this chapter, we formulated the mathematical basis for a comprehensive general-purpose model of fluid flow and heat transfer from the basic principles of conservation of mass, momentum, and energy. The governing equations are derived by the consideration of an infinitesimally small control volume to conserve mass and energy, and that the net force acting on the control volume is equivalent to the time rate of change of linear momentum. Although we employ the Newtonian model of viscous stresses to close the system of equations, the accommodation of fluids having non-Newtonian characteristics can be easily incorporated within the framework of these equations. The appearance of viscosity and other thermophysical properties, such as the density, thermal conductivity, and specific heat, in the governing equations may vary with local conditions, but accounting for these effects is a straightforward process.

The reader was also introduced to some aspects of turbulence modeling, which resulted in the derivation of the widely applied two-equation k-ε model. This turbulence model still comes highly recommended for general-purpose CFD computations and is the default model used in many commercial codes.

Whether the fluid flow is laminar or turbulent, there are significant commonalities between the conservation equations. This leads to the formulation of the generic form of the governing equations accompanied by a discussion of the physical boundary conditions commonly employed within the CFD framework to close the fluid-flow system. A road map to encapsulate all these aforementioned aspects, beginning from the conception of the fluid flow and ending with the system of equations to be solved, is shown in Figure 3.9.

The means of obtaining a solution to these governing equations are discussed in the next chapter. We present some of the basic computational techniques that can be employed to solve such partial differential equations. The equations

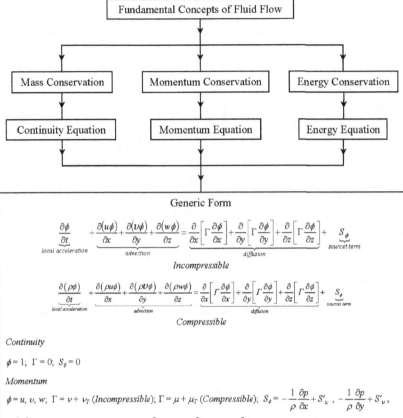

FIGURE 3.9   The road map for Chapter 3.

presented in this chapter are usually regarded as the starting point for the application of the many *discretization* procedures predicting the fluid-motion and heat-transfer processes, which are further elaborated in the next chapter.

REVIEW QUESTIONS

3.1 Simplify the general continuity equation:

$$\frac{\partial \rho}{\partial t} + \frac{\partial(\rho u)}{\partial x} + \frac{\partial(\rho v)}{\partial y} + \frac{\partial(\rho w)}{\partial z} = 0$$

for a two-dimensional constant-density case.

3.2 In a converging nozzle, the flow accelerates due to the narrowing geometry. Discuss the changes in the velocity gradients $\frac{\partial u}{\partial x}$ and $\frac{\partial v}{\partial y}$ during the flow (assume a constant density).

3.3 What is Newton's second law of motion?

3.4 Write a force balance equation for all the forces acting on a differential control volume.

3.5 For the momentum of a fluid property in the x direction, discuss how the local acceleration $\frac{\partial u}{\partial t}$ and the advection terms $u\frac{\partial u}{\partial x} + v\frac{\partial u}{\partial y}$ contribute to the overall transport of the fluid.

3.6 A simplified, one-dimensional, inviscid, incompressible, laminar flow is defined by the following momentum equation in the x direction:

$$\frac{\partial u}{\partial t} + u\frac{\partial u}{\partial x} = -\frac{1}{\rho}\frac{\partial p}{\partial x}$$

Name each term and discuss its contribution to the flow.

3.7 The momentum of a fluid in the y direction is given by the following equation:

$$\frac{\partial v}{\partial t} + u\frac{\partial v}{\partial x} + v\frac{\partial v}{\partial y} = -\frac{1}{\rho}\frac{\partial p}{\partial y} + v\frac{\partial^2 v}{\partial x^2} + v\frac{\partial^2 v}{\partial y^2} - g$$

Discuss the forces that act to transport the fluid.

3.8 What are the differences between the momentum equation in Question 3.7 and the following momentum equation?

$$\rho u\frac{\partial u}{\partial x} + \rho v\frac{\partial u}{\partial y} = -\frac{\partial p}{\partial x} + \mu\left(\frac{\partial^2 u}{\partial x^2} + \frac{\partial^2 u}{\partial y^2}\right)$$

3.9 The hydrodynamic length for a channel flow shown below is equal to L_e when air is used ($u_{in} = 0.03$ m/s, $\mu_1 = 1.65 \times 10^{-5}$ kg/m · s, and $\rho = 1.2$ kg/m³). To obtain the same hydrodynamic length for water, what inlet velocity is required ($\mu_1 = 1.003 \times 10^{-3}$ kg/m · s, $\rho = 1000$ kg/m³)?

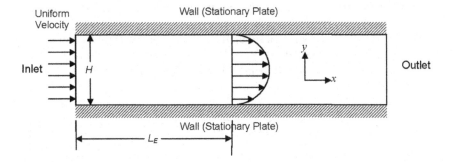

3.10 The Reynolds number is a ratio of two fluid properties. What are they?

3.11 If the Reynolds number is very high ($Re \gg 10,000$), what does this suggest? If it is very low ($Re \ll 100$), what does this imply?

3.12 Explain the first law of thermodynamics.

3.13 Name the sources of energy that contribute to the energy equation.

3.14 Write an equation for the energy balance using the sources of energy defined in Question 3.11 for the first law of thermodynamics.

3.15 Apply *Fourier's law of heat conduction* to obtain the heat flux in the x direction.

3.16 Write the equation that defines the substantial derivative for the transport of temperature in terms of the local acceleration derivative and the advection derivative of temperature, and explain why this is so.

3.17 If a car travels across a warmer environment, the car's body will experience a sudden rise in temperature. Is this an example of the local acceleration derivative or is an advection derivative at work?

3.18 In what type of situation can you simplify the general two-dimensional energy equation:

$$\frac{\partial T}{\partial t} + u\frac{\partial T}{\partial x} + v\frac{\partial T}{\partial y} = \frac{k}{\rho C_p}\frac{\partial^2 T}{\partial x^2} + \frac{k}{\rho C_p}\frac{\partial^2 T}{\partial y^2}$$

to reach the well-known Laplace's equation:

$$\frac{\partial^2 T}{\partial x^2} + \frac{\partial^2 T}{\partial y^2} = 0$$

3.19 Obtain the general analytical solution for Laplace's equation for a one-dimensional case.

3.20 The Prandtl number is a ratio of two fluid properties. What are they?

3.21 Fluids like oils have a high Prandtl number ($Pr \gg 1$). What does this suggest?

3.22 What is the significance of a Prandtl number equal to 1 in terms of entry lengths?

3.23 What is the *energy cascade* process in turbulence?

3.24 Why do large eddies tend to be anisotropic? Why are small-scale eddies isotropic?

3.25 The use of direct numerical simulation (DNS) remains a problem for engineering applications. Why?

3.26 Which profile indicated below is the laminar or turbulent velocity profile?

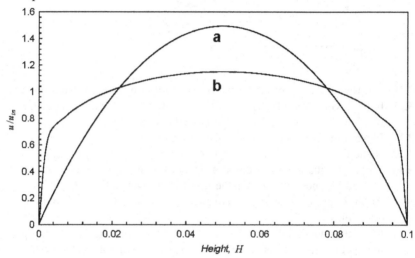

3.27 From Question 3.26, describe the shape of the turbulent velocity profile. Explain why the profile is so different from the laminar profile.

3.28 For incompressible flow, the non-dimensional transport equation can be simplified as

$$u\frac{\partial u}{\partial x} = \frac{1}{Re}\frac{\partial^2 u}{\partial x^2}$$

Provide an explanation why, for a laminar flow, the diffusion term (the right-hand side of the equation) will be more dominant than the convective term (the left-hand side of the equation). At a highly turbulent flow, the situation is the opposite. Why?

3.29 Explain the terms in the following general equation:

$$u\frac{\partial \phi}{\partial x} = \Gamma\frac{\partial^2 \phi}{\partial x^2} + S_\phi$$

3.30 The equation in Question 3.29 represents a transport process for the property ϕ. Under what circumstances can this equation be applied?

3.31 Spot the errors in the equation below. (Hint: It is a 3D x-momentum equation.)

$$\frac{\partial u}{\partial t} + \frac{\partial(uu)}{\partial x} + \frac{\partial(vu)}{\partial x} + \frac{\partial(uu)}{\partial z} = \frac{\partial}{\partial x}\left[(v+v_T)\frac{\partial u}{\partial x}\right] + \frac{\partial}{\partial y}\left[(v+v_T)\frac{\partial u}{\partial x}\right]$$

$$+ \frac{\partial}{\partial z}\left[(v+v_T)\frac{\partial u}{\partial x}\right] + \left(S_u = -\frac{1}{\rho}\frac{\partial p}{\partial x} + S'_u\right)$$

3.32 Referring to the diagram below, answer the following questions.
   (a) At the wall, what are the values of u and v?
   (b) At the symmetry plane, what are the values of u, v, and p?

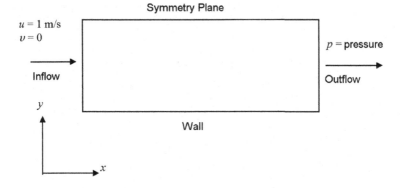

Chapter 4

CFD Techniques—The Basics

4.1 INTRODUCTION

Some of the basic computational techniques that are required to solve the governing equations of fluid dynamics are examined in this chapter. Here, the authors endeavor to demonstrate how these techniques are employed to obtain an approximate solution for the governing equations of flow problems with appropriate boundary conditions applied for the specific problem considered.

The process of obtaining the computational solution consists of two stages. The *first stage* involves the conversion of the partial differential equations and auxiliary (boundary and initial) conditions into a system of discrete algebraic equations. This stage is commonly known as the *discretization* stage.

In Chapter 3, some analytical solutions were derived for the partial differential Navier–Stokes equations to some simple one-dimensional flow problems. There, the closed-form expressions of u, v, w, p, etc., as functions of one of the spatial locations x, y, and z, were used to provide the desired values of the flow-field variables. However, real fluid flows are generally three-dimensional in nature; analytical relationships are not easily attainable. Even for a simplified three-dimensional flow applied to a two-dimensional problem, this can be difficult. We saw in Chapter 2 how CFD can be employed to solve a simple two-dimensional channel flow problem. Instead of closed-form expressions, the respective values of u, v, w, p, etc., were obtained at discrete locations within the flow domain by the CFD solver, and the original Navier–Stokes equations were approximated by algebraic derivatives. The partial differential equations, totally replaced by a system of algebraic equations, solved the discrete values of the flow-field variables. For such a solution, the original partial differential equations were considered to be discretized in order to yield the values at the discrete locations.

The process of *discretization* can be identified through some common methods that are still in use today, which is shown in the overview of the computational solution procedure in Figure 4.1. The two main headings of the methods constitute the most popular discretization approaches in CFD (and they are further elaborated upon in Section 4.2). It is also worth mentioning that there are other discretization methods available in CFD: the finite-element and spectral methods. In brief, the finite-element method has many similarities to the

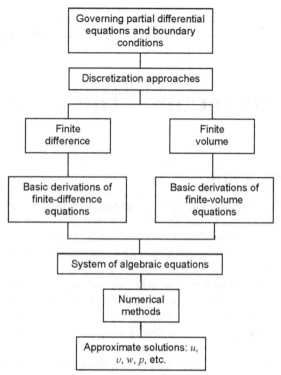

FIGURE 4.1   Overview of the computational solution procedure.

finite-volume method. The distinguishing feature is that it uses simple piecewise polynomial functions on local elements to describe the variations of the unknown flow variables. The concept of weighted residuals is introduced to measure the errors associated with the approximate functions, which are later minimized. A set of non-linear algebraic equations for the unknown terms of the approximating functions is solved, hence yielding the flow solution. The finite-element method has not enjoyed extensive use in CFD although there are a number of commercial and research codes available that employ it. A significant advantage of the finite-element method is the ability to handle arbitrary geometries. Nevertheless, it has generally been found that the finite-element method requires greater computational resources and computer processing power than the equivalent finite-volume method; therefore, its popularity has been limited. The spectral method employs the same general approach as the finite-difference and finite-element methods, where the unknowns of the governing equations are replaced with a truncated series. The difference is that, where the previous two methods employ local approximations, the spectral method uses a global approximation; that is, by means of Fourier series, Legendre polynomials, or Chebyshev polynomials for the entire flow domain. The discrepancy between the exact solution and the approximation is dealt with by using a weighted residuals concept similar to the finite-element method. The authors concentrate in detail on the basic derivations of the finite-difference and finite-volume methods in this book. A brief description of the finite-element and spectral methods is given in Sections 4.2.3 and 4.2.4.

The finite-difference method is illustrated because of its simplicity in formulating the algebraic equations and because it also forms the foundation for comprehending the essential basic features of discretization. The finite-volume method is employed in the majority of all commercial CFD codes today. We believe readers should familiarize themselves with this approach because of its ability to be applied not only to *structured* mesh but also to *unstructured* mesh, which is gaining in popularity and use in handling arbitrary geometrical shapes. Structured mesh is usually designated as a mesh containing cells having either a regularly shaped element with four-nodal corner points in two dimensions or a hexahedron-shaped element with eight-nodal corner points in three dimensions. Unstructured mesh commonly refers to a mesh overlay with cells that are in the form of either a triangle-shaped element in two dimensions or a tetrahedron-shaped element in three dimensions.

The *second stage* of the solution process involves the implementation of numerical methods to provide a solution to the system of algebraic equations. Appropriate methods for obtaining the numerical solution for the system of algebraic equations are discussed in Section 4.3.

Throughout this chapter, we concentrate more on systems of algebraic equations typically arising from the solution of steady-state flow problems. These governing equations contain spatial derivatives that can be discretized by employing either the finite-difference or the finite-volume method. Nevertheless, some basic approximations to the time derivatives for unsteady flow problems, which in practice are exclusively discretized using the finite-difference method, are briefly described in Section 4.2.1.

4.2   DISCRETIZATION OF GOVERNING EQUATIONS

4.2.1   Finite-Difference Method

The finite-difference method is the oldest of the methods for the numerical solution of partial differential equations. It is believed to have been developed by Euler in 1768 and was used to obtain numerical solutions to differential equations by hand calculation. At each nodal point of the grid used to describe the fluid-flow domain, the Taylor series expansions are used to generate finite-difference approximations to the partial derivatives of the governing equations. These derivatives, replaced by finite-difference approximations, yield an algebraic equation for the flow solution at each grid point. In principle, the finite-difference method can be applied to any type of grid system. However, the method is more commonly applied to structured grids since it requires a mesh having a high degree of regularity. The grid spacing between the nodal points need not be uniform, but there are limits on the amount of grid stretching or distortion that can be imposed, to maintain accuracy. Topologically, the finite-difference structured grids must conform to the constraints of general coordinate systems, such as Cartesian grids comprised of six-sided computational domains. However, use of intermediate coordinate mapping, such as the body-fitted coordinate system, allows this major geometrical constraint to be relaxed so that complex shapes can be modeled.

Finite-difference allows the incorporation of higher-order differencing approximations on regular grids, which provide a higher degree of accuracy in the solution. However, the main disadvantage is that the conservation property is not usually enforced unless special care is taken.

The first step toward obtaining a numerical solution involves the discretization of the geometric domain; i.e., a numerical grid must be defined. In the finite-difference method, the grid is usually taken to be locally structured, which means that each grid node may be considered the origin of a local coordinate system, whose axes coincide with the grid lines. These two gridlines also imply that they do not intersect elsewhere and that any pair of grid lines belonging to the different families intersects only once at the grid point. In three dimensions, three grid lines intersect at each node; none of the lines intersects any other at any other grid nodal point. Figure 4.2 illustrates examples of one-dimensional and two-dimensional uniformly distributed Cartesian grids commonly used in the finite-difference method. Within these two grid systems, each node is

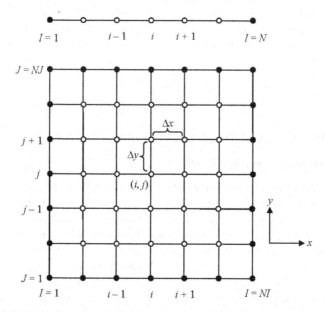

FIGURE 4.2   One-dimensional and two-dimensional uniformly distributed Cartesian grids for the finite-difference method (full symbols denote boundary nodes and open symbols denote computational nodes).

uniquely identified by a set of indices, which are indices of the grid lines that intersect at (i, j) in a two-dimensional grid and at (i, j, k) in a three-dimensional grid. The neighboring nodes are defined by increasing or reducing one of the indices by unity.

Analytical solutions of partial differential equations involve closed-form expressions that provide variation of the flow-field variables continuously throughout the domain. This is in contrast to numerical solutions, which provide answers only at discrete points in the geometric domain, for example, at the grid points (open symbols) shown in Figure 4.2. To illustrate the finite-difference method, let us conveniently assume that the spacing of the grid points in the x direction is uniform and is given by Δx, and that the spacing of the points in the y direction is also uniform and is given by Δy. The spacing of Δx or Δy need not necessarily be uniform. We could have easily dealt with totally unequal spacing in both directions. This uniform spacing does not have to correspond to physical x-y space. As is frequently handled in CFD, the numerical calculations can be performed in a transformed computational space that has uniform spacing in the transformed independent variables but still corresponds to a non-uniform spacing in physical space. In any event, we assume uniform spacing in each of the coordinate directions in describing the basic techniques of the finite-difference method.

The starting point for the representation of the partial derivatives in the governing equations is the Taylor series expansion. We encountered the Taylor series expansion in formulating the conservation equations in Chapter 3. Here, we are interested in replacing the partial derivative with a suitable algebraic difference quotient—a finite difference. For example, referring to Figure 4.2, if at the indices (i, j) there exists a generic flow-field variable ϕ, then the variable at point $(i+1, j)$ can be expressed in terms of a Taylor series expanded about the point (i, j) as

$$\phi_{i+1,j} = \phi_{i,j} + \left(\frac{\partial \phi}{\partial x}\right)_{i,j} \Delta x + \left(\frac{\partial^2 \phi}{\partial x^2}\right)_{i,j} \frac{\Delta x^2}{2} + \left(\frac{\partial^3 \phi}{\partial x^3}\right)_{i,j} \frac{\Delta x^3}{6} + \dots \quad (4.1)$$

Similarly, the variable at point $(i - 1, j)$ can also be expressed in terms of a Taylor series about points (i, j) as

$$\phi_{i-1,j} = \phi_{i,j} - \left(\frac{\partial \phi}{\partial x}\right)_{i,j} \Delta x + \left(\frac{\partial^2 \phi}{\partial x^2}\right)_{i,j} \frac{\Delta x^2}{2} - \left(\frac{\partial^3 \phi}{\partial x^3}\right)_{i,j} \frac{\Delta x^3}{6} + \dots \quad (4.2)$$

Equations (4.1) and (4.2) are mathematically exact expressions for the respective variables $\phi_{i+1,j}$ and $\phi_{i-1,j}$ if the number of terms is infinite, the series converges, and/or $\Delta x \to 0$. By subtracting the two equations, we obtain the approximation for the first-order derivative of ϕ:

$$\left(\frac{\partial \phi}{\partial x}\right) = \frac{\phi_{i+1,j} - \phi_{i-1,j} - (\Delta x^3/3)(\partial^3 \phi/\partial x^3)}{2\Delta x} + \dots$$

or

$$\left(\frac{\partial\phi}{\partial x}\right) = \frac{\phi_{i+1,j} - \phi_{i-1,j}}{2\Delta x} + \underbrace{O(\Delta x^2)}_{\text{Truncation error}} \quad \textit{Central difference} \quad (4.3)$$

The term $O(\Delta x^n)$ signifies the truncation error of the finite-difference approximation, which measures the accuracy of the approximation and determines the rate at which the error decreases as the spacing between points is reduced. Equation (4.3) is taken to be second-order accurate because the truncation error is of an order of 2. This is a major simplification, and its validity depends on the size of Δx. The smaller Δx is, the better the agreement. This equation is called *central difference* since it depends equally on values at both sides of the node at x. It is also possible to form other expressions for the first-order derivative by invoking Eqs. (4.1) and (4.2). These are

$$\left(\frac{\partial\phi}{\partial x}\right) = \frac{\phi_{i+1,j} - \phi_{i,j}}{\Delta x} + \underbrace{O(\Delta x)}_{\text{Truncation error}} \quad \textit{Forward difference} \quad (4.4)$$

and

$$\left(\frac{\partial\phi}{\partial x}\right) = \frac{\phi_{i,j} - \phi_{i-1,j}}{\Delta x} + \underbrace{O(\Delta x)}_{\text{Truncation error}} \quad \textit{Backward difference} \quad (4.5)$$

The above two equations are termed the *forward* and *backward differences*. They reflect their respective biases, and both of these finite-difference approximations are only first-order accurate. It is expected that they will be less accurate in comparison to the central difference for a given value of Δx.

Let's further explore the idea behind finite-difference approximations, which is borrowed directly from the definition of derivatives. A geometric interpretation of Eqs. (4.3), (4.4), and (4.5) is provided in Figure 4.3. The first-order derivative $\partial\phi/\partial x$ at the point i in the direction of x is the slope of the tangent to the curve $\phi(x)$ at that point and is the line marked "Exact" in the figure. Its slope can be approximated by the slope of a line passing through the neighboring points—$i+1$ and $i-1$—on the curve. The forward difference is evaluated by the slope BC between the points i and $i+1$, while the backward difference is achieved by the slope AB between the points $i-1$ and i. The line labeled *central* represents the approximation by a central difference that evaluates the slope AC. From the figure, it can be seen that some approximations are better than others. The line for the central difference, indicated by the slope AC, appears to be closer to the slope of the exact line; if the function $\phi(x)$ were a second-order polynomial and the points were equally spaced in the x direction, the slopes would match exactly. It also appears that the quality of approximation improves when additional points are made closer to a point; i.e., as the grid is refined, the approximation improves.

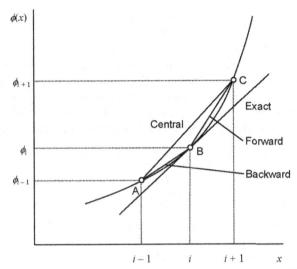

FIGURE 4.3   Finite-difference representation of the first-order derivative for $\partial\phi/\partial x$.

Differences for the y derivatives are obtained in exactly the same fashion. The results are directly analogous to the previous equations for the x derivatives. They are given by

$$\left(\frac{\partial\phi}{\partial y}\right) = \frac{\phi_{i,j+1} - \phi_{i,j}}{\Delta y} + \underbrace{O(\Delta y)}_{\text{Truncation error}} \quad \textit{Forward difference} \qquad (4.6)$$

$$\left(\frac{\partial\phi}{\partial y}\right) = \frac{\phi_{i,j} - \phi_{i,j-1}}{\Delta y} + \underbrace{O(\Delta y)}_{\text{Truncation error}} \quad \textit{Backward difference} \qquad (4.7)$$

$$\left(\frac{\partial\phi}{\partial y}\right) = \frac{\phi_{i,j+1} - \phi_{i,j-1}}{2\Delta y} + \underbrace{O(\Delta y^2)}_{\text{Truncation error}} \quad \textit{Central difference} \qquad (4.8)$$

The second-order derivative can also be obtained through the Taylor series expansion, as was applied to approximate the first-order derivative. By summing Eqs. (4.1) and (4.2), we have

$$\left(\frac{\partial^2\phi}{\partial x^2}\right) = \frac{\phi_{i+1,j} - 2\phi_{i,j} + \phi_{i-1,j}}{\Delta x^2} + \underbrace{O(\Delta x^2)}_{\text{Truncation error}} \qquad (4.9)$$

This equation represents the central finite-difference for the second-order derivative with respect to x evaluated at the point (i, j). The approximation is second-order accurate. An analogous expression can easily be obtained for the second-order derivative with respect to y, which results in

$$\left(\frac{\partial^2 \phi}{\partial y^2}\right) = \frac{\phi_{i,j+1} - 2\phi_{i,j} + \phi_{i,j-1}}{\Delta y^2} + \underbrace{O(\Delta y^2)}_{\text{Truncation error}} \qquad (4.10)$$

A Taylor series expansion for time derivatives can also be obtained similar to the Taylor series expansion for space, Eq. (4.1). Since the numerical solution is more likely to be marched continuously in a discrete time interval of Δt, the finite approximation derived for the first-order spatial derivatives applies equally for the first-order time derivative. For a forward-difference approximation in time,

$$\left(\frac{\partial \phi}{\partial t}\right) = \frac{\phi_{i,j}^{n+1} - \phi_{i,j}^{n}}{\Delta t} + \underbrace{O(\Delta t)}_{\text{Truncation error}} \quad \textit{Forward difference} \qquad (4.11)$$

The above equation introduces a truncation error of $O(\Delta t)$. More accurate approximations to the time derivative can be obtained through the consideration of additional discrete values of $\phi_{i,j}$ in time.

4.2.2   Finite-Volume Method

The finite-volume method discretizes the integral form of the conservation equations directly in physical space. It was initially introduced by researchers like McDonald (1971) and MacCormack and Paullay (1972) for the solution of two-dimensional time-dependent Euler equations, and it was later extended to three-dimensional flows by Rizzi and Inouye (1973). The computational domain is subdivided into a finite number of contiguous control volumes, where the resulting statements express the exact conservation of relevant properties for each of the control volumes. At the centroid of each of the control volumes, the variable values are calculated. Interpolation is used to express variable values at the control volume surface in terms of the center values, and suitable quadrature formulae are applied to approximate the surface and volume integrals. An algebraic equation for each of the control volumes can be obtained, in which a number of the neighboring nodal values appear.

As the finite-volume method works with control volumes and not grid intersection points, it has the capacity to accommodate any type of grid. Here, instead of structured grids, unstructured grids can be employed that allow a large number of options for the definition of the shape and location of the control volumes. Since the grid defines only the control volume boundaries, the method is conservative so long as the surface integrals that are applied at these boundaries are the same as the control volumes sharing the boundary. One disadvantage of this method compared to the finite-difference schemes is that higher-order differencing approximations greater than the second order are more difficult to develop in three dimensions. This is because of the requirement for two levels of approximation, which are interpolation and integration.

However, the finite-volume method has more advantages than disadvantages. One important feature of the method is that a "finite-element" type mesh can be used; the mesh can be formed by the combination of triangles or quadrilaterals in the case of two dimensions or tetrahedra and hexahedra in three dimensions. This type of unstructured mesh offers greater flexibility for handling complex geometries. Another attractive feature is that the method requires no transformation of the equations in terms of a body-fitted coordinate system, as is required in the finite-difference method.

As with the finite-difference method, a numerical grid must be initially defined to discretize the physical flow domain of interest. For the finite-volume method, we now have the flexibility of representing the grid by either structured or unstructured mesh. For purposes of illustrating the finite-volume method, we consider a typical representation of structured (quadrilateral) and unstructured (triangular) finite-volume elements in two dimensions (shown in Figure 4.4) for the discretization of the partial differential equations. The cornerstone of the finite-volume method is the *control volume integration*. In a control volume, the bounding surface areas of the element are directly linked to the discretization of the first- and second-order derivatives for ϕ (the generic flow-field

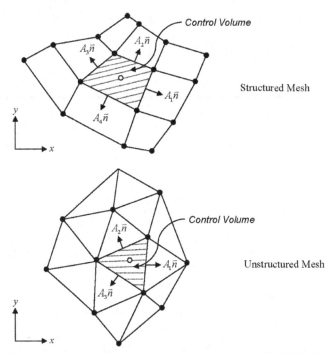

FIGURE 4.4   A representation of structured and unstructured mesh for the finite-volume method (full symbols denote element vertices and open symbols at the center of the control volumes denote computational nodes).

variable). Here, the surface areas in the normal direction ($\vec{n}$) to the volume surfaces, as indicated in Figure 4.4, are resolved with respect to the Cartesian coordinate directions to yield the projected areas A_i^x and A_i^y in the x and y directions, respectively. The projected areas are positive if their outward normal vectors from the volume surfaces are directed in the same directions as the Cartesian coordinate system; otherwise, they are negative.

By applying Gauss's divergence theorem to the volume integral, the first-order derivative of ϕ in two dimensions, for example, the term along the x direction represented in Eq. (3.47), can be approximated by

$$\left(\frac{\partial\phi}{\partial x}\right) = \frac{1}{\Delta V}\int_V \frac{\partial\phi}{\partial x}dV = \frac{1}{\Delta V}\int_A \phi dA^x \approx \frac{1}{\Delta V}\sum_{i=1}^{N}\phi_i A_i^x \qquad (4.12)$$

where ϕ_i represents the variable values at the elemental surfaces and N denotes the number of bounding surfaces on the elemental volume. Equation (4.12) applies for any type of finite-volume element that can be represented within the numerical grid. For a quadrilateral element in two dimensions with structured mesh, as seen in Figure 4.4, N has the value of four, since there are four bounding surfaces of the element. In three dimensions, for a hexagonal element, N becomes six. Similarly, the first-order derivative for ϕ in the y direction is obtained in exactly the same fashion, which can be written as

$$\left(\frac{\partial\phi}{\partial y}\right) = \frac{1}{\Delta V}\int_{\Delta V} \frac{\partial\phi}{\partial y}dV = \frac{1}{\Delta V}\int_A \phi dA^y \approx \frac{1}{\Delta V}\sum_{i=1}^{N}\phi_i A_i^y \qquad (4.13)$$

Example 4.1

Consider the conservation of mass described in Chapter 3. Determine the discretized form of the two-dimensional continuity equation, $\frac{\partial u}{\partial x}+\frac{\partial v}{\partial y}=0$, by the finite-volume method in a structured uniform grid arrangement.

Solution

An elemental control volume of the two-dimensional structured grid is shown in Figure 4.1.1. The centroid of the control volume is indicated by the point P, which is surrounded by the adjacent control volumes having their respective centroids indicated by these points: east, E; west, W; north, N; and south, S. The control volume face between points P and E is denoted by the area A_e^x. Subsequently, the rest of the control volume faces are, respectively, A_w^x, A_n^y and A_s^y.

We begin the analysis by introducing the control volume integration, which forms the key step of the finite-volume method. Applying Eqs. (4.12) and (4.13) yields the following expression, which is applicable to both structured and unstructured grids:

$$\frac{1}{\Delta V}\int_V \frac{\partial u}{\partial x}dV = \frac{1}{\Delta V}\int_A u dA^x \approx \frac{1}{\Delta V}\sum_{i=1}^{4}u_i A_i^x = \frac{1}{\Delta V}\left(u_e A_e^x - u_w A_w^x + u_n A_n^x - u_s A_s^x\right)$$

$$\frac{1}{\Delta V}\int_V \frac{\partial v}{\partial y}dV = \frac{1}{\Delta V}\int_A v dA^y \approx \frac{1}{\Delta V}\sum_{i=1}^{4}v_i A_i^y = \frac{1}{\Delta V}\left(v_e A_e^y - v_w A_w^y + v_n A_n^y - v_s A_s^y\right)$$

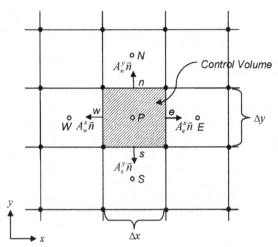

FIGURE 4.1.1   Control volume for the two-dimensional continuity equation problem.

For the structured uniform grid arrangement, the projection areas A_n^x and A_s^x in the x direction and the projection areas A_e^y and A_w^y in the y direction are zero. One important aspect demonstrated here by the finite-volume method is that it allows direct discretization in the physical domain (or in a body-fitted conformal grid) without the need for transforming the continuity equation from the physical domain to a computational domain.

Since the grid has been taken to be uniform, the face velocities u_e, u_w, v_n and v_s are located midway between each of the control volume centroids, which allows us to determine the face velocities from the values located at the centroids of the control volumes. Thus,

$$u_e = \frac{u_P + u_E}{2}; \quad u_w = \frac{u_P + u_W}{2}; \quad v_n = \frac{v_P + v_N}{2}; \quad v_s = \frac{v_P + v_S}{2}$$

After substitution of the above expressions into the discretized form of the velocity first-order derivatives, the final form of the discretized continuity equation becomes

$$\left(\frac{u_P + u_E}{2}\right)A_e^x - \left(\frac{u_P + u_W}{2}\right)A_w^x + \left(\frac{v_P + v_N}{2}\right)A_n^y - \left(\frac{v_P + v_S}{2}\right)A_s^y = 0$$

From Figure 4.1.1, $A_e^x = A_w^x = \Delta y$ and $A_n^y = A_s^y = \Delta x$; the above equation can then be expressed by

$$\left(\frac{u_P + u_E}{2}\right)\Delta y - \left(\frac{u_P + u_W}{2}\right)\Delta y + \left(\frac{v_P + v_N}{2}\right)\Delta x - \left(\frac{v_P + v_S}{2}\right)\Delta x = 0$$

and reduced to

$$\left(\frac{u_E - u_W}{2}\right)\Delta y + \left(\frac{v_N - v_S}{2}\right)\Delta x = 0$$

or, in another form,

$$\frac{u_E - u_W}{2\Delta x} + \frac{v_N - v_S}{2\Delta y} = 0$$

For a uniform grid arrangement, the distances between P and E and W and P are equivalent to Δx. Similarly, the distances between P and N and S and P are given by Δy. If the finite-difference method is applied to discretize the continuity equation through the central-difference scheme, we obtain the same discretized form of the equation at point P as compared to the form derived through the finite-volume method. The accuracy obtained is of second order, as inferred from the finite-difference central-difference scheme.

Discussion

The purpose of this example is to demonstrate the use of the finite-volume method to discretize the two-dimensional continuity equation and compare its form to the finite-difference approximation. We observe that the exact representation of the discretized form for the continuity equation can be obtained by applying either the finite-volume or finite-difference method for a uniform grid arrangement. Nevertheless, there are two major advantages that the finite-volume method holds over the finite-difference method. First, it has good conservation properties from the physical viewpoint, and second, it allows the accommodation of complicated physical domains to be discretized in a simpler way, rather than requiring the transformation of the equation to generalized coordinates in the computational domain.

From the above example, the first-order derivatives that appear in the continuity equation have been discretized through the finite-volume method. The discretization of the second-order derivatives is no different from that for the first-order derivatives. The second-order derivative along the x direction, for example, the diffusion terms represented in Eqs. (3.24) and (3.25), can be evaluated by

$$\left(\frac{\partial^2 \phi}{\partial x^2}\right) = \frac{1}{\Delta V}\int_{\Delta V}\frac{\partial^2 \phi}{\partial x^2}dV = \frac{1}{\Delta V}\int_A \frac{\partial \phi}{\partial x}dA^x \approx \frac{1}{\Delta V}\sum_{i=1}^{N}\left(\frac{\partial \phi}{\partial x}\right)_i A_i^x \quad (4.14)$$

An analogous expression can be easily obtained for the second-order derivative with respect to y, which is given by

$$\left(\frac{\partial^2 \phi}{\partial y^2}\right) = \frac{1}{\Delta V}\int_{\Delta V}\frac{\partial^2 \phi}{\partial y^2}dV = \frac{1}{\Delta V}\int_A \frac{\partial \phi}{\partial y}dA^y \approx \frac{1}{\Delta V}\sum_{i=1}^{N}\left(\frac{\partial \phi}{\partial y}\right)_i A_i^y \quad (4.15)$$

From Eqs. (4.14) and (4.15), it can be seen that in order to approximate the second-order derivatives, the respective first-order derivatives of $(\partial \phi/\partial x)_i$ and $(\partial \phi/\partial y)_i$ appearing in the equations are required to be evaluated at the elemental surfaces of the control volume.

The approximation of the first-order derivatives at the control volume faces is usually determined from the discrete ϕ values of the surrounding elements. For example, in a structured mesh arrangement, as shown in Figure 4.5, where the central control volume (shaded) is surrounded by only one adjacent control volume at each face, the first-order derivatives could be approximated by a piecewise-linear gradient profile between the central and adjacent nodes. If needed, higher-order quadratic profiles could also be employed to attain higher accuracy for the numerical solution, which requires more surrounding elemental volumes within the mesh system.

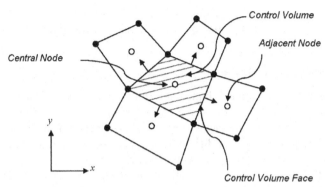

FIGURE 4.5   A representation of a structured-mesh arrangement (open symbols at the center of the control volumes denote computational nodes) for the evaluation of the face first-order derivatives.

4.2.3   Finite-Element Method

In many aspects, the finite-element method, which was developed initially as an ad hoc engineering procedure for determining the stress and strain displacement solutions in structural analysis, is similar to the finite-volume method. Both the finite-element and finite-volume methods are suitable for irregular computational domains, meaning that they can accommodate complex geometries.

Nevertheless, one distinguishing feature of the finite-element method is that the governing equations are first approximated by multiplication with the so-called shape functions before they are integrated over the entire computational domain. For the domain that is divided into a set of finite elements, the generic variable ϕ can be approximated by

$$\phi = \sum_{j=1}^{n} \phi_j \psi_j(x, y, z) \tag{4.16}$$

where n represents the number of discrete nodal unknowns ϕ_j and $\psi_j(x, y, z)$ are the shape functions. For the consideration of linear shape functions, they can be constructed simply from the values at the corners of the elements. As a general guide, the use of linear shape functions generates solutions of about the same accuracy as those of the second-order finite-difference method. This approximation is then substituted into the integral of the weighted residual over the computation domain that is taken to be equal to zero,

$$\iiint W_m(x, y, z) R \, dx \, dy \, dz = 0 \tag{4.17}$$

in order to generate a system of algebraic equations for ϕ_j, which normally can be solved via numerical methods. In Eq. (4.17), R is referred to as the equation residual, while W_m represents the weight functions. Different choices of W_m give rise to different methods. For example, if the Galerkin method is adopted, the weight functions are chosen to be the same as the shape functions. More details on the application of the finite-element method to fluid mechanics can be found in Thomasset (1981), Baker (1983), and Fletcher (1984).

4.2.4 Spectral Method

The finite-difference, finite-volume, and finite-element methods are generally considered *local* methods. Consequently, these methods can be identified in terms of discrete nodal unknowns. In contrast, the spectral method is a *global* method, which makes it more difficult to implement in practice. Nevertheless, the method allows approximate solutions with a high degree of accuracy; the number of grid points required to achieve the desired precision can therefore be very low.

The spectral method is actually built on the same principal ideas as the finite-element method discussed in the previous section. One main difference between the two methods is that the spectral method approximates the solution with a linear combination of continuous functions that are generally non-zero throughout the whole domain (Fourier series, Legendre polynomials, and Chebyshev polynomials), while the finite-element method approximates the solution of piecewise functions that are non-zero on localized subdomains. Thus, the spectral method works best when the solution is smooth.

As in the finite-element method, the starting point for the spectral method is the introduction of the approximate solution of ϕ, which can be assumed to be

$$\phi = \sum_{j=1}^{n} a_j(t)\, \varphi_j(x, y, z) \tag{4.18}$$

where $a_j(t)$ are the unknown coefficients and $\varphi_j(x, y, z)$ are the trial functions. The choice of trial functions is dependent on the particular consideration of the flow problem being solved. Fourier series are particularly suited for periodic boundary conditions, while Chebyshev polynomials are more appropriate for non-periodic boundary conditions. Substitution of Eq. (4.18) into the transport equation yields the residual of the equation. The unknown coefficients are now determined via the integration of the vanishing weighted residual over the computational domain, i.e., applying Eq. (4.17). Here again, if the Galerkin method is adopted, the weight functions are automatically chosen from the same family as the trial functions. More details on the application of the spectral method to fluid mechanics can be found in Fletcher (1984) and Canuto et al. (1987).

4.3 CONVERTING GOVERNING EQUATIONS TO ALGEBRAIC EQUATION SYSTEM

The basic derivations of the finite-difference and finite-volume methods and some basic descriptions of the finite-element and spectral methods have been provided. Primarily concentrating on the finite-difference and finite-volume methods, the next task is to consider the appropriate application of these various discretization techniques to numerically solving the governing partial differential equations. A numerical method is developed by considering the three transport processes associated with (1) pure diffusion in steady state, (2) steady convection-diffusion, and (3) unsteady convection-diffusion. We further simplify the problem by systematically working through a one-dimensional equation in order to demonstrate the application of the discretization techniques to finally attain the algebraic form of the governing equation. It will be seen later that the discretized form of the equation is easily extended and accommodated for two- and three-dimensional diffusion problems.

Pure diffusion process: Let us consider the steady-state diffusion of the generic variable ϕ in a one-dimensional domain. The equation that governs such a process is

$$\frac{\partial}{\partial x}\left(\Gamma\frac{\partial\phi}{\partial x}\right) + S_\phi = 0 \tag{4.19}$$

where Γ is the diffusion coefficient and S_ϕ is the source term. This equation is typical of a one-dimensional heat-conduction process, which we further investigate in detail through worked examples. For this problem, the applications of the finite-difference and finite-volume methods to discretize the equation are illustrated.

4.3.1 Finite-Difference Method

The use of the finite-difference method is explored here. The first step in developing a numerical solution for this method involves dividing the geometric domain into discrete nodal points. Let us consider a general nodal point P and its surrounding neighboring nodal points to the west and east, W and E, for the one-dimensional geometry as demonstrated in Figure 4.6. A uniform grid spanning the three nodal points W, P, and E is produced.

The next step in the discretization process is to discretize Eq. (4.19) around the nodal point P. To derive a suitable expression for the finite-difference method, Eq. (4.19) is required to be expanded into its non-conservative form, which is given by

$$\frac{\partial\Gamma}{\partial x}\frac{\partial\phi}{\partial x} + \Gamma\frac{\partial^2\phi}{\partial x^2} + S_\phi = 0 \tag{4.20}$$

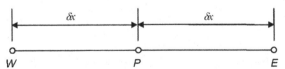

FIGURE 4.6 A schematic representation of the uniform grid spacing along the x direction for the one-dimensional geometry.

By applying the central differencing of the first-order derivative equation, Eq. (4.8), and the second-order derivative equation, Eq. (4.9), the discretized form of Eq. (4.20) is obtained as

$$\frac{(\Gamma_E - \Gamma_W)}{2\delta x}\frac{(\phi_E - \phi_W)}{2\delta x} + \Gamma_P \frac{(\phi_E - 2\phi_P + \phi_W)}{\delta x^2} + S_\phi = 0 \qquad (4.21)$$

After some rearrangement, Eq. (4.21) can be alternatively expressed as

$$\frac{2\Gamma_P}{\delta x^2}\phi_P = \left[\frac{(\Gamma_E - \Gamma_W)}{4\delta x^2} + \frac{\Gamma_P}{\delta x^2}\right]\phi_E + \left[-\frac{(\Gamma_E - \Gamma_W)}{4\delta x^2} + \frac{\Gamma_P}{\delta x^2}\right]\phi_W + S_\phi \quad (4.22)$$

Identifying the coefficients of ϕ_E and ϕ_W in Eq. (4.19) as a_E and a_W and the coefficient of ϕ_P as a_P, the equation can be written in a simple form:

$$a_P\phi_P = a_E\phi_E + a_W\phi_W + b \qquad (4.23)$$

where

$$a_E = \frac{(\Gamma_E - \Gamma_W)}{4\delta x^2} + \frac{\Gamma_P}{\delta x^2}; \quad a_W = -\frac{(\Gamma_E - \Gamma_W)}{4\delta x^2} + \frac{\Gamma_P}{\delta x^2}; \quad a_P = \frac{2\Gamma_P}{\delta x^2}; \quad b = S_\phi$$

4.3.2   Finite-Volume Method

When the finite-volume method is applied, we have to consider the physical domain as being divided into finite control volumes surrounding the nodal points W, P, and E. Figure 4.7 shows a control volume surrounding the nodal point P. The distances between the nodes W and P, and between nodes P and E, are identified by the respective notations δx_W and δx_E. For this one-dimensional case, the control volume width surrounding the nodal point P is Δx, since Δy and Δz have dimensions of unit length.

To apply the finite-volume discretization, the gradient term in Eq. (4.19) can be approximated by the use of Eq. (4.12). This gives

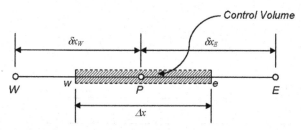

FIGURE 4.7 A schematic representation of a control volume around node P in a one-dimensional domain using the finite-volume method.

$$\frac{\partial}{\partial x}\left(\Gamma\frac{\partial\phi}{\partial x}\right) = \frac{1}{\Delta V}\int\limits_{\Delta V}\frac{\partial}{\partial x}\left(\Gamma\frac{\partial\phi}{\partial x}\right)dV = \frac{1}{\Delta V}\int\limits_{A}\left(\Gamma\frac{\partial\phi}{\partial x}\right)dA^x$$

$$\approx \frac{1}{\Delta V}\sum_{i=1}^{2}\left(\Gamma\frac{\partial\phi}{\partial x}\right)_i A_i^x \qquad (4.24)$$

Here, the projected areas A_i^x for the one-dimensional case are given by $A_1^x = -A_W$ and $A_2^x = A_E$. Equation (4.24) can thus be written as

$$\left(\Gamma\frac{\partial\phi}{\partial x}\right)_e A_E - \left(\Gamma\frac{\partial\phi}{\partial x}\right)_w A_w \qquad (4.25)$$

For the remaining term in the equation, the source term is approximated as

$$\frac{1}{\Delta V}\int\limits_{\Delta V} S_\phi dV = S_\phi \qquad (4.26)$$

where S_ϕ is assumed to be constant within ΔV, which is the finite control volume. The final form of the discretized equation becomes

$$\frac{1}{\Delta V}\left(\Gamma\frac{\partial\phi}{\partial x}\right)_e A_E - \frac{1}{\Delta V}\left(\Gamma\frac{\partial\phi}{\partial x}\right)_w A_w + S_\phi = 0 \qquad (4.27)$$

To express an algebraic form for Eq. (4.27) with the nodal points W, E, and P, approximations to the gradients $\partial\phi/\partial x$ at the west (w) and east (e) faces of the control volume are required. We will assume the piecewise-linear gradient profiles spanning the nodal points between W and P and between P and E to sufficiently approximate the first-order derivatives at w and e; the diffusive fluxes are evaluated as

$$\left(\Gamma\frac{\partial\phi}{\partial x}\right)_e A_E = \Gamma_e A_E\left(\frac{\phi_E - \phi_P}{\delta x_E}\right) \qquad (4.28)$$

$$\left(\Gamma\frac{\partial\phi}{\partial x}\right)_w A_W = \Gamma_w A_W\left(\frac{\phi_P - \phi_W}{\delta x_W}\right) \qquad (4.29)$$

Substitution of Eqs. (4.28) and (4.29) into Eq. (4.27) gives

$$\frac{\Gamma_e A_E}{\Delta V}\left(\frac{\phi_E - \phi_P}{\delta x_E}\right) - \frac{\Gamma_w A_W}{\Delta V}\left(\frac{\phi_P - \phi_W}{\delta x_W}\right) + S_\phi = 0 \qquad (4.30)$$

Equation (4.30) presents a very attractive feature of the finite-volume method. This discretized equation possesses a clear physical interpretation. It states that the difference between the diffusive fluxes of ϕ at the east and west faces of the control volume equal the generation of ϕ, and it constitutes a balance equation for ϕ over the control volume. Equation (4.30) can be re-arranged as

$$\frac{1}{\Delta V}\left(\frac{\Gamma_e A_E}{\delta x_E} + \frac{\Gamma_w A_W}{\delta x_W}\right)\phi_P = \frac{1}{\Delta V}\left(\frac{\Gamma_e A_E}{\delta x_E}\right)\phi_E + \frac{1}{\Delta V}\left(\frac{\Gamma_w A_W}{\delta x_W}\right)\phi_W + S_\phi \quad (4.31)$$

As above, by identifying the coefficients of ϕ_E and ϕ_W in Eq. (4.31) as a_E and a_W and the coefficient of ϕ_P as a_P, the algebraic form can be written as

$$a_P\phi_P = a_E\phi_E + a_W\phi_W + b \quad (4.32)$$

where

$$a_E = \frac{\Gamma_e A_E}{\Delta V \delta x_E}; \quad a_W = \frac{\Gamma_w A_W}{\Delta V \delta x_W}; \quad a_P = a_E + a_W; \quad b = S_\phi$$

Equation (4.32) represents the discretized form through the finite-volume method for Eq. (4.19). For the one-dimensional problem considered here, the face areas A_E and A_W are unity, since Δy and Δz have dimensions of unit length; the finite control volume ΔV is therefore the width Δx.

4.3.3 Comparison of the Finite-Difference and Finite-Volume Discretizations

Although the same algebraic form of equation for the steady-state one-dimensional diffusion process is obtained, different expressions for the coefficients of a_E, a_W, and a_P are derived, as seen in Eq. (4.23) for the finite-difference method and Eq. (4.32) for the finite-volume method. Nevertheless, let us consider a special case where the diffusion coefficient is spatially invariant and the mesh is uniformly distributed. The coefficients in Eq. (4.23) for the finite-difference method reduce to

$$a_E = \frac{\Gamma}{\delta x^2}; \quad a_W = \frac{\Gamma}{\delta x^2}; \quad a_P = a_E + a_W; \quad b = S_\phi \quad (4.33)$$

while for the finite-volume method the coefficients in Eq. (4.32) become

$$a_E = \frac{\Gamma}{\delta x^2}; \quad a_W = \frac{\Gamma}{\delta x^2}; \quad a_P = a_E + a_W; \quad b = S_\phi \quad (4.34)$$

where the control volume is $\Delta x = \delta x$ (uniform grid and Δy, Δz is unity). From the example above for the discretization of the continuity equation, the resultant algebraic equations are again the same for either the finite-difference or the finite-volume discretization.

It should be noted that the finite-difference method generally requires a uniformly distributed mesh in order to apply the first- and second-order derivative approximations to the governing equation. For a non-uniform grid distribution, some mathematical manipulation (e.g., transformation functions) is required to transform Eq. (4.20) into a computational domain in generalized coordinates before applying the finite-difference approximations. However, this requirement is not a prerequisite for the finite-volume method. Because of the availability of different control volume sizes, any non-uniform grid can therefore be easily accommodated.

In comparison to the finite-difference method, which is mathematically derived from the Taylor series, the finite-volume method ensures that the property is conserved, and it retains this physical significance throughout the discretization process. Almost all commercial CFD codes adopt the finite-volume discretization of the Navier–Stokes equation to obtain numerical solutions for complex fluid-flow problems because the mesh is not restricted to structured-type elements but can include a variety of unstructured-type elements of different shapes and sizes.

Example 4.2

Consider the problem of steady heat conduction in a large brick plate with uniform heat generation. The faces A and B, as shown in Figure 4.2.1, are maintained at constant temperatures. The governing equation is of the generic form presented in Eq. (4.15). The diffusion coefficient Γ governing the heat-conduction problem becomes the thermal conductivity k of the material. For a given thickness $L = 2$ cm with constant thermal conductivity $k = 5$ W/m$^2 \cdot$ K, determine the steady-state distribution in the plate. Temperatures at T_A and T_B are, respectively, 100°C and 400°C, and heat generation q is 500 kW/m^3.

Solution

Assuming that the dimensions in the y direction and z direction are so large that the temperature gradients are only significant in the x direction, we can reduce

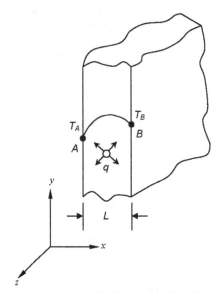

FIGURE 4.2.1   Schematic representation of the large brick plate with heat generation.

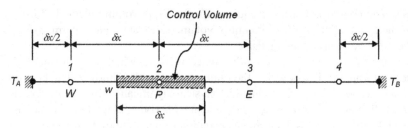

FIGURE 4.2.2   Finite-volume discretization of the large brick plate domain.

the problem to a one-dimensional analysis. We apply the finite-volume method to obtain the solution of this simple heat-conduction problem.

Let us divide the domain into four control volumes (see Figure 4.2.2) giving $\delta x = 0.005$ m. There are four nodal points, each representing the central location for the four control volumes. For illustration purposes, a unit area is considered in the y-z plane.

Since the thermal conductivity is constant, we can define $k = k_e = k_w$; and the volumetric source term S_ϕ equal to q, the general discretized form of the equation at point P (node 2), as from Eq. 4.29, is given as

$$a_P T_P = a_E T_E + a_W T_W + b$$

where the coefficients are

$$a_E = \frac{kA}{\delta x^2}; \quad a_W = \frac{kA}{\delta x^2}; \quad a_P = a_E + a_W; \quad b = q$$

This algebraic equation is also valid for the control volume of node 3.

For the control volumes of nodes 1 and 4, we apply a linear approximation for the temperatures between the boundary points and adjacent nodal point of the control volume. At the west face of the control volume at the fixed end, the temperature is known by T_A (see Figure 4.2.3). We begin by re-visiting Eq. (4.24).

For a constant thermal conductivity and uniform heat generation, the equation becomes

$$k\left(\frac{\partial T}{\partial x}\right)_e - k\left(\frac{\partial T}{\partial x}\right)_w + q\delta x = 0$$

Introducing linear approximations to the gradients at the east and west faces of control volume 1 gives

$$kA\left(\frac{T_E - T_P}{\delta x}\right) - kA\left(\frac{T_P - T_A}{\delta x/2}\right) + q\delta x = 0$$

The above equation can be re-arranged to yield the discretized equation for node 1:

$$a_P T_P = a_E T_E + a_W T_W + b$$

where the coefficients are

$$a_E = \frac{kA}{\delta x}; \quad a_W = 0; \quad a_P = a_E + a_W + \frac{2kA}{\delta x}; \quad b = q\delta x + \frac{2kA}{\delta x} T_A$$

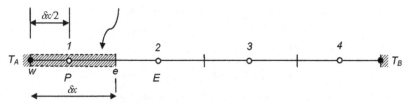

FIGURE 4.2.3   Finite-volume discretization of the large brick plate domain, showing the first control volume.

The area $a_W = 0$, since it is a fixed end. At node 4, the temperature of the east face of the control volume is known. The node is similarly treated, and we obtain

$$k\left(\frac{\partial T}{\partial x}\right)_e - k\left(\frac{\partial T}{\partial x}\right)_w + q\delta x = 0$$

and

$$kA\left(\frac{T_B - T_P}{\delta x/2}\right) - kA\left(\frac{T_P - T_W}{\delta x}\right) + q\delta x = 0$$

with the discretized equation for node 4 given as

$$a_P T_P = a_E T_E + a_W T_W + b$$

where

$$a_E = 0; \quad a_W = \frac{kA}{\delta x}; \quad a_P = a_E + a_W + \frac{2kA}{\delta x}; \quad b = q\delta x + \frac{2kA}{\delta x}T_B$$

Substitution of numerical values for the thermal conductivity $k = 5$ W/m$^2 \cdot$ K, heat generation $q = 500$ kW/m^3, $\delta x = 0.005$ m, and unit area $A = 1$ m^2 throughout provides the coefficients of the discretized equation summarized in Table 4.2.1.

The resulting set of algebraic equations for this example is

$$\begin{aligned}
3000\,T_1 &= 1000\,T_2 + 2000\,T_A + 2500 \\
2000\,T_2 &= 1000\,T_1 + 1000\,T_3 + 2500 \\
2000\,T_3 &= 1000\,T_2 + 1000\,T_4 + 2500 \\
3000\,T_4 &= 1000\,T_3 + 2000\,T_B + 2500
\end{aligned}$$

TABLE 4.2.1 The coefficients at each node of the control volumes

Node	a_E	a_W	a_P	b
1	1000	0	3000	$2500 + 2000\,T_A$
2	1000	1000	2000	2500
3	1000	1000	2000	2500
4	0	1000	3000	$2500 + 2000\,T_B$

This set of equations can be re-arranged in matrix form as

$$\begin{bmatrix} 3000 & -1000 & 0 & 0 \\ -1000 & 2000 & -1000 & 0 \\ 0 & -1000 & 2000 & -1000 \\ 0 & 0 & -1000 & 3000 \end{bmatrix} \begin{bmatrix} T_1 \\ T_2 \\ T_3 \\ T_4 \end{bmatrix} = \begin{bmatrix} 2000T_A + 2500 \\ 2500 \\ 2500 \\ 2000T_B + 2500 \end{bmatrix}$$

The above set of equations yields the steady-state temperature distribution for the given situation.

Discussion

For a one-dimensional steady heat-conduction process, we obtained the algebraic equations in matrix form. Because the problem involves only a small number of nodes, the matrix can be solved directly with a software package like MATLAB (*The Student Edition of MATLAB*, The Math Works Inc., 1992). Analytically, the matrix can be solved by methods like Gaussian elimination. This matrix algorithm is discussed in the next section.

In most CFD problems, however, the complexity and size of a set of equations depend on the dimensionality of the problem and the number of grid nodes, where the system can consist of up to 100,000 or 1 million equations. For such problems, available computer resources are a powerful constraint and the Gaussian elimination may not be the most economical way to obtain the solution of the discretized equations. We explore other algorithms that are currently available to efficiently obtain the solution to the algebraic equations in Section 4.3.

Steady convection-diffusion process: In the absence of sources, the equation governing the steady convection-diffusion process of a property ϕ in a given one-dimensional flow field u from Eq. (3.53) simplifies to

$$\frac{\partial}{\partial x}(\rho u \phi) = \frac{\partial}{\partial x}\left(\Gamma \frac{\partial \phi}{\partial x}\right) \qquad (4.35)$$

The mass conservation for the one-dimensional convection-diffusion process is enforced through

$$\frac{\partial(\rho u)}{\partial x} = 0 \qquad (4.36)$$

By applying the finite-volume method based on the elemental volume described in Figure 4.7, the algebraic form of Eq. (4.35) can be expressed as

$$(\rho u \phi)_e A_E - (\rho u \phi)_w A_W = \left(\Gamma \frac{\partial \phi}{\partial x}\right)_e A_E - \left(\Gamma \frac{\partial \phi}{\partial x}\right)_w A_W \qquad (4.37)$$

Assuming piecewise-linear gradient profiles spanning the nodal points between W and P and between P and E, the first-order derivatives at w and e of the diffusive fluxes can be similarly approximated using the expressions formulated in Eqs. (4.28) and (4.29) to yield

$$(\rho u \phi)_e A_E - (\rho u \phi)_w A_W = \Gamma_e A_E \left(\frac{\phi_E - \phi_P}{\delta x_E} \right) - \Gamma_w A_W \left(\frac{\phi_P - \phi_W}{\delta x_W} \right) \quad (4.38)$$

In hindsight, it seems sensible for the convective term to be approximated in a similar fashion. Using the Taylor series expansion, the interface values ϕ_w and ϕ_e at the respective cell faces w and e can be obtained via linear interpolation between the neighboring points as

$$\phi_w = \frac{1}{2}(\phi_W + \phi_P) \quad (4.39)$$

$$\phi_e = \frac{1}{2}(\phi_P + \phi_E) \quad (4.40)$$

The above approximation is second-order accurate and the scheme does not exhibit any bias on the flow direction. Using these relationships, Eq. (4.38) thus becomes

$$(\rho u)_e A_E \frac{1}{2}(\phi_P + \phi_E) - (\rho u)_w A_W \frac{1}{2}(\phi_W + \phi_P) =$$
$$\Gamma_e A_E \left(\frac{\phi_E - \phi_P}{\delta x_E} \right) - \Gamma_w A_W \left(\frac{\phi_P - \phi_W}{\delta x_W} \right) \quad (4.41)$$

Equation (4.41) can be re-arranged as

$$\left(\frac{\Gamma_e A_E}{\delta x_E} + \frac{(\rho u)_e A_E}{2} + \frac{\Gamma_w A_W}{\delta x_W} - \frac{(\rho u)_w A_W}{2} \right) \phi_P =$$
$$\left(\frac{\Gamma_e A_E}{\delta x_E} - \frac{(\rho u)_e A_E}{2} \right) \phi_E + \left(\frac{\Gamma_w A_W}{\delta x_W} + \frac{(\rho u)_w A_W}{2} \right) \phi_W \quad (4.42)$$

By identifying the coefficients of ϕ_E and ϕ_W in Eq. (4.42) as a_E and a_W and the coefficient of ϕ_P as a_P, the algebraic form can be written as

$$a_P \phi_P = a_E \phi_E + a_W \phi_W \quad (4.43)$$

where

$$a_E = \frac{\Gamma_e A_E}{\delta x_E} - \frac{(\rho u)_e A_E}{2}; \quad a_W = \frac{\Gamma_w A_W}{\delta x_W} + \frac{(\rho u)_w A_W}{2}$$
$$a_P = a_E + a_W + (\rho u)_e A_E - (\rho u)_w A_w$$

One well-known inadequacy of the central differencing scheme in a strongly convective flow is its inability to identify the flow direction. It is well recognized that the above treatment usually results in large "undershoots" and "overshoots," eventually causing the numerical procedure to diverge. Increasing the mesh resolution for the flow domain with very small grid spacing can probably overcome the problem, but such an approach usually precludes practical flow calculations from being carried out robustly and effectively.

Consider for a moment the flow moving from the upstream point W (left) to the downstream point E (right). Through the central differencing approximation, the interface values of ϕ are always assumed to be equally weighted by the influence of the available variables at the neighboring grid nodal points. This implies that the downstream values of ϕ_P and ϕ_E are prevailing during the evaluation of ϕ_w and ϕ_e. In the majority of flow cases, these values are not known a priori; the remedy is therefore to design a numerical solution to recognize the direction of the flow by exerting an unequal weighting influence based on the available variables located at the surrounding grid nodal points to appropriately determine the interface values. This is essentially the hallmark of the *upwind* or *donor-cell* concept. The first-order upwind scheme is described here. With reference to Figure 4.7, if the interface velocities $u_w > 0$ and $u_e > 0$, the interface values ϕ_w and ϕ_e, according to the donor-cell concept, are now approximated according to their upstream neighboring counterparts as

$$\phi_w = \phi_W \tag{4.44}$$

$$\phi_e = \phi_P \tag{4.45}$$

Using the above relationships, Eq. (4.38) can be expressed as

$$(\rho u)_e A_E \phi_P - (\rho u)_w A_W \phi_W = \Gamma_e A_E \left(\frac{\phi_E - \phi_P}{\delta x_E} \right) - \Gamma_w A_W \left(\frac{\phi_P - \phi_W}{\delta x_W} \right) \tag{4.46}$$

and the coefficients a_E, a_W, and a_P as described by Eq. (4.43) are given by

$$a_E = \frac{\Gamma_e A_E}{\delta x_E}; \quad a_W = \frac{\Gamma_w A_W}{\delta x_W} + (\rho u)_w A_W; \quad a_P = a_E + a_W + (\rho u)_e A_E - (\rho u)_w A_w$$

Similarly, if the interface velocities $u_w < 0$ and $u_e < 0$, the interface values ϕ_w and ϕ_e are conversely evaluated by

$$\phi_w = \phi_P \tag{4.47}$$

$$\phi_e = \phi_E \tag{4.48}$$

Using the above relationships, Eq. (4.38) can be written as

$$(\rho u)_e A_E \phi_E - (\rho u)_w A_W \phi_P = \Gamma_e A_E \left(\frac{\phi_E - \phi_P}{\delta x_E} \right) - \Gamma_w A_W \left(\frac{\phi_P - \phi_W}{\delta x_W} \right) \tag{4.49}$$

and the coefficients a_E, a_W, and a_P are now given by

$$a_E = \frac{\Gamma_e A_E}{\delta x_E} - (\rho u)_e A_E; \quad a_W = \frac{\Gamma_w A_W}{\delta x_W}; \quad a_P = a_E + a_W + (\rho u)_e A_E - (\rho u)_w A_w$$

A more in-depth look at the different discretization schemes based on upwind and central differencing schemes can be described by three fundamental properties: *conservativeness*, *boundedness*, and *transportiveness*.

Conservativeness concerns the fluxes being represented in a reliable manner. The upwind and central differencing schemes use consistent expressions to evaluate the advective and diffusive fluxes at the control volume faces. This means that fluxes cancel out when summed and overall conservation is satisfied.

Boundedness concerns the condition whereby the resulting matrix of the coefficients is diagonally dominant to ensure numerical convergence. It can be demonstrated that the matrix coefficients for the central differencing scheme can become negative, which violates the requirement of boundedness and may lead to unphysical solutions. For the coefficient a_E, it is possible that it can become negative if convection dominates. For a_E to be positive, it must satisfy the following condition:

$$\frac{(\rho u)_e}{\Gamma_e/\delta x_E} = Pe_e < 2 \qquad (4.50)$$

If the Peclet number Pe_e is greater than 2, a_E becomes negative and this violates the requirement of boundedness and may result in physically impossible solutions. However, all coefficients in the upwind differencing scheme are positive and the coefficient matrix is diagonally dominant, which satisfies the requirement of boundedness.

Transportiveness concerns the direction of the flow. The central differencing scheme introduces the influence at node P from all directions from its neighboring nodes to calculate the advective and diffusive fluxes. Hence, this particular scheme does not recognize the direction or the strength of advection relative to diffusion and does not possess the transportiveness property. Nevertheless, the upwind differencing scheme accounts for flow direction and satisfies the transportiveness property.

It should be noted that, despite the favorable properties of the upwind differencing scheme in satisfying the above fundamental properties and promoting numerical stability, it is widely known that it generally causes unwanted numerical diffusion in space. In order to reduce these numerical errors, high-order approximations, such as the second-order upwind and third-order QUICK scheme, which are widely applied in many CFD problems, have been proposed. The formulation of these schemes is described in Appendix B.

Unsteady convection-diffusion process: The equation governing the unsteady convection and diffusion process of a property ϕ in a given one-dimensional flow field u is

$$\frac{\partial(\rho\phi)}{\partial t} + \frac{\partial}{\partial x}(\rho u \phi) = \frac{\partial}{\partial x}\left(\Gamma\frac{\partial\phi}{\partial x}\right) + S_\phi \qquad (4.51)$$

For the purpose of illustration, let us assume that the fluid is incompressible (i.e., density is constant) and Eq. (4.46) can be rearranged as

$$\frac{\partial \phi}{\partial t} + \frac{\partial}{\partial x}(u\phi) = \frac{1}{\rho}\frac{\partial}{\partial x}\left(\Gamma \frac{\partial \phi}{\partial x}\right) + \frac{S_\phi}{\rho} \qquad (4.52)$$

The mass conservation becomes

$$\frac{\partial u}{\partial x} = 0 \qquad (4.53)$$

Assuming central differencing for the convective and diffusive terms, the partial algebraic form of Eq. (4.51) via the finite-volume method is given by

$$\frac{\partial \phi}{\partial t} + u_e A_E \frac{1}{2}(\phi_W + \phi_P) - u_w A_W \frac{1}{2}(\phi_P + \phi_E) =$$
$$\Gamma_e A_E \frac{1}{\rho}\left(\frac{\phi_E - \phi_P}{\delta x_E}\right) - \Gamma_w A_W \frac{1}{\rho}\left(\frac{\phi_P - \phi_W}{\delta x_W}\right) + \frac{S_\phi}{\rho}\Delta V \qquad (4.54)$$

In the majority of cases, the time derivative can be approximated by applying the first-order forward difference scheme as above:

$$\frac{\partial \phi}{\partial t} = \frac{\phi^{n+1} - \phi^n}{\Delta t} = \frac{\phi_P^{n+1} - \phi_P^n}{\Delta t} \qquad (4.55)$$

where Δt is the incremental time step and the superscripts n and $n+1$ denote the previous and current time levels, respectively. It should be clear from the above that a suitable time-marching procedure is needed to appropriately update the property ϕ at the central point P and the neighboring points through time. Let us first examine the implication due to an *explicit* approach. For this approach, Eq. (4.54) can be written as

$$\frac{\phi_P^{n+1} - \phi_P^n}{\Delta t} + u_e^n A_E \frac{1}{2}(\phi_W^n + \phi_P^{n+1}) - u_w^n A_W \frac{1}{2}(\phi_P^{n+1} + \phi_E^n) =$$
$$\Gamma_e^n A_E \frac{1}{\rho}\left(\frac{\phi_E^n - \phi_P^{n+1}}{\delta x_E}\right) - \Gamma_w^n A_W \frac{1}{\rho}\left(\frac{\phi_P^{n+1} - \phi_W^n}{\delta x_W}\right) + \frac{S_\phi^n}{\rho}\Delta V \qquad (4.56)$$

Here again, by identifying the coefficients of ϕ_E and ϕ_W in Eq. (4.56) as a_E and a_W and the coefficient of ϕ_P as a_P, the algebraic form can be written as

$$a_P \phi_P^{n+1} = a_E \phi_E^n + a_W \phi_W^n + b \qquad (4.57)$$

where

$$a_E = \frac{\Gamma_e^n A_E}{\rho \delta x_E} - \frac{u_e^n A_E}{2}; \quad a_W = \frac{\Gamma_w^n A_W}{\rho \delta x_W} + \frac{u_w^n A_W}{2}$$

$$a_P = a_E + a_W + u_e^n A_E - u_w^n A_W + \frac{1}{\Delta t}; \quad b = \frac{S_\phi}{\rho}\Delta V + \frac{1}{\Delta t}\phi_P^n$$

Casting our attention to Eq. (4.57) and the sketch in Figure 4.8, we see a straightforward mechanism for evaluating the unknown ϕ_P^{n+1} (indicated by the square at time level $n+1$) that is calculated directly from all the values obtained from the indicated circles of the property ϕ at the previous time level n. By definition, in an *explicit* approach, each difference equation contains only one unknown and therefore can be solved *explicitly* for this unknown in a simple manner.

On the other hand, let us consider a time-matching procedure that requires the solution for all the variables at the time level $n+1$. Equation (4.54) can be rewritten as

$$\frac{\phi_P^{n+1} - \phi_P^n}{\Delta t} + u_e^{n+1} A_E \frac{1}{2}\left(\phi_W^{n+1} + \phi_P^{n+1}\right) - u_w^{n+1} A_W \frac{1}{2}\left(\phi_P^{n+1} + \phi_E^{n+1}\right) =$$
$$\Gamma_e^{n+1} A_E \frac{1}{\rho}\left(\frac{\phi_E^{n+1} - \phi_P^{n+1}}{\delta x_E}\right) - \Gamma_w^{n+1} A_W \frac{1}{\rho}\left(\frac{\phi_P^{n+1} - \phi_W^{n+1}}{\delta x_W}\right) + \frac{S_\phi^{n+1}}{\rho}\Delta V$$

$$(4.58)$$

In its algebraic form, Eq. (4.58) can be recast as

$$a_P \phi_P^{n+1} = a_E \phi_E^{n+1} + a_W \phi_W^{n+1} + b \qquad (4.59)$$

where

$$a_E = \frac{\Gamma_e^{n+1} A_E}{\rho \delta x_E} - \frac{u_e^{n+1} A_E}{2}; \quad a_W = \frac{\Gamma_w^{n+1} A_W}{\rho \delta x_W} + \frac{u_w^{n+1} A_W}{2}$$

$$a_P = a_E + a_W + u_e^{n+1} A_E - u_w^{n+1} A_W + \frac{1}{\Delta t}; \quad b = \frac{S_\phi}{\rho}\Delta V + \frac{1}{\Delta t}\phi_P^n$$

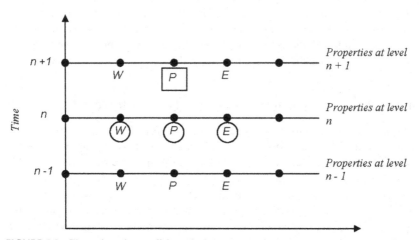

FIGURE 4.8 Illustration of an explicit method.

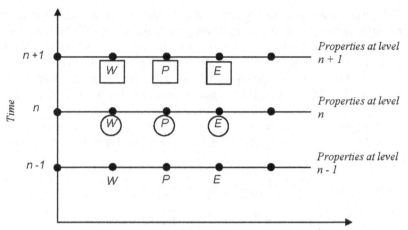

FIGURE 4.9 Illustration of an implicit method.

As sketched in Figure 4.9, the property ϕ at the central point P as well as the properties at the neighboring points are required to be solved simultaneously. This property ϕ may also need to be coupled with other flow variables, such as pressure and temperature, appearing in the source term $S\phi$ at the current time level within the same difference equation. This represents an example of the fully implicit approach. By definition, an *implicit* approach is one where the unknowns must be obtained by means of a simultaneous solution of the difference equations applied at *all* grid nodal points at a given time level. Implicit methods usually involve the manipulation of large matrices because of the need to solve large systems of algebraic equations.

The above explicit and implicit schemes are generally considered as methods having first order in time. Similar to the first order in space, these methods may also cause unwanted numerical diffusion in time. In order to reduce these numerical errors, second-order approximations, such as the explicit Adams-Bashford and semi-implicit Crank-Nicolson schemes, have been proposed. The formulation of these schemes is described in Appendix C.

4.4 NUMERICAL SOLUTIONS TO ALGEBRAIC EQUATIONS

The various discretization methods for the partial differential equations have been described. Through this process, we obtain a system of linear or non-linear algebraic equations that need to be solved by some numerical methods. The complexity and size of this set of equations depend on the dimensionality and geometry of the physical problem. Whether the equations are linear or nonlinear, efficient and robust numerical methods are required to solve the system. There are essentially two families of numerical methods: *direct methods* and *iterative methods*.

Previously, we highlighted a direct method, such as Gaussian elimination, which can be used to solve the resultant matrix form of a simple one-dimensional steady heat-conduction process. It is noted that there are other direct methods that may be employed, such as Cramer's rule for matrix inversion and the Thomas algorithm (tri-diagonal matrix algorithm). For finite-difference or finite-volume discretization on structured mesh, the resultant matrix of the algebraic equations is typically sparse; most of the elements are zero and the non-zero terms are close to the diagonal. For one-dimensional situations, the Thomas algorithm or tri-diagonal matrix algorithm is actually a direct method that takes advantage of this particular matrix structure. It is computationally inexpensive and requires a minimum amount of storage in the core memory and as a result is extensively used in CFD programs. (The Thomas algorithm is described in the next section.)

Iterative methods are, however, based on repeated applications of an algorithm, leading to its eventual convergence after a number of repetitions. Iterative methods are generally much more economical, and only non-zero terms of the algebraic equations are required to be stored in the core memory. For non-linear problems, they are used out of necessity, but they are just as valuable for sparse linear systems. Well-known point-by-point methods, such as the Jacobi and Gauss-Siedel methods, are described in Section 4.4.2 in order to provide the reader with a basic understanding of iterative methods. Other variants from these two iterative methods are also described, particularly those algorithms that are used in solving CFD problems.

4.4.1 Direct Methods

One of the most basic methods for solving linear systems of algebraic equations is *Gaussian elimination*. The algorithm derives from the systematic reduction of large systems of equations to smaller ones. Let us suppose that the systems of equations can be written in the form

$$A\phi = B \tag{4.60}$$

where ϕ is the unknown nodal variables. Matrix A contains non-zero coefficients of the algebraic equations as illustrated below:

$$A = \begin{bmatrix} A_{11} & A_{12} & A_{13} & \cdots & A_{1n} \\ A_{21} & A_{22} & A_{23} & \cdots & A_{2n} \\ A_{31} & A_{32} & A_{33} & \cdots & A_{3n} \\ \vdots & \vdots & \vdots & \ddots & \vdots \\ A_{n1} & A_{n2} & A_{n3} & \cdots & A_{nn} \end{bmatrix} \tag{4.61}$$

while B comprises known values of ϕ, for example, that are given by the boundary conditions or source/sink terms.

It can be observed that the diagonal coefficients of matrix A are represented by the entries $A_{11}, A_{22}, \ldots, A_{nn}$. The heart of the algorithm is to eliminate the entries below the diagonal to yield a lower triangle of zeros. This means eliminating the elements $A_{21}, A_{31}, A_{32}, \ldots, A_{nn-1}$ by replacing them with zeros. We begin the elimination process by considering the first column of elements $A_{21}, A_{31}, \ldots$, A_{n1} in matrix A. If we multiply the first row of the matrix by A_{21}/A_{11} and subtract these values from the second row, all the elements in the second row are subsequently modified, which includes the terms in B on the right-hand side of the equations. The other elements $A_{31}, A_{41}, \ldots, A_{n1}$ in the first column of matrix A are treated similarly, so that repeating this process down the first column reduces all the elements below A_{11} to zero. The same procedure is then applied in the second column (for all elements below A_{22}), and so forth, until the process reaches the $n-1$ column.

After this process is complete, the original matrix A becomes an *upper triangular matrix* that is given by

$$
U = \begin{bmatrix}
A_{11} & A_{12} & A_{13} & \cdots & A_{1n} \\
0 & A_{22} & A_{23} & \cdots & A_{2n} \\
0 & 0 & A_{33} & \cdots & A_{3n} \\
\vdots & \vdots & \vdots & \ddots & \vdots \\
0 & 0 & 0 & \cdots & A_{nn}
\end{bmatrix}
\tag{4.62}
$$

All the elements in the matrix U except the first row differ from those in the original matrix A. It is therefore more efficient to store the modified elements in place of the original ones. This process is called the *forward elimination* process. The upper triangular system of equations can now be solved by the *back substitution* process. It is observed that the entry of matrix U contains only one variable, ϕ_n, and is solved here:

$$
\phi_n = \frac{B_n}{U_{nn}}
\tag{4.63}
$$

The entry in matrix U just above Eq. (4.63) contains only ϕ_{n-1} and ϕ_n and, once ϕ_n is known, it can be solved for ϕ_{n-1}. By proceeding in an upward fashion, each of the variables ϕ_i is solved in turn. The general form of the equation for ϕ_i can be expressed as

$$
\phi_i = \frac{B_i - \sum_{j=i+1}^{n} A_{ij}\phi_j}{A_{ii}}
\tag{4.64}
$$

It is not difficult to see that the bulk of the computational effort is in the forward elimination process; the back substitution process requires fewer arithmetic operations and is thus much less costly. Gaussian elimination can be expensive, especially for a full matrix containing a large number of unknown variables to be solved, but it is as good as any other methods that are currently available.

As observed in Example 4.2, we obtained a matrix of the form typically found from the application of the finite-difference or finite-volume method. This special form of matrix can be solved by the Thomas algorithm. We observed that the non-zero elements (neighboring entries) lie close to the main diagonal entries; it is useful to consider variants of Gaussian elimination that take advantage of this particular banded structure of the matrix (tri-diagonal) to maximize computational resources and to reduce the arithmetic operations. Let us consider the tri-diagonal form of a system of algebraic equations as

$$
\begin{bmatrix}
A_{11} & A_{12} & & & & & \\
A_{21} & A_{22} & A_{23} & & & & \\
& \cdots & \cdots & \cdots & & & \\
& & A_{ii-1} & A_{ii} & A_{ii+1} & & \\
& & & \cdots & \cdots & \cdots & \\
& & & & A_{nn-2} & A_{n-1n-1} & A_{n-1n} \\
& & & & & A_{nn-1} & A_{nn}
\end{bmatrix}
\begin{bmatrix}
\phi_1 \\ \phi_2 \\ \cdots \\ \phi_i \\ \cdots \\ \phi_{n-1} \\ \phi_n
\end{bmatrix}
=
\begin{bmatrix}
B_1 \\ B_2 \\ \cdots \\ B_i \\ \cdots \\ B_{n-1} \\ B_n
\end{bmatrix}
$$

$$(4.65)$$

The *Thomas algorithm,* like Gaussian elimination, solves the system of equations above in two parts: forward elimination and back substitution. For the forward elimination process, the neighboring banded entries are eliminated below the diagonal to yield zero entries. This means replacing the elements $A_{21}, A_{32}, A_{43}, \ldots, A_{nn-1}$ with zeros. For the first row, the diagonal entry A_{11} is normalized to unity and the neighboring entry A_{12} and matrix B term B_1 are modified according to

$$
A'_{12} = \frac{A_{12}}{A_{11}}, \quad B'_1 = \frac{B_1}{A_{11}}
\tag{4.66}
$$

As in Gaussian elimination, multiplying the first row of the matrix by A_{21} and subtracting it from the second row makes all the elements in the second row subsequently modified; these elements also include the terms in B on the right-hand side of the equations. Applying the same procedure to the rest of the rows of the matrix, the neighboring element entries and the matrix B terms in general form are

$$
A'_{ii+1} = \frac{A_{ii+1}}{A_{ii} - A_{ii-1}A'_{i-1i}}, \quad B'_i = \frac{B_i - A_{ii-1}B'_{i-1}}{A_{ii} - A_{ii-1}A'_{i-1i}}
\tag{4.67}
$$

The matrix containing the non-zero coefficients is therefore manipulated into

$$
\begin{bmatrix}
1 & A'_{12} \\
 & 1 & A'_{23} \\
 & & \cdots & \cdots \\
 & & & 1 & A'_{ii+1} \\
 & & & & \cdots & \cdots \\
 & & & & & 1 & A'_{n-1n} \\
 & & & & & & 1
\end{bmatrix}
\begin{bmatrix}
\phi_1 \\ \phi_2 \\ \cdots \\ \phi_i \\ \cdots \\ \phi_{n-1} \\ \phi_n
\end{bmatrix}
=
\begin{bmatrix}
B'_1 \\ B'_2 \\ \cdots \\ B'_i \\ \cdots \\ B'_{n-1} \\ B_n
\end{bmatrix}
\tag{4.68}
$$

The second stage simply involves the back substitution process, which involves evaluating

$$
\phi_n = B'_n \quad \text{and} \quad \phi_i = B'_i - \phi_{i+1} A'_{ii+1}
\tag{4.69}
$$

It can be seen that the Thomas algorithm is more economical than Gaussian elimination because of the absence of arithmetic operations (multiplication and division) in obtaining ϕ_i during back substitution. Nevertheless, in order to prevent ill-conditioning (and hence round-off error) for the two direct methods, it is necessary that

$$
|A_{ii}| > |A_{ii-1}| + |A_{ii+1}|
\tag{4.70}
$$

This means that the diagonal coefficients are required to be much larger than the sum of the neighboring coefficients.

Example 4.3

A steady heat-conduction problem in a large brick plate with a uniform heat generation is presented here. With the boundary temperatures of T_A and T_B given, respectively, by 100°C and 400°C, determine the discrete nodal temperatures across the brick plate using the Thomas algorithm.

Solution

In this example, we illustrate the arithmetic operations of the Thomas algorithm to solve the resultant system of equations for the one-dimensional steady heat-conduction process. The system of equations in matrix form that was previously derived is given as

$$
\begin{bmatrix}
3000 & -1000 & 0 & 0 \\
-1000 & 2000 & -1000 & 0 \\
0 & -1000 & 2000 & -1000 \\
0 & 0 & -1000 & 3000
\end{bmatrix}
\begin{bmatrix} T_1 \\ T_2 \\ T_3 \\ T_4 \end{bmatrix}
=
\begin{bmatrix} 2000 T_A + 2500 \\ 2500 \\ 2500 \\ 2000 T_B + 2500 \end{bmatrix}
$$

Substituting the temperatures $T_A = 100°C$ and $T_B = 400°C$ into the right-hand side of the forcing terms, we have

$$
\begin{bmatrix}
3000 & -1000 & 0 & 0 \\
-1000 & 2000 & -1000 & 0 \\
0 & -1000 & 2000 & -1000 \\
0 & 0 & -1000 & 3000
\end{bmatrix}
\begin{bmatrix} T_1 \\ T_2 \\ T_3 \\ T_4 \end{bmatrix}
=
\begin{bmatrix} 202500 \\ 2500 \\ 2500 \\ 802500 \end{bmatrix}
$$

In the Thomas algorithm, the first step of the forward elimination process involves eliminating the lower triangular coefficients below the diagonal coefficients to yield zero entries. By applying Eqs. (4.36) and (4.37), the matrix is reduced to

$$
\begin{bmatrix}
1 & -0.333 & 0 & 0 \\
0 & 1 & -0.6 & 0 \\
0 & 0 & 1 & -0.714 \\
0 & 0 & 0 & 1
\end{bmatrix}
\begin{bmatrix}
T_1 \\ T_2 \\ T_3 \\ T_4
\end{bmatrix}
=
\begin{bmatrix}
67.5 \\ 42.0 \\ 31.8 \\ 365.0
\end{bmatrix}
$$

The second stage of the Thomas algorithm simply involves the back substitution process. Using Eq. (4.39), the solution to the above system is

$$
\begin{bmatrix}
T_1 \\ T_2 \\ T_3 \\ T_4
\end{bmatrix}
=
\begin{bmatrix}
140.0 \\ 217.4 \\ 292.4 \\ 365.0
\end{bmatrix}
$$

Discussion
For this simple problem, we could have obtained the solution using Gaussian elimination instead of the Thomas algorithm. For such a small matrix, the additional arithmetic operations required for Gaussian elimination to perform on the zero entries may not be as significant in comparison to the Thomas algorithm. Nevertheless, this is not true when a number of grid points are used to better predict the temperature distribution across the plate. This is because of the additional and more cumbersome numerical computations (multiplication and division) that have to be performed on the matrix entries. The algorithm degenerates and becomes inefficient once the order of the matrix becomes higher (>10).

4.4.2 Iterative Methods

Direct methods, such as Gaussian elimination, can be employed to solve any system of equations. Unfortunately, in most CFD problems that usually result in a large system of non-linear equations, the cost of using direct methods is generally quite high. It has been demonstrated through the worked example in the previous section that the Thomas algorithm is particularly economical in obtaining the solution for a one-dimensional steady-state heat-conduction problem because of the inherent banded matrix structure (tri-diagonal). For multi-dimensional situations, however, the nature of the solver cannot be readily extended to solve such problems.

This therefore leaves the option of employing iterative methods. In an iterative method, one guesses the solution and uses the equation to systematically improve the solution until it reaches some level of convergence. If the number of iterations needed to achieve convergence is small, an iterative solver may cost less to use than a direct method. This is usually the case for CFD problems.

Jacobi and Gauss-Siedel methods: The simplest method from the various classes of iterative methods is the Jacobi method. Let us revisit the system of equations, $A\phi = B$, as described in the previous section. The general form of the algebraic equation for each unknown nodal variables of ϕ can be written as

$$\sum_{j=1}^{i-1} A_{ij}\phi_j + A_{ii}\phi_i + \sum_{j=i+1}^{n} A_{ij}\phi_j = B_i \tag{4.71}$$

In Eq. (4.71), the Jacobi method requires that the nodal variables ϕ_j (nondiagonal matrix elements) are assumed to be known at iteration step k and the nodal variables ϕ_i are treated as unknown at iteration step $k+1$. Solving for ϕ_i, we have

$$\phi_i^{(k+1)} = \frac{B_i}{A_{ii}} - \sum_{j=1}^{i-1} \frac{A_{ij}}{A_{ii}} \phi_j^{(k)} - \sum_{j=i+1}^{n} \frac{A_{ij}}{A_{ii}} \phi_j^{(k)} \tag{4.72}$$

The iteration process begins with an initial guess of the nodal variables ϕ_j ($k=0$). After repeated application of Eq. (4.72) to all the n unknowns, the first iteration, $k=1$, is completed. We proceed to the next iteration step, $k=2$, by substituting the iterated values at $k=1$ into Eq. (4.71) to estimate the new values at the next iteration step. This process is continuously repeated for as many iterations as required to converge to the desired solution.

A more immediate improvement to the Jacobi method is provided by the Gauss-Siedel method, in which the updated nodal variables $\phi_j^{(k+1)}$ are immediately used on the right-hand side of Eq. (4.71) as soon as they are available. In such a case, the previous values of $\phi_j^{(k)}$ that appear in the second term of the right-hand side of Eq. (4.72) are replaced by the current values of $\phi_j^{(k)}$, for which the equivalent of Eq. (4.72) becomes

$$\phi_i^{(k+1)} = \frac{B_i}{A_{ii}} - \sum_{j=1}^{i-1} \frac{A_{ij}}{A_{ii}} \phi_j^{(k+1)} - \sum_{j=i+1}^{n} \frac{A_{ij}}{A_{ii}} \phi_j^{(k)} \tag{4.73}$$

Comparing the above two iterative procedures, the Gauss-Siedel iteration is typically twice as fast as the Jacobi iteration. After repeated applications of Eqs. (4.72) and (4.73), convergence can be gauged in a number of ways. One convenient condition to terminate the iteration process is to ensure that the maximum difference $\phi_j^{(k+1)} - \phi_j^{(k)}$ falls below some predetermined value of acceptable error. The smaller the acceptable error, the more accurate the solution will be, but it is noted that this is achieved at the expense of a larger number of iterations.

Example 4.4

Based on the same worked example of the steady heat-conduction problem in a large brick plate with a uniform heat generation, determine the discrete nodal temperatures across the brick plate using

(a) The Jacobi method
(b) The Gauss-Siedel method

Solution

(a) To illustrate the Jacobi method, the resulting set of algebraic equations, as previously derived in Example 4.2, is rewritten

$$3000T_1 + 1000T_2 + 0 \times T_3 + 0 \times T_4 = 2500 + 2000T_A$$
$$1000T_1 + 2000T_2 + 1000T_3 + 0 \times T_4 = 2500$$
$$0 \times T_1 + 1000T_2 + 2000T_3 + 1000T_4 = 2500$$
$$0 \times T_1 + 0 \times T_2 + 1000T_3 + 3000T_4 = 2500 + 2000T_B$$

Substituting the boundary temperatures $T_A = 100°C$ and $T_B = 400°C$, we have

$$3000T_1 + 1000T_2 + 0 \times T_3 + 0 \times T_4 = 202,500$$
$$1000T_1 + 2000T_2 + 1000T_3 + 0 \times T_4 = 2500$$
$$0 \times T_1 + 1000T_2 + 2000T_3 + 1000T_4 = 2500$$
$$0 \times T_1 + 0 \times T_2 + 1000T_3 + 3000T_4 = 802,500$$

The above set of equations can be reorganized so that the required variable is on the left-hand side of the equation.

$$T_1 = 0.333T_2 + 0 \times T_3 + 0 \times T_4 + 67.5$$
$$T_2 = 0.5T_1 + 0.5T_3 + 0 \times T_4 + 1.25$$
$$T_3 = 0 \times T_1 + 0.5T_2 + 0.5T_4 + 1.25$$
$$T_4 = 0 \times T_1 + 0 \times T_2 + 0.333T_3 + 267.5$$

By employing initial guesses: $T_1^{(0)} = T_2^{(0)} = T_3^{(0)} = T_4^{(0)} = 100$, the nodal temperatures for the first iteration are determined as

$$T_1^{(1)} = 0.333(100) + 67.5 = 100.8$$
$$T_2^{(1)} = 0.5(100) + 0.5(100) + 1.25 = 101.25$$
$$T_3^{(1)} = 0.5(100) + 0.5(100) + 1.25 = 101.25$$
$$T_4^{(1)} = 0.333(100) + 267.5 = 300.8$$

The above first iteration values of $T_1^{(1)} = 100.8$, $T_2^{(1)} = 101.25$, $T_3^{(1)} = 101.25$, and $T_4^{(1)} = 300.8$ are substituted back into the system of equations; the second iteration yields

$$T_1^{(2)} = 0.333(101.25) + 67.5 = 101.2$$
$$T_2^{(2)} = 0.5(100.8) + 0.5(101.25) + 1.25 = 102.3$$
$$T_3^{(2)} = 0.5(101.25) + 0.5(300.8) + 1.25 = 202.3$$
$$T_4^{(2)} = 0.333(101.25) + 267.5 = 301.2$$

After repeated applications of the iterative process up to 10 and 20 iterations, the nodal temperatures have advanced to

$$\begin{bmatrix} T_1^{(10)} \\ T_2^{(10)} \\ T_3^{(10)} \\ T_4^{(10)} \end{bmatrix} = \begin{bmatrix} 135.2 \\ 207.9 \\ 282.2 \\ 360.5 \end{bmatrix} \text{ and } \begin{bmatrix} T_1^{(20)} \\ T_2^{(20)} \\ T_3^{(20)} \\ T_4^{(20)} \end{bmatrix} = \begin{bmatrix} 139.7 \\ 217.1 \\ 292.0 \\ 364.7 \end{bmatrix}$$

In Example 4.3, we obtained the exact direct solution by the Thomas algorithm, which is

$$\begin{bmatrix} T_1^{exact} \\ T_2^{exact} \\ T_3^{exact} \\ T_4^{exact} \end{bmatrix} = \begin{bmatrix} 140.0 \\ 217.4 \\ 292.4 \\ 365.0 \end{bmatrix}$$

It is observed that the nodal temperatures after 20 iterations are edging closer toward the exact nodal temperature values.

(b)   Let us now employ the iterative Gauss-Siedel method to the system of algebraic equations. We begin as in the Jacobi method with the set of equations

$$T_1 = 0.333T_2 + 67.5$$
$$T_2 = 0.5T_1 + 0.5T_3 + 1.25$$
$$T_3 = 0.5T_2 + 0.5T_4 + 1.25$$
$$T_4 = 0.333T_3 + 267.5$$

Employing the same initial guesses, the first iteration yields

Immediate substitution

$$T_1^{(1)} = 0.333(100) + 67.5 = 100.8$$
$$T_2^{(1)} = 0.5T_1^{(1)} + 0.5(100) + 1.25 = 0.5(100.8) + 0.5(100) + 1.25 = 101.7$$
$$T_3^{(1)} = 0.5T_2^{(1)} + 0.5(100) + 1.25 = 0.5(101.7) + 0.5(100) + 1.25 = 102.1$$
$$T_4^{(1)} = 0.5T_3^{(1)} + 267.5 = 0.333(102.1) + 267.5 = 301.5$$

After 10 iterations have been performed, the nodal temperatures have advanced to

$$\begin{bmatrix} T_1^{(10)} \\ T_2^{(10)} \\ T_3^{(10)} \\ T_4^{(10)} \end{bmatrix} = \begin{bmatrix} 139.4 \\ 216.7 \\ 291.9 \\ 364.7 \end{bmatrix}$$

The temperature values obtained through the Gauss-Siedel method at this present stage are comparable to the values obtained by the Jacobi method at 20 iterations.

Discussion

We can infer from this example that the Gauss-Siedel iteration is twice as fast as the Jacobi iteration. Convergence is achieved more quickly by the Gauss-Siedel method because of the *immediate substitution* of the temperatures to the right-hand side of the equations whenever they are made available. Thus far, we have not discussed the issue of terminating the iteration process for this particular problem. The degree to which you wish convergence to be achieved is entirely up to you. If the absolute maximum difference $|\phi_j^{(k+1)} - \phi_j^{(k)}|$ is chosen as the condition for the termination process, the accuracy of the solution depends on the targeted number of significant figures you wish to obtain for the temperatures. The smaller the acceptable error, the higher the number of iterations, but this will achieve greater accuracy.

Sometimes convergence to a solution can be enhanced by utilizing the numerical technique called *successive over-relaxation*. The idea behind this technique is an extrapolation procedure where the intermediate nodal variables $\phi_j^{(k+1)}$ are further advanced by a weighted average of the current values of $\phi_j^{(k+1)}$ with the previous values of $\phi_j^{(k)}$. The extrapolated values for $\phi_j^{(k+1)}$ are obtained as follows:

$$\bar{\phi}_i^{(k+1)} = (1 - \lambda)\phi_i^{(k)} + \lambda\phi_i^{(k+1)} \tag{4.74}$$

These extrapolated values are continuously used in the system of equations as the iteration process progresses. In the above equation, λ is a relaxation factor whose value is usually found by trial-and-error experimentation for a given problem. Generally, the value of λ is bounded between $0 < \lambda < 2$ in order to ensure convergence. If successive over-relaxation is used in conjunction with the Gauss-Siedel method, for a value of λ between $1 < \lambda < 2$, a significant improvement to the nodal temperatures obtained is realized at each iteration step and hence convergence is achieved at a faster rate.

Until now, we have discussed the application of iterative methods only to one-dimensional problems. For multi-dimensional problems, with a larger number of grid points, and thus a larger number of equations to be solved, the Jacobi and Gauss-Siedel methods, despite their simplicity, may prove expensive, especially since they generally require a large number of iterations to reach convergence. Successive over-relaxation, though, provides a way of accelerating the iteration process; however, the difficulty in determining the optimum values of λ precludes its wide application in tackling CFD problems. Nevertheless, the primary aim of this section is to introduce the reader to some basic understanding of iterative methods and demonstrate with some worked examples their numerical computations.

Other methods: A practice often applied to multi-dimensional problems is the use of iterative matrices that correspond to lower-dimensional problems. One commonly used method is the ADI (alternating direction implicit) method, introduced by Peaceman and Rachford (1955), which is used to reduce multi-dimensional problems, whether they are two-dimensional or three-dimensional, to a sequence of one-dimensional problems. The resulting matrix is tri-diagonal, and the Thomas algorithm is applied in each of the coordinate directions; this procedure solves the nodal variables for the lines in one direction and repeats for the lines in other directions. Another iterative method for solving multi-dimensional discretization equations, particularly for structured mesh, is the strongly implicit procedure (SIP) proposed by Stone (1968). The basic idea of this method involves approximating the matrix A, Eq. (4.61), by an incomplete LU (lower-upper) factorization to yield an iteration matrix M. Unlike other methods, SIP is a good iterative technique in its own right. It has been used in some commercial CFD codes as the standard solver for non-linear equations. It also provides a good basis for acceleration techniques, such as the conjugate-gradient methods and multi-grid methods.

4.5   PRESSURE–VELOCITY COUPLING—"SIMPLE" SCHEME

The incompressible form of the conservation equations governing fluid flow are derived in Chapter 3 and are summarized in Section 3.6. Because of the incompressible assumption, the solution of the governing equations is complicated by the lack of an independent equation for pressure. In each of the momentum equations, fluid flow is driven by the contribution of the pressure gradients. With the additional equation provided by the continuity equation, this system of equations is self-contained; there are four equations for four dependents, u, v, w, and p, but no independent transport equation for pressure. The implication here is that the continuity and momentum equations are all that are required to solve for the velocity and pressure fields in an incompressible flow. For such a flow, the continuity equation is a kinematic constraint on the velocity field rather than a dynamic equation. In order to link the pressure with the velocity for an incompressible flow, one possible way is to construct the pressure field so as to guarantee conservation of the continuity equation.

In this section, we describe the basis of one of the most popular schemes of pressure–velocity coupling for an incompressible flow. It belongs to the class of iterative methods that is embodied in a scheme called SIMPLE, where the acronym stands for Semi-Implicit Method for Pressure-Linkage Equations. The SIMPLE scheme was developed for practical engineering solutions by Patankar and Spalding (1972). Ever since their pioneering work, it has found widespread application in the majority of commercial CFD codes. In the SIMPLE scheme, a guessed pressure field is used to solve the momentum equations. A pressure correction equation, deduced from the continuity equation, is then solved to obtain a pressure correction field, which in turn is used to update the velocity and pressure fields. These guessed fields are progressively improved through the iteration process until convergence is achieved for the velocity and pressure fields. The salient features of the SIMPLE scheme and the assembly of the complete iterative procedure are discussed later.

Variable arrangement on the grid: Before describing the SIMPLE scheme, the choice of arrangement on the grid requires some consideration. Among the many arrangements, the two most popular that have gained wide acceptance are the *staggered* and *non-staggered* grid arrangements.

The aim of having a *staggered* grid arrangement for CFD computations is to evaluate the velocity components at the control volume faces while the rest of the variables governing the flow field, such as the pressure, temperature, and turbulent quantities, are stored at the central node of the control volumes. A typical arrangement is depicted in Figure 4.10 for a structured finite-volume grid, and it can be demonstrated that the discrete values of the velocity component, u, from the x-momentum equation are evaluated and stored at the east, e, and west, w, faces of the control volume. Evaluation of the other velocity components using the y-momentum and z-momentum equations on the rest of the control volume faces allows a straightforward evaluation of the mass fluxes that

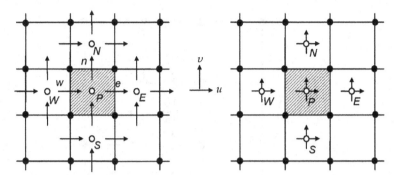

FIGURE 4.10 Staggered and co-located arrangements of velocity components on a finite-volume grid (full symbols denote element vertices and open symbols at the center of the control volumes denote computational nodes for the storage of other governing variables).

are used in the pressure correction equation. This arrangement therefore provides a strong coupling between the velocities and pressure, which helps to avoid some types of convergence problems and oscillations in the pressure and velocity fields. Historically, staggered grid arrangement enjoyed its dominance within the CFD framework between the 1960s and 1980s. However, as the use of non-orthogonal grids became commonplace because of the need to handle complex geometries, alternative grid arrangements had to be explored because of some inherent difficulties in the staggered approach. In particular, if the staggered approach is used in generalized coordinates, curvature terms are required to be introduced into the equations that are usually difficult to treat numerically and may create non-conservative errors when the grid is not smooth.

Nowadays, the alternative grid arrangement that is frequently adopted in many commercial CFD codes is the *non-staggered* grid arrangement. Here, all the flow-field variables including the velocities are stored at the same set of nodal points. For the finite-volume grid shown in Figure 4.11, they are stored at the central node of the control volumes (open symbols). The co-located arrangement offers significant advantages in complicated domains, especially the capability of accommodating slope discontinuities or boundary conditions that may be discontinuous. Furthermore, if multi-grid methods are used, the co-located arrangement allows the ease of transfer of information between various grid levels for all the variables. This grid arrangement was out of favor for incompressible flow computation for a substantial period because of the difficulties faced in coupling the pressure with the velocity and the occurrence of oscillations in the pressure. Nevertheless, widespread use of the co-located grid arrangement became prominent once again through significant developments of the pressure–velocity coupling algorithms, such as the well-known Rhie and Chow (1983) interpolation scheme. This scheme, which has provided physically sensible solutions on structured co-located meshes, generated much

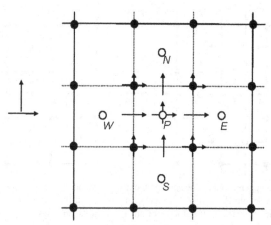

FIGURE 4.11   Arrangement of velocity components on a control volume element of a structured grid at the central node, element faces, and element vertices.

interest for unstructured meshing applications. For the vast majority of general flows, this treatment ties together the pressure fields to yield smooth solutions, while only minimally affecting the mass fluxes. More details of this interpolation scheme are left to the interested reader.

Pressure correction equation and its solution: The SIMPLE scheme is essentially a guess-and-correct procedure for the calculation of pressure through the solution of a pressure correction equation. The method is illustrated by considering a two-dimensional steady laminar flow problem in a structured grid, as shown in Figure 4.11.

The SIMPLE scheme provides a robust method of calculating the pressure and velocities for an incompressible flow. When coupled with other governing variables, such as temperature and turbulent quantities, the calculation needs to be performed sequentially, since it is an iterative process. The sequence of operations in a typical CFD iterative process that embodies the SIMPLE scheme is given in Figure 4.12, with more details of each iterative step elaborated below.

Step 1: The iterative SIMPLE calculation process begins by guessing the pressure field, p^*. During the iterative process, the discretized momentum equations are solved using the guessed pressure field. Applying the finite-volume method, the equations for the x-momentum and y-momentum that yield the velocity components, u^* and v^*, can be expressed in the same algebraic form as previously derived in Eq. (4.29), which can be recast into

$$a_P^u u_P^* = \sum a_{nb}^u u_{nb}^* - \frac{\partial p^*}{\partial x} \Delta V + b' \qquad (4.75)$$

$$a_P^v v_P^* = \sum a_{nb}^v v_{nb}^* - \frac{\partial p^*}{\partial y} \Delta V + b' \qquad (4.76)$$

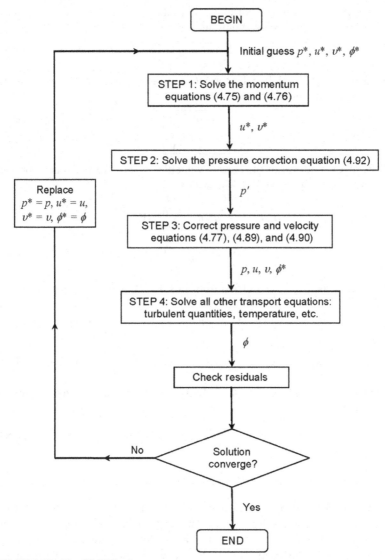

FIGURE 4.12   The SIMPLE scheme.

where ΔV is the finite-control volume. Here, we simplify the above expressions by introducing a_{nb} to represent the presence of the neighboring coefficients and u_{nb}^* and v_{nb}^* to denote the neighboring nodal velocity components. The pressure gradient terms appearing in the above two equations are taken from the original source term b of the momentum equations while the other terms governing the fluid flow are left in the source term b'.

Step 2: If we define the correction p' as the difference between the correct pressure field and the guessed pressure field p^*, we get

$$\boxed{p = p^* + p'} \tag{4.77}$$

Similarly, we can also define the corrections u' and v' to relate the correct velocities u and v to the guessed velocities u^* and v^*:

$$\boxed{u = u^* + u'} \tag{4.78}$$

$$\boxed{v = v^* + v'} \tag{4.79}$$

The algebraic form of the correct velocities u and v can also be expressed similarly to Eqs. (4.75) and (4.76), so that

$$a_P^u u_p = \sum a_{nb}^u u_{nb} - \frac{\partial p}{\partial x} \Delta V + b' \tag{4.80}$$

$$a_p^v v_p = \sum a_{nb}^v v_{nb} - \frac{\partial p}{\partial y} \Delta V + b' \tag{4.81}$$

Subtracting Eqs. (4.75) and (4.76) from Eqs. (4.80) and (4.81), we obtain

$$a_P^u \left(u_p - u_p^* \right) = \sum a_{nb}^u \left(u_{nb} - u_{nb}^* \right) - \frac{\partial (p - p^*)}{\partial x} \Delta V \tag{4.82}$$

$$a_P^u \left(v_p - v_p^* \right) = \sum a_{nb}^u \left(v_{nb} - v_{nb}^* \right) - \frac{\partial (p - p^*)}{\partial y} \Delta V \tag{4.83}$$

Using the correction formulae (4.77–4.79), the above equations may be rewritten as follows:

$$a_P^u u_p' = \sum a_{nb}^u u_{nb}' - \frac{\partial p'}{\partial x} \Delta V \tag{4.84}$$

$$a_P^v v_p' = \sum a_{nb}^v v_{nb}' - \frac{\partial p'}{\partial y} \Delta V \tag{4.85}$$

The SIMPLE scheme approximates Eqs. (4.84) and (4.85) by the omission of the terms $\sum a_{nb}^u u_{nb}'$ and $\sum a_{nb}^v v_{nb}'$. The reader is reminded that this scheme is an iterative approach, and there is no reason why the formula designed to predict p' needs to be physically correct. Hence, we are allowed to construct a formula for p' that is simply a numerical artifice with the aim of expediting the convergence of the velocity field to a solution that satisfies the continuity equation. This is the essence of the algorithm. Once the pressure correction field is known, the correct velocities u and v can be obtained through the guessed velocities u^* and v^* from simplified Eqs. (4.84) and (4.85):

$$u_p = u_P^* - D^u \frac{\partial p'}{\partial x} \tag{4.86}$$

$$v_p = v_P^* - D^v \frac{\partial p'}{\partial y} \tag{4.87}$$

where

$$D^u = \frac{\Delta V}{a_p^u} \quad \text{and} \quad D^v = \frac{\Delta V}{a_p^v} \tag{4.88}$$

Although Eqs. (4.86) through (4.88) have been developed to correct the velocities from the guessed velocities at the central node of the control volume, these correction formulae can also be generally applied to any location where the velocity components reside within the computational grid (as shown in Figure 4.9, the velocities may be located at central node P or at the control volume faces or at the vertices of the control volume). The general form of the velocity correction formulae, by removing the subscript P, can be expressed as

$$\boxed{u = u^* - D^u \frac{\partial p'}{\partial x}} \tag{4.89}$$

$$\boxed{v = v^* - D^v \frac{\partial p'}{\partial y}} \tag{4.90}$$

The derivation of a pressure correction equation utilizes the above two equations. Differentiating Eq. (4.89) by the Cartesian direction x and Eq. (4.90) by the Cartesian direction y and summing them together yields

$$-\frac{\partial}{\partial x}\left(D^u \frac{\partial p'}{\partial x}\right) - \frac{\partial}{\partial y}\left(D^v \frac{\partial p'}{\partial y}\right) + \underbrace{\frac{\partial u^*}{\partial x} + \frac{\partial v^*}{\partial y}}_{\text{guessed velocity gradients}} = \underbrace{\frac{\partial u}{\partial x} + \frac{\partial v}{\partial y}}_{\text{correct velocity gradients}}{}^{=0} \tag{4.91}$$

By invoking the continuity equation, it is shown that the term represented by the source term of the right-hand side of Eq. (4.91) is zero, and Eq. (4.91) can be rearranged as

$$\boxed{\frac{\partial}{\partial x}\left(D^u \frac{\partial p'}{\partial x}\right) + \frac{\partial}{\partial y}\left(D^v \frac{\partial p'}{\partial y}\right) = \underbrace{\left(\frac{\partial u^*}{\partial x} + \frac{\partial v^*}{\partial y}\right)}_{\text{mass residual}}} \tag{4.92}$$

Interestingly, Eq. (4.92) behaves like a steady-state diffusion process in a two-dimensional domain. It is a Poisson equation—one of the well-known equations from classical physics and mathematics. The solution to this Poisson equation can be achieved through some efficient numerical solvers (conjugate-gradient and multi-grid methods), as previously discussed, to accelerate its convergence.

Step 3: Once the pressure correction p' field is obtained, the pressure and velocity components are subsequently updated through the correction formulae of Eqs. (4.77), (4.89), and (4.90). If the solution concerns only a laminar CFD flow problem, the iteration process proceeds directly to check the convergence of the solution. If the solution is not converged, the process is repeated by returning to Step 1. The source term appearing in the pressure correction equation, Eq. (4.92), commonly known as the *mass residual*, is normally used in CFD computations as a criterion for terminating the iteration procedure. As the mass residual continues to diminish, the pressure correction p' will be zero, thereby yielding a converged solution of $p^* = p$, $u^* = u$, and $v^* = v$.

Step 4: This step is executed if the CFD flow problem is turbulent or if it involves the transfer of heat or mass exchanges between different flow phases. Additional transport equations governing such a flow system need to be solved before convergence is checked. If the solution is not converged, the iterative process returns to Step 1 and repetitive calculations are carried out until convergence is reached.

The application of this SIMPLE scheme is best illustrated by solving the Chapter 2 CFD problem of a steady two-dimensional incompressible laminar flow in a channel, which is described in Example 4.5.

Example 4.5

Consider the case of a steady, two-dimensional, incompressible, laminar flow between two stationary parallel plates, as in Chapter 2. By obtaining the solution from a CFD code using the finite-volume method, track the progress of the intermediate values of u, v, p, p' and the mass residual during the iterative process at a computational nodal point at the center of the channel (Figure 4.5.1).

Solution

The problem is described as follows: To demonstrate the robustness of the SIMPLE scheme, the iterative process begins by employing the initial guesses: $p^* = 0$, $u^* = 0$ and $v^* = 0$. The discretized equations governing the momentum and pressure correction are solved using the default iterative solvers provided in the commercial CFD code. The inlet, outlet, and wall conditions remain the same as applied in Chapter 2.

Based on Eqs. (4.75), (4.76), (4.77), (4.89), (4.90), and (4.92), the calculated values of the pressure p, pressure correction p', velocities u and v, and mass residual for the first iteration at the *monitoring point* are

$$\begin{bmatrix} p_{monitor}^{(1)} \\ p'^{(1)}_{monitor} \\ u_{monitor}^{(1)} \\ v_{monitor}^{(1)} \end{bmatrix} = \begin{bmatrix} 0.02043 \\ 0.06812 \\ 0.01033 \\ -0.1246 \times 10^{-4} \end{bmatrix} \qquad \boxed{\begin{array}{l} \textit{mass residual} \Rightarrow \\ 1.2 \times 10^{-4} \ kg/s \end{array}}$$

The solution of the first iteration from above is subsequently used as intermediate values for the next iteration step; the second iteration yields

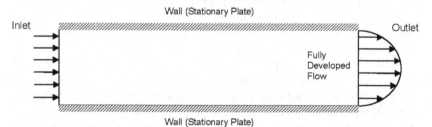

FIGURE 4.5.1   Two-dimensional laminar flow in a channel with a monitoring point located at the center of the channel.

$$
\begin{bmatrix} p^{(2)}_{monitor} \\ p'^{(2)}_{monitor} \\ u^{(2)}_{monitor} \\ v^{(2)}_{monitor} \end{bmatrix} = \begin{bmatrix} 4.774 \times 10^{-3} \\ -0.0522 \\ 0.01181 \\ 1.104 \times 10^{-5} \end{bmatrix}
$$

$$\boxed{mass\ residual \Rightarrow 2.0494 \times 10^{-4}\ kg/s}$$

After repeated applications of the iterative process, the respective values for the pressure p, pressure correction p', velocities u and v, and mass residual after 10 and 20 iterations are

$$
\begin{bmatrix} p^{(10)}_{monitor} \\ p'^{(10)}_{monitor} \\ u^{(10)}_{monitor} \\ v^{(10)}_{monitor} \end{bmatrix} = \begin{bmatrix} 6.83 \times 10^{-4} \\ -4.949 \times 10^{-4} \\ 0.01419 \\ -2.963 \times 10^{-6} \end{bmatrix}
$$

$$\boxed{mass\ residual \Rightarrow 1.8639 \times 10^{-5}\ kg/s}$$

and

$$
\begin{bmatrix} p^{(20)}_{monitor} \\ p'^{(20)}_{monitor} \\ u^{(20)}_{monitor} \\ v^{(20)}_{monitor} \end{bmatrix} = \begin{bmatrix} 6.626 \times 10^{-4} \\ 3.793 \times 10^{-6} \\ 0.01482 \\ 2.444 \times 10^{-7} \end{bmatrix}
$$

$$\boxed{mass\ residual \Rightarrow 5.061 \times 10^{-7}\ kg/s}$$

From a theoretical viewpoint, the vertical velocity v is zero at the monitoring point and the iterative process confirms the trend of the prediction toward the zero value. It is seen during the iterative process that the intermediate values of this velocity are much smaller than the rest of the other governing variables; the convergence history plot of this velocity component is therefore omitted, since no quantitative comparison can be realized against the other variable convergence histories. The convergence histories for the rest of the governing variables, which include the pressure p, pressure correction p', horizontal velocity u, and mass residual are illustrated in Figures 4.5.2 through 4.5.5.

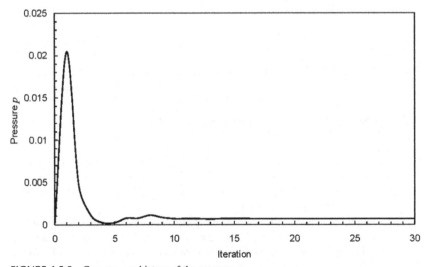

FIGURE 4.5.2   Convergence history of the pressure p.

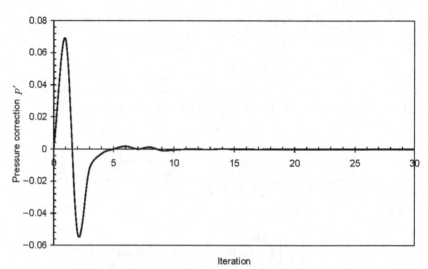

FIGURE 4.5.3   Convergence history of the pressure correction p'.

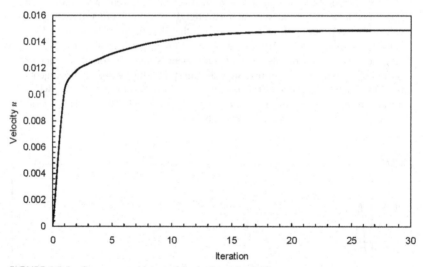

FIGURE 4.5.4   Convergence history of the horizontal velocity u.

Discussion

From this worked example of a channel flow, the SIMPLE scheme provides an efficient iterative procedure for obtaining the velocity and pressure fields for an incompressible flow. The SIMPLE scheme is a robust method that produces rapid stabilization of the velocity and pressure, as seen by their respective convergence histories after 5 iterations. The mass residual that appears as a source term in the pressure correction equation, Eq. (4.80), continues to diminish during the iteration process, thus reaffirming conservation in the continuity equation. Subsequently, the

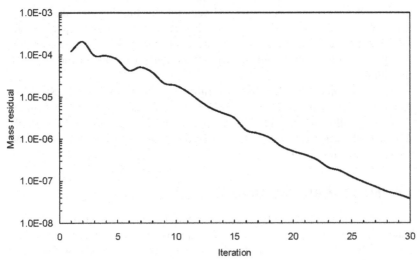

FIGURE 4.5.5   Convergence history of the mass residual.

pressure correction p' is seen to be approaching zero. Hence, the corrections that are required to update the velocity field are also approaching zero. The trend of the convergence histories favors the likelihood of a converged steady-state solution.

The reader should be aware of other types of pressure–velocity coupling algorithms that employ a philosophy similar to the SIMPLE algorithm and that are employed by CFD users or are adopted in commercial CFD codes. These variant SIMPLE algorithms have been formulated with the aim of improving the robustness and convergence rate of the iterative process. We do not intend to provide the reader with all the details of the available algorithms, but we will briefly indicate and describe the modifications made to the original SIMPLE algorithm.

The SIMPLEC (SIMPLE-Consistent) algorithm by Van Doormal and Raithby (1984) follows the same iterative steps as in the SIMPLE algorithm. The main difference between SIMPLEC and SIMPLE is that the discretized momentum equations are manipulated so that the SIMPLEC velocity correction formulae omit terms that are less significant than those omitted in SIMPLE. Another pressure correction procedure that is also commonly employed is the PISO (Pressure Implicit with Splitting of Operators) algorithm proposed by Issa (1986). This pressure–velocity calculation procedure was originally developed for non-iterative computation of unsteady compressible flows. Nevertheless, it has been adapted successfully for the iterative solution of steady-state problems. PISO is simply an extension of SIMPLE with an additional corrector step that involves an additional pressure correction equation to enhance the convergence. The SIMPLER (SIMPLE-Revised) algorithm developed by Patankar (1980) also falls within the framework of two corrector steps, as in PISO. Here, a discretized equation for the pressure provides the intermediate pressure field before the discretized momentum equations are solved. A pressure correction is later solved where the velocities are corrected through the correction formulae, derived similarly to those in the SIMPLE algorithm.

There are other SIMPLE-like algorithms, such as SIMPLEST (SIMPLE-ShorTened) of Spalding (1980) or SIMPLEX of Van Doormal and Raithby (1985) or SIMPLEM (SIMPLE-Modified) of Archarya and Moukalled (1989), that share the same essence in their derivations. More details of all the above pressure–velocity coupling algorithms are left to interested readers.

4.6   MULTI-GRID METHOD

The multi-grid method can be categorized into two types: geometric and algebraic. The former, also known as the full approximation scheme (FAS) multi-grid, involves a hierarchy of meshes (cycling between fine and coarse grids) and the discretized equations are evaluated on every level, while in the latter, the coarse level equations are generated without any geometry or re-discretization on the coarse levels—a feature that makes algebraic multi-grid particularly attractive for use on unstructured meshes.

The multi-grid method is ideal for solving the Poisson-like pressure or pressure-correction equation, such as the SIMPLE method, which is discussed further in the next section. Conceptually, the multi-grid method can be described in the following. Focusing on the system of equations, $A\phi = B$, intermediate solutions of ψ are obtained if this system is solved with an iterative method after some unspecified number of iterations. The residuals R can be defined as

$$A\psi = B - R \tag{4.93}$$

By also defining the errors as the difference between the true and intermediate solutions,

$$E = \phi - \psi \tag{4.94}$$

and subtracting Eq. (4.94) from Eq. (4.93), the relationship between the errors and residuals is

$$AE = R \tag{4.95}$$

During the multi-grid cycle, the matrix of coefficients of A and the residuals described by Eq. (4.95) are transferred from a finer grid to a coarser grid through volume-weighted *restrictions*. After obtaining the converged solutions of the errors, *prolongations* of computed corrections on the coarse grids are transferred to the next fine grids through tri-linear interpolations. The simplest choice of a multi-grid cycle can be described by the typical V-cycle, with five different grid levels, as shown in Figure 4.13 (see also Section 8.2.4 and

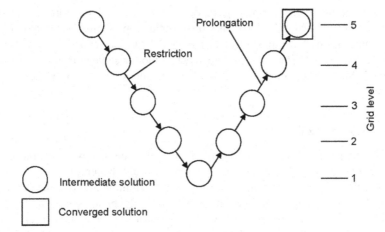

FIGURE 4.13   Schematic representation of a multi-grid method using a V-cycle.

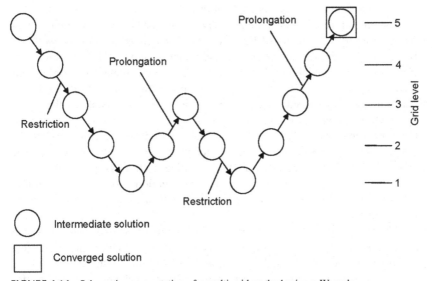

FIGURE 4.14   Schematic representation of a multi-grid method using a W-cycle.

references given within). Another strategy, for example, a W-cycle, as depicted in Figure 4.14, may also be used for cycling between coarse and fine grids. Efficiency of the rate of convergence may be improved by the decision to switch from one grid to another through the combination of V-W cycles or other possible combinations. The optimal choice of parameters is problem dependent but their effect on performance is not as dramatic as for the single-grid method.

In theory, the advantage of geometric multi-grid over algebraic multi-grid is that the former should perform better for non-linear problems, since non-linear properties in the system are propagated down to the coarse levels through the re-discretization, while for the latter, once linearization is performed on the system of equations, the non-linear properties are not "felt" by the solver until the fine-level operator is updated. The multi-grid approach is more a strategy than a particular method. More details of these acceleration techniques are left to the interested reader.

4.7   SUMMARY

Let's review some of the basic computational techniques that were examined in this chapter to solve the governing equations of fluid dynamics.

The first stage of obtaining the computational solution involves the conversion of the governing equations into a system of algebraic equations. This is usually known as the *discretization stage*. We discussed some of the *discretization tools*, such as the finite-difference and finite-volume methods, which form the foundation of understanding the basic features of *discretization*. Both of these methods abound in CFD applications.

The second stage involves numerically solving the system of algebraic equations, which can be achieved by either *direct methods* or *iterative methods*. Basic direct methods, such as Gaussian elimination and the Thomas algorithm, were described, of which the latter is exceedingly economical for a tri-diagonal matrix system and is a standard algorithm for the solution of fluid-flow equations in a structured mesh. Simple iterative methods, such as the point-by-point Jacobi and Gauss-Siedel methods, were also described. Nevertheless, CFD problems are generally multi-dimensional and are comprised of a large system of equations to be solved. Efficient iterative methods, such as the ADI or Stone's SIP, are often applied to solve such a system of equations. To further enhance the convergence of the computational solution, precondition conjugate-gradient methods or multi-grid methods are employed to accelerate the iteration process.

The reader may return to Figure 4.1 to view how these two stages fit within the process of the computational solution procedure. Within the block that comprises numerical methods, an iterative algorithm for the calculation of pressure and velocity fields based on the SIMPLE scheme is presented for an incompressible flow. The basic philosophy behind this popular scheme is to initially guess a pressure field in the discretized momentum equations to yield the intermediate velocities. The continuity equation in the form of a pressure correction is subsequently solved, and then is used to correct the velocity and pressure fields. These guessed fields are continuously improved until convergence is reached. The reader may refer to Figure 4.10 for a more exhaustive description of the iterative steps that are involved within the SIMPLE scheme.

Finally, we did not discuss in depth the assessment of *convergence*. In practice, the algebraic equations that result from the discretization process yield the flow solution at each nodal point on a finite-grid layout. It is expected that, from the truncation errors given in Section 4.2.1, more accurate solutions can be obtained through refining the grid. For an unsteady problem, this can be achieved by employing smaller time intervals. However, for a given required *solution accuracy*, it may be more economical to solve higher-order approximations of the first- and second-order derivative equations governing the fluid flow on a coarse grid rather than using a low-order approximation on a finer grid. This leads to the concept of *computational efficiency*. Other issues, such as the *solution consistency* and *stability* of the numerical procedure, are also important considerations for the *convergence* of the computational solution. All of these are investigated in the next chapter.

REVIEW QUESTIONS

4.1 What are the differences between solving a fluid-flow problem analytically and solving it numerically? What are the advantages and disadvantages of each method?

4.2 What are the main advantages and disadvantages of discretization of the governing equations through the finite-difference method?

4.3 Is finite-difference more suited for structured or for unstructured mesh geometries? Why?

4.4 Consider the following finite-difference formulation for a simplified flow:

$$\frac{\phi_{i-1} - 2\phi_i + \phi_{i+1}}{\Delta x^2} = 0$$

Is the flow steady or transient? Is it one-, two- or three-dimensional? Is the nodal spacing constant or variable?

4.5 Using finite-difference, show that the steady, one-dimensional, heat-conduction equation,

$$k\frac{\partial^2 T}{\partial x^2} = 0$$

can be expressed as

$$\frac{T_{i-1} - 2T_i + T_{i+1}}{\Delta x^2} = 0$$

4.6 What is the second term in the central-difference approximation for a first-order derivative (given below) called? What does it measure?

$$\left(\frac{\partial \phi}{\partial x}\right) = \frac{\phi_{i+1,j} - \phi_{i-1,j}}{2\Delta x} + O(\Delta x^2)$$

4.7 Which of the following, *forward difference*, *backward difference*, and *central difference*, is most accurate and why?

4.8 What are the main advantages and disadvantages of discretization of the governing equations through the finite-volume method?

4.9 Is the finite-volume method more suited for structured or for unstructured mesh geometries? Why?

4.10 What is the significance of the integration of the governing equations over a control volume during the finite-volume discretization?

4.11 For the control volume below, show how the one-dimensional steady-state diffusion term $\frac{\partial}{\partial x}\left(\Gamma\frac{\partial\phi}{\partial x}\right)$ is discretized to obtain its discretized equation $\left(\Gamma\frac{\partial\phi}{\partial x}\right)_e A_E - \left(\Gamma\frac{\partial\phi}{\partial x}\right)_w A_w$ for central grid nodal point P?

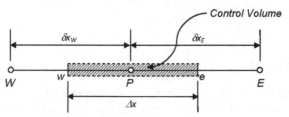

4.12 In a finite-difference scheme, data are resolved at nodal points. How is this different from the finite-volume scheme?

4.13 How is a steady convective-diffusion process different from a pure diffusion process?

4.14 Why are upwind schemes important for strongly convective flow?

4.15 Why are higher-order upwind schemes more favorable than the first-order upwind scheme?

4.16 For the unsteady convection-diffusion process, what is the difference between *explicit* and *implicit* time-marching approaches?

4.17 What is the difference between a direct method and an iterative method for solving the discretized equations?

4.18 Is the direct method or the iterative method more suitable in solving for a large system of non-linear equations?

4.19 Why does the Gauss-Siedel iterative method converge to a solution quicker than the Jacobi method?

4.20 What is the technique associated with *successive over-relaxation* and why is it used?

4.21 Where are the flow-field variables located in co-located grids? How is this different from the locations in a staggered grid?

4.22 Write down the formulation of the central-difference scheme for u velocity in the x direction. What is its truncation error in terms of Δx? Also, state the order of this discretization scheme.

4.23 What is the purpose of the SIMPLE scheme? Does it give us a direct solution or depend on the iterative concept?

4.24 What is the Gaussian elimination method based on? Can this method be used to solve a system of non-linear algebraic equations?

4.25 Solve the following set of equations by Gaussian elimination:

$$
\begin{bmatrix}
100 & 100 & 0 \\
200 & 100 & - \\
300 & - & - \\
- & 200 & 300
\end{bmatrix}
\begin{bmatrix}
T_1 \\
T_2 \\
T_3
\end{bmatrix}
=
\begin{bmatrix}
400 \\
100 \\
-300 \\
400
\end{bmatrix}
$$

4.26 Solve the following set of equations by the Thomas algorithm:

$$
\begin{bmatrix}
100 & - & 200 & - \\
200 & - & 300 & - \\
100 & 100 & 100 & 0 \\
100 & - & 400 & 300
\end{bmatrix}
\begin{bmatrix}
T_1 \\
T_2 \\
T_3 \\
T_4
\end{bmatrix}
=
\begin{bmatrix}
800 \\
-2000 \\
-200 \\
400
\end{bmatrix}
$$

4.27 Solve the following set of equations using the Gauss-Seidel method:

$$
\begin{bmatrix}
-1000 & -100 & 200 \\
-100 & -1100 & -100 \\
200 & -100 & 1000 \\
0 & 300 & -100
\end{bmatrix}
\begin{bmatrix}
T_1 \\
T_2 \\
T_3
\end{bmatrix}
=
\begin{bmatrix}
600 \\
-2500 \\
1100 \\
1500
\end{bmatrix}
$$

4.28 For the same matrix given in Question 4.27, use the Jacobi method to solve the set of equations. Compare the number of iterations for convergence between the Jacobi method and the Gauss-Seidel method.

4.29 Solve the following set of equations using the Gauss-Seidel method:

$$
\text{(a)} \quad
\begin{aligned}
3x_1 &- x_2 + 3x_3 = 0 \\
-x_1 &+ 2x_2 + x_3 = 3 \\
2x_1 &- x_2 - x_3 = 2
\end{aligned}
$$

$$
\text{(b)} \quad
\begin{aligned}
10x_1 &- x_2 + 2x_3 & &= 6 \\
-x_1 &+ 11x_2 - x_3 + 3x_4 &= 25 \\
2x_1 &- x_2 - 10x_3 - x_4 &= -11 \\
&\ 3x_2 - x_3 + 8x_4 &= 15
\end{aligned}
$$

4.30 Following the grid arrangement below, derive the following expression:

$$
\left(\frac{\partial^2 u}{\partial x \partial y}\right)_{i,j} = \frac{u_{i+1,j+1} - u_{i+1,j-1} - u_{i-1,j+1} + u_{i-1,j-1}}{4\Delta x \Delta y}
$$

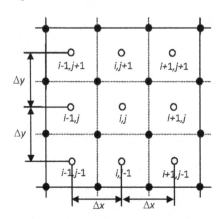

Chapter 5

CFD Solution Analysis— Essentials

5.1 INTRODUCTION

Analyzing a *computational solution* is an integral part of the use of CFD. In Chapter 4, some basic discretization techniques are introduced to allow the reader to become familiar with common methodologies of converting the governing partial differential equations into a system of algebraic equations. This system of algebraic equations is subsequently solved through numerical methods to provide *approximate solutions* to the governing equations. It is these *approximate solutions* that we can interchangeably refer to as *computational solutions*.

In the context of CFD, some of the primary concerns about the computational solutions are whether the solution can be guaranteed to approach the exact solution of the partial differential equations and, if so, under what circumstances. This can be (superficially) achieved by forcing the computational solution to converge to an exact solution as the finite quantities shrink to zero. We recall from Chapter 4 that the finite quantities, in time Δt and in space Δx, Δy, and Δz, are prevalent in the system of algebraic equations as a result of the discretization of the partial differential equations. Nevertheless, *convergence* can be neither straightforward nor directly established. Indirect considerations of convergence need, however, to be implicated from aspects such as *consistency* and *stability*. First, it is required that the formulation of the system of algebraic equations through the discretization process should be consistent with the original partial differential equations. The implication of consistency here is the recovery of the governing equations by reversing the discretization process through a Taylor series expansion. Second, for any chosen numerical algorithm adopted to solve the algebraic equations, stability shares a part of the platform and, with the consistency criteria, ensures convergence.

The *accuracy* of the computational solutions can be affected by errors and uncertainties in the numerical calculations. These errors and uncertainties can be generated either in the conceptual modeling or during the computational design phase, and need to be measured or bounded. The credibility of the computational solutions is strongly dependent on whether the errors and uncertainties are identified and qualified, irrespective of their sources. Systematic reduction of errors and uncertainties leads to better representation of real physical flow problems and thus increases confidence in the use of the CFD simulation code. We provide a pragmatic approach for estimating these errors and uncertainties in CFD through the *verification* and *validation* procedures.

To close this chapter, indicative case studies are selected to demonstrate the relevance and credibility of the computational solutions by addressing the essentials of consistency, stability, convergence, and accuracy that concern a CFD solution analysis. Discussing these essentials will enable the reader to realize and to appreciate the various numerical aspects that are involved in solving the particular flow problems. Each of the problems has its own physical significance; it is assumed that the reader already has some basic knowledge of fluid-flow and heat-transfer processes in order to better understand the physical considerations of the numerical simulations.

5.2 CONSISTENCY

The property of *consistency* can appear rhetorical. Nevertheless, it is an important property and concerns the discretization of the partial differential equations where the approximation performed should diminish or become exact if the finite quantities, such as the time step Δt and mesh spacing Δx, Δy, and Δz, tend to zero. In Section 4.2.1, on the finite-difference method, *truncation error* measures the discrete approximation obtained through a Taylor series expansion about a single nodal point. Essentially, the truncation error represents the difference between the discretized equation and the exact one. As a result, the original partial differential equation is recovered by the addition of a remainder, the truncation error. This error basically measures the accuracy of the approximation and determines the rate at which the error decreases as the time step and/or mesh spacing is reduced.

For any numerical method to be *consistent*, the truncation error must become zero when the time step $\Delta t \to 0$ and/or mesh spacing Δx, Δy, and $\Delta z \to 0$. This error is usually proportional to an nth power for the finite quantities. If the most important term is proportional to $(\Delta t)^n$ or $(\Delta x_i)^n$, the numerical method results in an nth-order approximation for $n > 0$. Ideally, all terms in the governing equations should be discretized with the approximation of the same order of accuracy. However, in practice, some terms (for example, advection terms for high-Reynolds-number flows) may be particularly dominant and a high-order approximation may be required to treat them with more accuracy than others.

Some basic ideas of consistency are further elucidated through the illustrative worked examples below.

Example 5.1

Consider the discretized form of the incompressible, steady-state, two-dimensional continuity equation $\frac{\partial u}{\partial x} + \frac{\partial v}{\partial y} = 0$ in a structured uniform grid arrangement, as in Example 4.1. Discuss the remainder or truncation error associated with the original form of the partial differential equation.

Solution

An elemental control volume of the two-dimensional structured grid is shown in Figure 5.1.1. The centroid of the control volume is indicated by the point P, which is surrounded by adjacent control volumes having their respective centroids indicated by the points: east, E; west, W; north, N; and south, S. The control volume having its centriod at P has respective faces indicated by east, e; west, w; north, n; and south, s.

The discretized form obtained through the finite-volume method is expressed by

$$\frac{u_e - u_w}{\Delta x} + \frac{v_n - v_s}{\Delta y} = 0$$

The face velocities u_e, u_w, v_n, and (v_s) are located midway between each of the control volume centroids, which allows us to determine the face velocities through interpolation of the centroid values. Thus,

$$u_e = \frac{u_P + u_E}{2}; \; u_w = \frac{u_P + u_W}{2}; \; v_n = \frac{v_P + v_N}{2}; \; v_s = \frac{v_P + v_S}{2}$$

By substituting the above expressions into the discretized form of the velocity first- order derivatives, we get

$$\frac{u_E - u_W}{2\Delta x} + \frac{v_N - v_S}{2\Delta y} = 0$$

To recover the original form of the partial differential equation, the above equation can be rewritten in terms of the truncation errors obtained through Taylor series expansion (described in Section 4.2.1) of Eqs. (4.3) and (4.8) as

$$\frac{u_E - u_W}{2\Delta x} + \frac{v_N - v_S}{2\Delta y} + \left[\underbrace{O\left(\Delta x^2, \Delta y^2\right)}_{Truncation\ error} \right]_P = 0$$

The numerical method results in an nth-order approximation of 2. The approximation is therefore second-order accurate at the grid nodal point P.

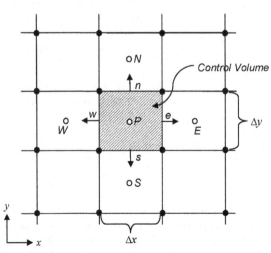

FIGURE 5.1.1 Control volume for the two-dimensional continuity equation problem.

Discussion

The original partial differential equation is recovered (satisfying consistency) from the discretized equation that includes the truncation error, as the mesh spacing Δx and $\Delta y \rightarrow 0$. For the above second-order scheme, halving the mesh spacing results in a reduction in the truncation errors by a factor of four.

Example 5.2

Consider the discretized form of the one-dimensional, transient diffusion equation. Assuming that the thermal diffusivity $\alpha\ (= k/\rho C_p)$ is constant, discuss the remainder or truncation error associated with the DuFort-Frankel (DuFort and Frankel, 1953) differencing of the transient heat-conduction equation.

Solution

The one-dimensional transient diffusion equation can be expressed as

$$\frac{\partial \phi}{\partial t} - \alpha \frac{\partial^2 \phi}{\partial x^2} = 0 \qquad (5.2\text{-A})$$

Applying the DuFort-Frankel differencing to the above equation yields

$$\frac{\phi_i^{n+1} - \phi_i^{n-1}}{2\Delta t} - \frac{\alpha}{\Delta x^2}\left(\phi_{i+1}^n - \phi_i^{n+1} - \phi_i^{n-1} + \phi_{i-1}^n\right) = 0$$

The left-hand term of Eq. (5.2-A), which represents the time derivative about the ith node, is analogous to the truncation error of the second-order central differencing in space; that is,

$$\frac{\partial \phi}{\partial t} = \frac{\phi_i^{n+1} - \phi_i^{n-1}}{2\Delta t} + \underbrace{O(\Delta t^2)}_{\text{Truncation error}}{}_i$$

The right-hand term of Eq. (5.2-A) can be represented according to the second-order derivative in space of Eq. (4.9), yielding the following truncation error:

$$\alpha \frac{\partial^2 \phi}{\partial x^2} = \frac{\alpha}{\Delta x^2}\left(\phi_{i+1}^n - \phi_i^{n+1} - \phi_i^{n-1} + \phi_{i-1}^n\right) + \underbrace{O(\Delta x^2)}_{\text{Truncation error}}{}_i$$

A Taylor series expansion of the exact solution substituted into the one-dimensional transient heat-conduction equation, neglecting higher-order terms, is thus given by

$$\left[\frac{\partial \phi}{\partial t} - \alpha \frac{\partial^2 \phi}{\partial x^2} + \alpha \left(\frac{\Delta t}{\Delta x}\right)^2 \frac{\partial^2 \phi}{\partial t^2}\right]_i + \left[\underbrace{O(\Delta t^2, \Delta x^2)}_{\text{Truncation error}}\right]_i = 0$$

Discussion

As demonstrated in Fletcher (1991), $\Delta t/\Delta x$ must $\rightarrow 0$ at the same rate as $\Delta t, \Delta x \rightarrow 0$ to achieve consistency. It is also required that $\Delta t \ll \Delta x$ for consistency or else the scheme becomes inaccurate (i.e., if $(\Delta t/\Delta x)^2$ is large). From a practical viewpoint, there is effectively a restriction on the size of Δt when using the DuFort-Frankel scheme.

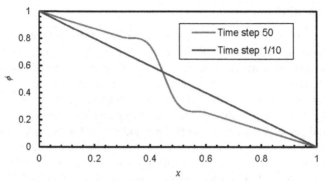

FIGURE 5.2.1   Steady-state solution for $\Delta t = 1/10$ and $\Delta t = 50$ with a fixed grid-step size $\Delta x = 1/10$.

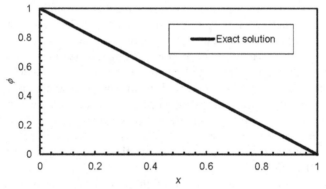

FIGURE 5.2.2   The exact solution for Eq. (5.2-A).

Consider the Dirichlet boundary conditions set at the opposite ends as ϕ $(t, x=0)=1$ and ϕ $(t, x=1)=0$. With a constant thermal diffusivity of $\alpha = 0.5$ and $\Delta x = 1/10$, the steady-state solutions to Eq. (5.2-A) subjected to time step $\Delta t = 1/10$ and time step $\Delta t = 50$ are shown in Figure 5.2.1. The exact solution is given in Figure 5.2.2. It is clearly seen that the solution for large $(\Delta t/\Delta x)$ is inconsistent with the exact solution when compared with the result obtained for a smaller $(\Delta t/\Delta x)$.

From the worked examples above, it is clear that the property of *consistency* is necessary if the approximate solution is to converge to the solution of the partial differential equation. Nevertheless, this property alone is not a sufficient condition. Even though the discretized equations might be equivalent to the partial differential equation as the finite quantities shrink to zero, it may not necessarily mean that the solution of the discretized equations follows the exact solution of the partial differential equation. The latter is evidenced by the inaccurate solution as a result of large $(\Delta t/\Delta x)$ in Example 5.2.

5.3   STABILITY

In addition to consistency, another property that also strongly governs the numerical solution method is *stability*. This property concerns the growth or decay of errors introduced at any stage during the computation. It is noted that the errors being referred to here are not those produced by incorrect logic but those that occur because of rounding-off at every step of computation due to the finite number of significant figures the computer hardware can accommodate as well as a poor initial guess. A numerical solution method is therefore considered to be stable if it does not magnify the errors that appear in the course of the numerical solution process. Stability in temporal problems guarantees that the method yields a bounded solution whenever the exact solution is bounded. Stability in the context of iterative methods ensures that the solution does not diverge.

The property of stability can be difficult to investigate. The problem is further exacerbated when boundary conditions and non-linearities are present. For this reason, the stability aspect of a numerical method is commonly investigated with constant coefficients without boundary conditions. The results obtained in this way can often be applied to more complicated problems, albeit with some notable exceptions. However, when solving complex, non-linear, and coupled equations with complex boundary conditions, there are few stability studies we can infer. In this circumstance, we have to rely on experience and intuition to ensure stability of the numerical procedure. A number of solution schemes require that the time step be set below a certain limit or promote under-relaxation in the system of algebraic equations. We discuss these issues and provide guidelines for the appropriate selection of time-step size and suitable under-relaxation factors in Section 5.4.3.

For linear problems, the two most common methods of stability analysis are the matrix method and the von Neumann method. Both methods are based on predicting whether there will be a growth of error between the true solution of the numerical method and the actual computed solution, which also includes the round-off contamination. Example 5.3 demonstrates the application of the von Neumann stability method to a convection-type equation. Other related stability issues are also demonstrated by additional numerical examples below.

Example 5.3

Consider the one-dimensional convection-type equation $\frac{\partial \phi}{\partial t} + u \frac{\partial \phi}{\partial x} = 0$. Demonstrate the use of the von Neumann stability method to analyze the stability properties of the linear partial differential equation.

Solution

The exact form of the stability criterion depends on the particular differencing approximation applied to the equation. Using the finite-difference method, let us approximate the time and spatial derivatives with forward and central differences, where the discretized form of the convection-type equation becomes

$$\frac{\phi_i^{n+1} - \phi_i^n}{\Delta t} + u\frac{\phi_{i+1}^n - \phi_{i-1}^n}{2\Delta x} = 0 \qquad (5.3\text{-A})$$

where u is the velocity. To analyze the stability of the above equation, consider the errors introduced at every grid point as

$$\xi_i^n = \phi_i^n - {}^*\phi_i^n \qquad (5.3\text{-B})$$

where ϕ_i^n is the true solution of the numerical method and ${}^*\phi_i^n$ is the actually computed solution. For the discretized equation, we are actually calculating

$$\frac{{}^*\phi_i^{n+1} - {}^*\phi_i^n}{\Delta t} + u\frac{{}^*\phi_{i+1}^n - {}^*\phi_{i-1}^n}{2\Delta x} = 0$$

Substituting Eq. (5.3-B) into the above, followed by the application of Eq. (5.3-A), yields

$$\frac{\xi_i^{n+1} - \xi_i^n}{\Delta t} + u\frac{\xi_{i+1}^n - \xi_{i-1}^n}{2\Delta x} = 0 \qquad (5.3\text{-C})$$

For a linear computational algorithm, the error ξ_i^n in the majority of textbooks (Fletcher, 1991; Anderson, 1995) deals with just one term of the finite complex Fourier series, which is given as

$$\xi_i^n = e^{at}e^{ik_m x} \qquad (5.3\text{-D})$$

where a is a constant and k_m is the wave number. Substituting the above into Eq. (5.3-C), we obtain

$$\frac{e^{a(t+\Delta t)}e^{ik_m x} - e^{at}e^{ik_m x}}{\Delta t} + u\frac{e^{at}e^{ik_m(x+\Delta x)} - e^{at}e^{ik_m(x-\Delta x)}}{2\Delta x} = 0$$

After some arithmetic manipulation and applying trigonometric identities, the above equation reduces to

$$e^{a\Delta t} = 1 - iC\ sin(k_m\Delta x)$$

where $C = u\Delta t/\Delta x$. For von Neumann stability, the following requirement needs to be satisfied for the amplification factor: $e^{a\Delta t} \leq 1$; therefore, the criterion for C is $C^2 \leq 2\Gamma$ (diffusion coefficient). Based on the latter criterion, the discretized Eq. (5.3-A), since being inviscid ($\Gamma = 0$), will lead to an unstable solution no matter what values of Δt are in parameter C. It is therefore classified as *unconditionally unstable*.

Alternatively, let us use the time variable ϕ_i^n in Eq. (5.3-A) as an average value between grid points $i+1$ and $i-1$; i.e.,

$$\phi_i^n = \frac{1}{2}\left(\phi_{i-1}^n + \phi_{i+1}^n\right)$$

Substituting the above into Eq. (5.3-A), the discretized form becomes

$$\frac{\phi_i^{n+1} - \frac{1}{2}\left(\phi_{i-1}^n + \phi_{i+1}^n\right)}{\Delta t} + u\frac{\phi_{i+1}^n - \phi_{i-1}^n}{2\Delta x} = 0 \qquad (5.3\text{-E})$$

We get a similar error equation in the form of Eq. (5.3-C) as

$$\xi_i^{n+1} = \frac{\xi_{i-1}^n + \xi_{i+1}^n}{2} - C\frac{\xi_{i+1}^n - \xi_{i-1}^n}{2}$$

By substituting Eq. (5.3-D) into the above, and after some arithmetic manipulation, the amplification factor becomes

$$e^{a\Delta t} = cos(k_m\Delta x) - iCsin(k_m\Delta x)$$

Here, the von Neumann stability requirement of $e^{a\Delta t} \leq 1$ is met as long as the parameter C is ≤ 1.

Discussion
The von Neumann stability analysis performed on a simple linear equation provided some fundamental insights into the application of various differencing schemes to achieve stability. The forward differencing in time employed in Eq. (5.3-A) fails to satisfy the stability requirement of $e^{a\Delta t} \leq 1$. However, by cleverly replacing the time derivative with a first-order difference where the variable $\phi(t)$ is represented by an average value between neighboring grid points, as illustrated in Eq. (5.3-E), the stability requirement of $e^{a\Delta t} \leq 1$ can be met for $C \leq 1$. The differencing used to represent the time derivative is called the *Lax method* (named after the mathematician Peter Lax, who first proposed it). The recurring parameter C in this example is commonly called the Courant number. It means that $\Delta t \leq \Delta x/u$ for the numerical solution to be stable. Moreover, it is also commonly called the *Courant-Friedrichs-Lewy* condition, generally written as the CFL condition. It is an important stability criterion for convection-type equations.

Example 5.4
Consider again the one-dimensional transient diffusion equation described in Example 5.2. Apply the finite-difference discretization to the equation and discuss the stability of the numerical solution using the explicit Euler method at two different time-step sizes of $\Delta t = 1/100,000$ and $\Delta t = 1/1000$ with a fixed grid-step size of $\Delta x = 1/100$. Approximate the time and spatial derivatives according to the first-order forward and second-order central differences. The initial condition is set according to $\phi(t=0, x) = 1 - x + sin(2\pi x)$, where $0 \leq x \leq 1$. The Dirichlet boundary conditions are $\phi(t, x=0) = 1$ and $\phi(t, x=1) = 0$. The thermal diffusivity α is also assumed to be constant, with a value of 0.5.

Solution
The finite-difference discretized form of the diffusion equation Eq. (5.2-A) using the explicit Euler method in this example can be expressed as

$$\phi_i^{n+1} = \phi_i^n + \alpha\frac{\Delta t}{\Delta x^2}\left(\phi_{i+1}^n - 2\phi_i^n + \phi_{i-1}^n\right)$$

The transient results for ϕ are shown in Figures 5.4.1 and 5.4.2.

Discussion
The sensitivity of the time step Δt to the time-advancement procedure is demonstrated for the explicit Euler method of a diffusion-type equation. For this particular example, the condition for stability is given by $\Delta t \leq \Delta x^2$. In Figure 5.4.1, where the time step Δt is much smaller than the grid size Δx, the numerical procedure marches in a stable fashion and shows signs of convergence tendencies. After 1000 iterations, the intermediate result is gradually approaching the exact solution

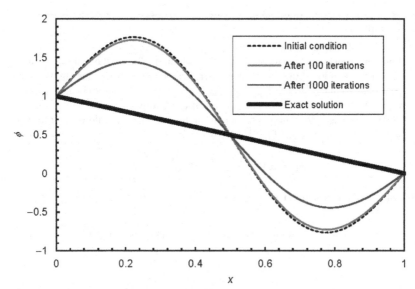

FIGURE 5.4.1 Time advancement with $\Delta t = 1/100{,}000$ and $\Delta x = 1/100$.

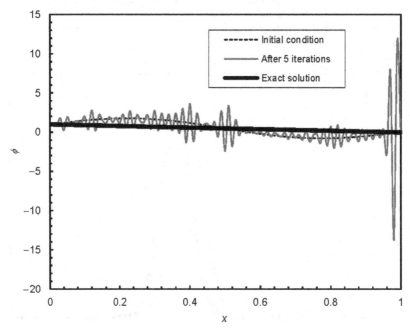

FIGURE 5.4.2 Time advancement with $\Delta t = 1/1000$ and $\Delta x = 1/100$.

profile. Nevertheless, for a larger time step, such as that employed in Figure 5.4.2 (note the difference in scale along the vertical axis for ϕ), when the time step Δt is much greater than the grid size Δx, the numerical procedure exhibits strong signs of instability. This is evidenced after only 5 iterations.

Example 5.5

Consider again the convection-type equation described in Example 5.3. Using the finite-difference method to discretize the equation, discuss the stability of the numerical solution using the explicit Euler method with a fixed time-step size of $\Delta t = 1$ at two different grid-step sizes of $\Delta x = 1$ and $\Delta x = 1/2$. Approximate the time and spatial derivatives according to the first-order forward and backward differences. The initial condition is set according to a Gaussian profile (similar to the initial value proposed in Hu et al., 1996):

$$\phi(t = 0, x) = \exp\left[-\ln(2)\left(\frac{x - 50}{3}\right)^2\right]$$

with the Dirichlet boundary condition of $\phi(t, x=0) = 0$.

Solution

The finite-difference discretization of the convection-type equation using the explicit Euler method is given by

$$\phi_i^{n+1} = \phi_i^n - u\frac{\Delta t}{\Delta x}\left(\phi_i^n - \phi_{i-1}^n\right)$$

For stability, the CFL number must be less than or equal to unity, i.e., $C \leq 1$. Assuming a constant CFL number of unity, $\Delta t \leq \Delta x/u$. If the velocity u is taken to be 1 m/s, the approximation is stable when $\Delta t \leq \Delta x$. The transient results for ϕ are shown in Figures 5.5.1 and 5.5.2.

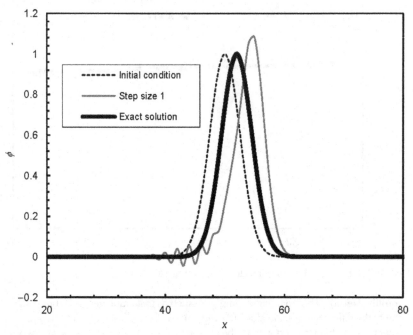

FIGURE 5.5.1 Time advancement with $\Delta t = 1$ and $\Delta x = 1$.

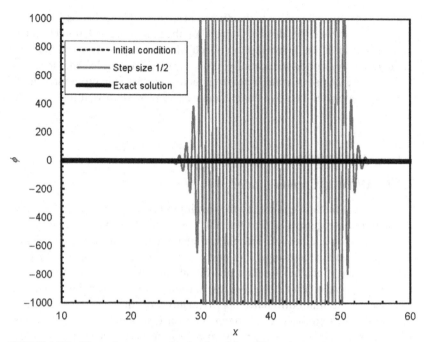

FIGURE 5.5.2   Time advancement with $\Delta t = 1$ and $\Delta x = 1/2$.

Discussion

The sensitivity of the grid size Δx to the time-advancement procedure, while maintaining a fixed time step Δt, is demonstrated for the explicit Euler method for a convection-type equation. This example illustrates the influence of the important CFL number on the stability of the numerically explicit marching procedure. For the case in Figure 5.5.1, where the condition is $\Delta t = \Delta x$, the numerical procedure is relatively stable, though some unfavorable wiggles are attained during the numerical computations. The removal of these unwanted wiggles can be overcome by marching with smaller time-step sizes. However, when $\Delta t > \Delta x$, as for the case depicted in Figure 5.5.2 (note again the difference in scale along the vertical axis for ϕ), the time-advancing numerical procedure is evidently unstable. The case that is solved in Figure 5.5.2 clearly accentuates the violation of the CFL number for explicit time-marching methods.

We have thus far discussed stability by predominantly focusing on explicit types of numerical procedures in the worked examples above. Explicit-type procedures can be considered *conditionally stable*, since they are strongly influenced by the temporal resolution. Implicit-type procedures are usually *unconditionally stable*. This is because allowance is provided for the variable to be continuously updated within the time step instead of calculating from the previous time-step values. The majority of commercial CFD codes employ implicit-type procedures due to their inherent stability. Nevertheless, instability that arises in these codes does not depend on the temporal resolution but rather on the adoption of the segregated approach where calculations of the transport variables are performed sequentially in the iterative process. In order to ensure *convergence*, the use of under-relaxation factors can assist in promoting stability of the segregated iterative computations. These are discussed further in the next section.

5.4 CONVERGENCE

5.4.1 What Is Convergence?

If a numerical method can satisfy the two important properties of consistency and stability, we generally find that the numerical procedure is convergent. *Convergence* of a numerical process can therefore be defined as the solution of the system of algebraic equations approaching the true solution of the partial differential equations having the same initial and boundary conditions as the refined grid system (*grid convergence*). For initial-value (marching) problems governed by the finite-difference approximations of linear partial differential equations, Lax's equivalence theorem is given here without proof. It states: "Given a property initial-valued problem and a finite-difference approximation, *consistency* and *stability* are the necessary and sufficient conditions that need to be satisfied for *convergence*"; i.e., *consistency + stability = convergence*. We might add that most computational work for non-linear partial differential equations, as used in CFD, proceeds as though this theorem applies, although it has not been proven directly for this general category of equations.

In the majority of commercial CFD codes, the system of algebraic equations is usually solved iteratively. When dealing with these codes, there are three important aspects to abide by for *iterative convergence*. First, all the discretized equations (momentum, energy, etc.) are deemed to be converged when they reach a specified tolerance at every nodal location. Second, the numerical solution no longer changes with additional iterations. Third, overall mass, momentum, energy, and scalar balances are obtained. During the numerical procedure, the imbalances (errors) of the discretized equations are monitored, and these defects are commonly referred to as the *residuals* of the system of algebraic equations; i.e., they measure the extent of imbalances arising from the equations and terminate the numerical process when a specified tolerance is reached. For satisfactory convergence, the residuals should diminish as the numerical process progresses. In the likelihood that the imbalances grow, as reflected by increasing residual values, the numerical solution is classified as being unstable (divergent). It is noted that iterative convergence is not the same as *grid convergence*. *Grid convergence* seeks a grid-independent solution, which means approaching the exact solution. We discuss this further later. In addition, in Section 5.4.2, the concepts of residuals and *convergence tolerance* are discussed in the context of attaining a numerical solution.

Example 5.6

Based on the explicit Euler method described in Example 5.4 for the one-dimensional transient diffusion equation with a time-step size of $\Delta t = 1/100{,}000$ and a grid-step size of $\Delta x = 1/100$, discuss the aspect of convergence for the numerical solution attained with identical initial and boundary conditions. The thermal diffusivity α is also assumed to be constant, with a value of 0.5.

Solution

The computational results showing the transient development for the variable ϕ are given in Figure 5.6.1.

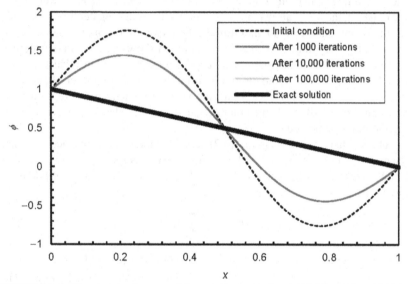

FIGURE 5.6.1 Transient development of the variable ϕ with $\Delta t = 1/100{,}000$ and $\Delta x = 1/100$.

Discussion

The purpose of this simple example is to illustrate the condition of *consistency + stability = convergence*. The first aspect concerns stability; it is observed that no signs of instability are experienced during the course of the numerical calculations. The second aspect concerns convergence; after 100,000 cycles, the numerical result converges and collapses to the exact solution profile. Since the difference between the discretized equation and the exact one is negligible and thus the remainder or truncation error diminishes, consistency prevails.

5.4.2 Residuals and Convergence Tolerance

For any transport variable ϕ, the discretized form of the partial differential equation (see also Eqs. (4.63) and (4.64) in Section 4.3.3) can be specifically written as

$$a_P \phi_P = \sum a_{nb} \phi_{nb} + b_P \qquad (5.1)$$

In Eq. (5.1), the central coefficient a_P and neighboring coefficients a_{nb} normally depend on the solution of other flow-field variables, including the time- and spatial-varying fluid-flow properties. These coefficients are updated consecutively during the iterative procedure. At the start of each iteration step, the equality in Eq. (5.1) will not hold. We can therefore rewrite the above equation by introducing an imbalance variable called residual R_p, where Eq. (5.1) can be re-expressed as

$$R_P = \sum a_{nb}\phi_{nb} + b_P - a_P\phi_p \qquad (5.2)$$

From the above equation, we introduce the concept of residuals as applied to each discretized equation of the system of transport equations. For a well-posed formulation, the residuals become *negligible* with increasing iterations. In CFD, residuals are employed to monitor the behavior of the numerical process. Importantly, they implicate whether the solution shows a trend of convergence or divergence. It is noted that the concept of *mass residual* introduced in Section 4.3.3 and used in Example 4.5 is different from the concept of residuals defined herein. The former is a source term appearing in the pressure-correction equation, Eq. (4.80), while the latter pertains to the imbalances in Eq. (5.2) during the iterative procedure.

The residual that arises in Eq. (5.2) actually depicts the imbalance (error) at the nodal point P for one cell volume. For practical purposes, a *global* residual R, taken as the sum of each *local* residual R_p over all the grid nodal points, is monitored:

$$R = \sum_{\text{grid points}} |R_P| \qquad (5.3)$$

Convergence is deemed to be achieved for the discretized Eq. (5.1) so long as the global residual $R \le \varepsilon$ or $\sum_{\text{grade points}} |R_P| \le \varepsilon$. The variable ε is usually referred to as the *convergence tolerance* for the system of algebraic equations. There is some practical guidance in selecting appropriate values for the convergence tolerance. It is noted that specifying appreciably small tolerance values will incur a large number of iteration steps in reaching convergence. On the other hand, large tolerance values constitute an early termination of the iteration process for which the numerical solution of the algebraic equations is considered to be rather coarse or not sufficiently converged. By default, the monitored residuals are usually scaled. Generally, a decrease of the residual by three orders of magnitude during the iteration process indicates at least *qualitative convergence*. Here, the major flow features are considered to be sufficiently established. Nevertheless, stricter convergence consideration is required for transport variables like energy and scalar species. It is recommended that, to achieve energy and species balance, the scaled energy residual decrease to a recommended convergence tolerance of 10^{-6}, while the scaled scalar species may need to decrease only to a convergence tolerance of 10^{-5}. For *quantitative convergence*, changes are monitored for all considered flow-field variables. During the monitoring of these residuals, the reader is also advised to ensure that property conservation is satisfied.

Example 5.7

Consider the two-dimensional CFD case of the incompressible laminar flow between two stationary parallel plates with the dimensions of height $H=0.1$ m and length $L=0.5$ m. Demonstrate the convergence behavior through monitoring the residuals of the transport variables u and v in their algebraic form represented in Eq. (5.2) as well as the pressure correction p' from Eq. (4.80), where air (density $\rho = 1.2$ kg/m^3 and viscosity 4×10^{-5} kg/m$\cdot$s) is the working fluid. The inlet velocity is fixed at $u_{in}=0.01$ m/s. The outlet and wall conditions remain the same as applied in Chapter 2.

Solution

The schematic diagram of the channel flow is identical to the one that is used in Example 4.5. Here, the discretized equations governing the momentum and pressure correction are solved using default iterative solvers provided by an in-house research CFD code.

Convergence histories of the velocities u and v, and pressure correction p' are shown in Figure 5.7.1. The residuals for each of the transport variables are not scaled and the convergence tolerance ε has been set at 10^{-7} to terminate the numerical simulation.

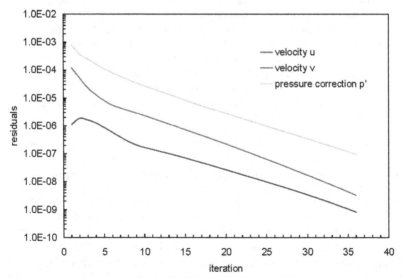

FIGURE 5.7.1 Convergence histories of the horizontal velocity u, vertical velocity v, and pressure correction p'.

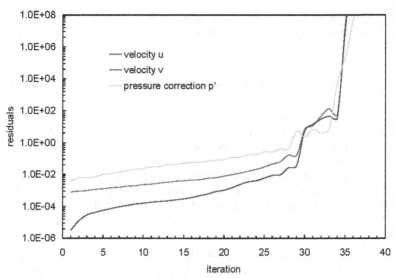

FIGURE 5.7.2   Divergence histories of the horizontal velocity u, vertical velocity v, and pressure correction p'.

Discussion

In Example 5.6, convergence is demonstrated and ascertained for an *explicit* methodology. Here, convergence is illustrated for an *implicit* methodology. The well-posed behavior of the numerical solution is evident by the diminishing values for the residuals of the velocities u and v, and pressure correction p'. Convergence is achieved, thereby satisfying the condition of Eq. (5.3), when all the residuals fall below the convergence tolerance of $\varepsilon = 1 \times 10^{-7}$. Nevertheless, there are circumstances where divergence can occur during the simulation of a channel flow problem. Such behavior is typically exhibited by the ascending trend of the residuals, leading to large, catastrophic values during the iterative process, as shown in Figure 5.7.2. This ill-posed (opposite to well-posed) numerical solution, which is particularly designed for the purpose of illustration, is due to the incorrect usage of the under-relaxation factors to control the numerical calculations. More discussion of the importance of under-relaxation factors generally required for implicit methods is elaborated in the next section.

5.4.3   Convergence Difficulty and Using Under-Relaxation

In consideration of the channel flow example discussed in previous chapters and herein, the residuals for the continuity (pressure-correction) and velocity components represent useful indicators that can be progressively evaluated to ascertain the convergence trend of the numerical calculations. Whether these residuals are being tracked locally at some grid nodal point within the flow system or globally through the sum of the local residuals, a convergence trend showing a diminishing residual value ensures the satisfaction of the conservation laws of mass and momentum (see Example 5.7). Since these laws are physical statements of fluid dynamics, the removal of any imbalances is imperative. In some flow cases, the local solution may be known, and the converged solution can be gauged and assessed by directly comparing the computed results against the available solution values. There are also practical considerations during the numerical calculations, where convergence can be assessed through the evaluation of some physical variables. For example, calculation of the drag or lift coefficients for a flow of air over an airfoil is useful for determining convergence of the flow system and thus determining when to terminate the numerical procedure.

In any numerical calculation, numerical instabilities can occur while solving the discretized equations. Poorly constructed mesh, improper solver settings, non-physical boundary conditions, and selection of inappropriate models are typically some of the factors that cause the ill-convergence of the numerical calculations. These are usually exhibited and amplified by the increasing (diverging) or "plateau" residual values throughout the iteration process (as seen in Example 5.7). Diverging residuals clearly imply an increase of imbalances in the conservation equations, and therefore that the physical laws of the fluid flow are being violated. Computational results that are not converged are misleading.

There are, however, some practical corrective steps that can be undertaken to overcome difficulties in achieving convergence. For a poor-quality mesh, the flow region can be re-meshed by increasing the number of grid nodal points. Such a strategy is particularly adopted to keep the grid from having large aspect ratios or highly skewed cells. In cases where high-order approximations are required, it may be sensible to initially compute the solution using low-order approximations. The diffusive nature of the low-order approximations allows large imbalances to dissipate quickly and promotes stability of the numerical procedure. Once the flow is established, a more accurate numerical solution can be attained by switching to the high-order approximations.

Another strategy to promote convergence is through the use of *under-relaxation factors*. Poor initial guesses or unresolved steep gradients in the flow field may cause divergence in the iterative process. The incorporation of under-relaxation factors into the system of algebraic equations can significantly moderate the iteration process by limiting the change in each of the transport variables from one step to the next. With the introduction of an under-relaxation factor α, the change in value of the transport variable ϕ at the central node of the cell volume between subsequent iteration steps can be expressed as

$$\phi_P^{New} = \phi_P^{Old} + \alpha \left(\phi_p^{New} - \phi_p^{Old} \right) \tag{5.4}$$

In CFD, under-relaxation is often introduced to stabilize the numerical calculations of the governing equations, which are generally non-linear, and where the equation of one transport variable is dependent on the others, for example, temperature affecting the velocities in buoyancy flows. The under-relaxation factor α in Eq. (5.4) controls the advancement of the transport variable ϕ_P during the iteration process. For a specified under-relaxation factor of 0.5, a restriction of 50% change is implied for ϕ_P from the value determined at the previous iteration to the current iteration step. The advancement for ϕ_P through the iteration process will therefore be increasingly impeded as smaller under-relaxation factors are employed. For the numerical solution in Example 5.7, the advancement of pressure has been under-relaxed by a factor of 0.3, which is typically employed when applying the SIMPLE scheme (Ferziger and Perić, 1999) to ensure stability and convergence of the iterative process.

In the majority of commercial CFD codes, the default settings of the under-relaxation factors are generally applicable to a wide range of problems. It is usually recommended to employ default factors in beginning the numerical calculations. Nevertheless, a more aggressive approach may be warranted by tightening the transport variable advancement through smaller under-relaxation factors to aid convergence. CFD users still constantly face many challenges in ascertaining optimal under-relaxation factors, usually not known a priori, to solve CFD problems. Settings of appropriate under-relaxation factors remain best learned from practical experience and application of CFD methodologies.

5.4.4 Accelerating Convergence

There are some practical guidelines for attaining quicker convergence of the computational solution. In the majority of cases handling flow problems, the supply of good initial or starting conditions is insurmountable, which leads to beneficial consequences in the iteration process. This can be achieved either by knowledge of similar physical conditions that can be imposed or by beginning the iteration from a previously converged solution. Inappropriate initial conditions generally lead to slower convergence, but may result in some untenable situations that promote divergence tendencies. Good initial or starting conditions promote computational efficiency and reduce computational efforts and resources. In the previous section, the use of under-relaxation factors was employed to promote stability during the iteration process. Default settings of the under-relaxation factors or Courant number in commercial CFD codes are generally applicable to a wide range of flow problems. Nevertheless, there are special circumstances where, depending on the flow problem, the under-relaxation factors or Courant number can be plausibly increased to accelerate the convergence. However, excessively high under-relaxation values can lead to unwanted instabilities. The reader may wish to adopt a strategy for storing intermediate solutions through incrementally increasing the under-relaxation factors. This can be carried out periodically before carrying out subsequent calculations. Such a procedure allows computations to be kept at a minimum and, more importantly, eliminates the need to re-compute the problem from the initial state. In Chapter 4, a number of accelerating techniques that are gaining prominence in solving CFD problems were described. In many commercial CFD codes, multi-grid solvers—a procedure to solve the algebraic equations by employing a combination of iterative solvers, such as Jacobi, Gauss-Siedel, or SIP, and direct solver cycling through different levels of grid densities (see sections 5.6 and 8.2.4 for more detailed description)—are offered as the default solvers in accelerating the convergence for the iteration process. In this case, provisions are given for users to change, at their discretion, the settings of the multi-grid solver. More often than not, the default settings provided are sufficiently robust and they do not necessarily need to be altered.

5.5   ACCURACY

The previous discussion of convergence, consistency, and stability has been primarily concerned with the solution behavior where the finite quantities, such as the time step Δt and mesh spacing Δx, Δy, and Δz, diminish. Since the discretized forms of the transport equations governing the flow and energy transfer are always solved numerically on a finite-grid layout and the effects of turbulence are generally modeled through approximate theories, the solution obtained is always approximate. The corresponding issue of *accuracy* therefore becomes another important consideration.

In Section 5.2, the determination of consistency produces an explicit expression for the truncation error. As mentioned, the truncation error represents the difference between the discretized equation and the exact one, and it provides a means of evaluating the accuracy of the solution for the partial differential equations. The order of the truncation error coincides with the order of the solution error if the grid spacings are sufficiently small and if the initial and auxiliary boundary conditions are sufficiently smooth. It is commonly implied that an improvement in accuracy (from the truncation error) of high-order approximations can be achieved for a sufficiently fine grid. Refining the grid will often produce a superior accuracy for high-order approximations over low-order approximations. However, at an absolute accuracy level, justification for more expensive computations may not demonstrate the desired superior accuracy due to limited computing capacity.

One method whereby accuracy can be assessed for a particular algorithm on a finite-grid is to apply it to a related but simplified problem that possesses an exact solution. However, accuracy is usually problem-dependent; an algorithm that is accurate for one model problem may not necessarily be as accurate for another, more complicated problem. Another way to assess accuracy is to obtain solutions on successively refined grids (*grid convergence*) and to check that, with successive refinements, the solution is not changing and is therefore satisfying some predetermined accuracy. This technique assumes that the approximate solutions will converge to the exact solution as the finite quantities diminish, and then the approximate solution on the finest grid can be used in place of the exact solution; a grid-independent solution is thus achieved. Assuming that the accuracy of this approximate solution can be assessed, it is important to consider the related question of how accuracy may be improved. At a specific level, the use of high-order approximation or grid refinement would be expected to produce more accurate solutions. Nevertheless, such choices are meaningful only if they are considered in conjunction with execution time and computational efficiency.

It is important that the reader be aware that *a converged solution does not necessarily mean an accurate solution*. Some possible sources of solution errors resulting from numerical calculations of the algebraic equations require analysis and are discussed in the next section. If these errors are to be minimized, some systematic steps in numerical analysis, such as grid independence and verification and validation of numerical models, are necessary.

5.5.1 Source of Solution Errors

Not only should the reader be aware of the existence of errors in *computational solutions,* but, more important, the reader must attempt to distinguish one error from another. This section addresses the possible sources of errors that the reader is likely to encounter in applying CFD methodologies. Errors are introduced because the numerical solutions of the fluid-flow and heat-transfer problems are only *approximate solutions*. Some prevalent sources of errors in dealing with numerical solutions include

- *Discretization error*
- *Round-off error*
- *Iteration or convergence error*
- *Physical modeling error*
- *Human error*

Before we elaborate on these errors in CFD, we should establish a clear and logical distinction between *error* and *uncertainty*; the distinction we use is based on the publication *AIAA Guide for the Verification and Validation of Computational Fluid Dynamics Simulations* (American Institute of Aeronautics and Astronautics, 1998). Error can be defined as *a recognizable deficiency that is not due to lack of knowledge*, while uncertainty can be defined as *a potential deficiency that is due to lack of knowledge*. Although these definitions appear to be philosophical, their practical application will become clearer as the origin of errors in CFD is explored below.

5.5.1.1 Discretization Error

Discretization errors are due to the difference between the exact solution of the modeled equations and a numerical solution with a limited time and space resolution. They arise because an exact solution to the equation being solved is not obtained but is numerically approximated. For a consistent discretization of the algebraic equations, the computed results are expected to become closer to the exact solution of the modeled equations as the number of grid cells is increased. However, the results are strongly affected by the density of the mesh and distribution of the grid nodal points.

We identify two types of discretization errors: local and global (or accumu-lated). To have an idea about the *local error* and *global error*, consider the finite-difference formulation of the derivatives for the transport variable ϕ in space and time at a specified grid nodal point expressed through the Taylor series expansion:

$$\left(\frac{\partial \phi}{\partial x}\right) = \frac{\phi_{i+1,j} - \phi_{i,j}}{\Delta x} + \underbrace{O(\Delta x)}_{\text{Truncation error}} \quad \textit{Spatial derivative} \quad (5.5)$$

$$\left(\frac{\partial \phi}{\partial t}\right) = \frac{\phi_{i,j}^{n+1} - \phi_{i,j}^{n}}{\Delta t} + \underbrace{O(\Delta t)}_{\text{Truncation error}} \quad \textit{Time derivative} \quad (5.6)$$

Termination of the Taylor series expansion in Eqs. (5.5) and (5.6) results in the so-called truncation errors involved in the approximation.

The local error is the formulation associated with a single step and provides an idea of the accuracy of the method used. For this error, the accuracy of the numerical solution concerns mainly the approximation of the spatial deriv-ative. The solution accuracy for a transient problem, however, focuses on the advancement of the transport variable ϕ through time, and it is usually charac-terized by global error. Local error and global error are illustrated in Figure 5.1. We can observe that the smaller the mesh size or time step in transient problems, the smaller the error and thus the more accurate the approximation.

Example 5.8 illustrates the significance of the discretization error.

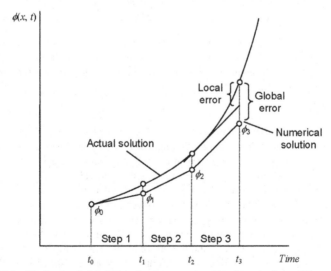

FIGURE 5.1   The local and global discretization errors resulting from the finite-difference method at a specified grid nodal point.

Example 5.9

Consider the transient, one-dimensional, convection-type equation to further illustrate the aspect of discretization error. A fourth-order central difference is employed for the spatial derivative to attain higher accuracy. Using the Euler explicit method, demonstrate the discretization error that is associated with the numerical solution obtained through the first- and second-order approximations to the time derivative at a fixed time-step size of $\Delta t = 1/128$ accompanied by the variation of two different grid-step sizes of $\Delta x = 1$ and $\Delta x = 1/2$.

Solution

Through consideration of additional grid nodal points along the spatial direction x and applying the Taylor series expansion, the fourth-order finite-difference approximation can be obtained as

$$\left(\frac{\partial \phi}{\partial x}\right) = \frac{-\phi_{i+2} + 8\phi_{i+1} - 8\phi_{i-1} + \phi_{i-2}}{12\Delta x} + \underbrace{O(\Delta x^4)}_{\text{Truncation error}}$$

Substituting the above approximation along with the first-order forward-difference to the time derivative yields the following algebraic equation:

$$\frac{\phi_i^{n+1} - \phi_i^n}{\Delta t} - u\left(\frac{-\phi_{i+2} + 8\phi_{i+1} - 8\phi_{i-1} + \phi_{i-2}}{12\Delta x}\right) = 0$$

For the second-order approximation of the time derivative, the central difference is employed; in other words,

$$\frac{\phi_i^{n+1} - \phi_i^{n-1}}{2\Delta t} - u\left(\frac{-\phi_{i+2} + 8\phi_{i+1} - 8\phi_{i-1} + \phi_{i-2}}{12\Delta x}\right) = 0$$

This newly developed formula is similar to the well-known *Leap Frog* method.

The computed results compared with the exact solution for the first-order and second-order time approximations against two different grid sizes are illustrated in Figures 5.8.1 and 5.8.2.

Discussion

By increasing the approximation of the time derivative from first order to second order, the numerical simulation for the higher-order time approximation at a grid size of $\Delta x = 1$ is shown to be less sensitive to the time-step changes with the suppression of wiggles through time (compare the solutions in Figures 5.7.1 and 5.7.2). By halving the grid size to $\Delta x = 1/2$, the numerical simulation is stabilized. The solution for the second-order time derivatives converges to the exact solution profile, but the first-order solution retains some oscillatory wiggles around the exact solution profile and is still far from its converged state. In this example, the systematic reduction of the grid-step sizes and the use of higher approximation for the time derivative characterize the diminishing contribution of the respective local and global discretization errors.

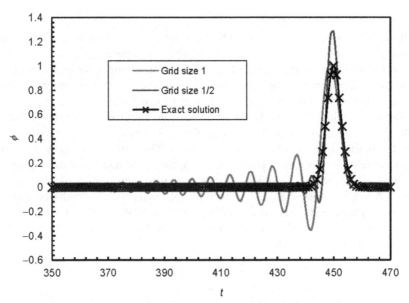

FIGURE 5.8.1 Fourth-order grid and first-order time approximations of convection-type equation for the variable ϕ with a fixed time step of $\Delta t = 1/128$ and two grid sizes of $\Delta x = 1$ and $\Delta x = 1/2$.

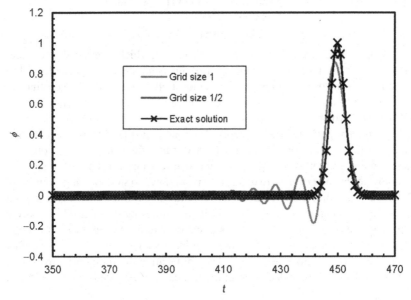

FIGURE 5.8.2 Fourth-order grid and second-order time approximations of convection-type equation for the variable ϕ with a fixed time step of $\Delta t = 1/128$ and two grid sizes of $\Delta x = 1$ and $\Delta x = 1/2$.

5.5.1.2   Round-Off Error

Round-off errors exist due to the difference between the machine accuracy of a computer and the true value of a variable. Every computer represents numbers that have a finite number of significant digits. The default value of the number of significant digits for many computers is 7, and this is commonly referred to as *single precision*. However, calculations can also be performed using 15 significant digits, which is referred as *double precision*. The error due to retention of a limited number of computer digits available for storage of a given physical value is therefore called the *round-off error*. This error is naturally random and there is no easy way to predict it. It depends on the number of calculations, rounding-off method, rounding-off type, and even sequence of calculations.

Consider the case of a simple arithmetic operation performed with a computer in single precision. Given that $a = 8888888, b = -8888887,$ and $c = 0.3333341$, let us evaluate the operations of $D = a + b + c$ and $E = a + c + b$. The arithmetic calculation for D proceeds as

$$D = 8888888 - 8888887 + 0.3333341$$
$$= 1 + 0.3333341$$
$$= 1.333334 \text{ (Correct result)}$$

while E performs the following operations:

$$E = 8888888 + 0.3333341 - 8888887$$
$$= 8888888 - 8888887$$
$$= 1.000000 \text{ (In error by } 25\%)$$

In algebra, we learned that $a + b + c = a + c + b$, which is reasonably accepted as a mathematically proven statement. But this is not necessarily true for calculations performed in a computer, as demonstrated above. It is noted that the sequence of calculations in single-precision mode results in a solution error of 25% in just two operations. Imagine that thousands or even millions of such operations are to be performed sequentially: The rounding-off error would accumulate and lead to serious error without any prior warning signs. If computer round-off errors are suspected of being significant, one test that can be performed is to employ double precision or to use a computer known to store floating-point numbers at a higher precision. An attempt to continuously refine a coarse grid solution to achieve a solution of diminishing finite quantities may be feasible; however, this may not be possible for more complex algorithms. Nevertheless, it is noted that round-off errors are usually not a dominant source of solution errors when compared to discretization errors. The implication of round-off errors in a CFD problem is expounded upon in the test case investigated in Section 5.7.1.

5.5.1.3   Iteration or Convergence Error

Iteration or convergence errors occur due to the difference between a fully converged solution of a finite number of grid points and a solution that has not fully achieved convergence. The majority of commercial CFD codes solve the discretized equations iteratively for steady-state solution methodologies. For procedures requiring an accurate intermediate solution at a given time step, the equations are solved iteratively in transient methods. It is expected that progressively better estimates of the solution are generated as the iteration step proceeds and ideally satisfies the imposed boundary conditions and equations in each local grid cell and globally over the whole domain. However, if the iterative process is terminated prematurely, then errors arise. Convergence errors therefore can occur either because of the user's being too impatient to allow the solution algorithm to complete its progress to the final converged solution or because of the user's applying too large convergence tolerances to halt the iteration process when the CFD solution may still be considerably far from its converged state.

5.5.1.4 Physical Modeling Error

Physical modeling errors are those due to uncertainty in the formulation of the mathematical models and deliberate simplifications of the models. Here, we reinforce the definition of uncertainty, where the Navier–Stokes equations can be considered to be exact and solving them is impossible for most flows of engineering interest because of lack of sufficient knowledge to model them. The sources of uncertainty in physical models are

1. The phenomenon is not thoroughly understood.
2. Parameters employed in the model are known to possess some degree of uncertainty.
3. Appropriate models are simplified, and thus uncertainty is introduced.
4. Experimental confirmation of the models is not possible or is incomplete.

Modeling is often required for turbulence, which places huge demands on computational resources if it is to be simulated directly. Other phenomena, such as combustion, multi-phase flow, chemical processes, etc., are difficult to describe exactly and inevitably require the introduction of approximate models. Even Newton's and Fourier's laws are themselves models, although they are solidly based on experimental observations for many fluids. In addition, the underlying mathematical model is nearly exact, but some fluid properties may not be exactly known. They depend strongly on temperature, species concentration, and, possibly, pressure; this dependence is ignored, thus introducing modeling errors (for example, the use of Boussinesq approximation for natural convection). In some situations, a simplified model may be adopted within the CFD code for the convenience of a more efficient computation, even though a physical process is known to a high level of accuracy. Physical modeling errors are examined by performing validation studies that focus on certain models. The conceptual idea and definition of validation are expounded upon in Section 5.5.3.

5.5.1.5 Human Error

In CFD, there are essentially two types of human error. First, computer programming errors involve human mistakes made in programming, which are the direct responsibility of the programmers. These errors can be removed by systematically performing verification studies of subprograms of the computer code and the entire code, reviewing the details inserted into the code, and performing validation studies of the code. (We review the concept of verification in Section 5.5.3.) Second, usage errors are due to application of the code in a less-than-accurate or improper manner. Inexperience in handling CFD codes may result from either incorrect computational domains (such as improper geometry construction or grid generation) or inappropriate setting of boundary conditions. Selection of bad numerical schemes or computational models to simulate certain flow problems compounds the undesirable usage errors. The reader should note that the potential for usage errors increases with the increased level of options available in CFD. Nevertheless, usage errors can be minimized and controlled through proper training and analysis and the accumulation of experience. Some practical guidelines are discussed in the next chapter.

5.5.2 Controlling the Solution Errors

It is good engineering practice for the CFD user to examine any potential pitfalls when employing numerical methodologies. Numerical errors are primarily discretization and round-off errors. They have the tendency to accumulate through computational processes and may yield unphysical CFD solutions. Results that at first seem reasonable in hindsight may be in considerable error at certain locations within the flow domain. Controlling solution errors therefore represents a crucial step toward obtaining reliable and meaningful CFD solutions.

We begin by focusing on the contribution of the discretization and round-off errors obtained through numerical methods. As the mesh or time-step size decreases (see Figure 5.2), the discretization error decreases while the round-

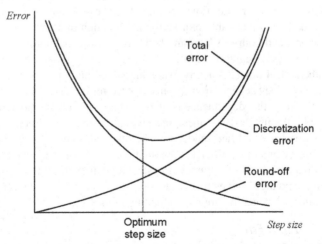

FIGURE 5.2 The discretization, round-off, and total errors as a function of mesh and/or time-step size.

off error increases. For the total error, which is taken as the sum of the discretization and round-off errors, it is evident that continually decreasing the step size does not necessarily mean that more accurate results are attained. The opposite is true at small step sizes, where less accurate results are obtained because of the quicker increase in the round-off error. In order to contain this error, we should therefore avoid a large number of computations with very small numbers.

In practice, we will not be able to determine the magnitude of the error involved in the numerical method. As shown above, the knowledge of discretization error alone is meaningless without a true estimate of the round-off error. To better assess the accuracy of the results obtained, some practical guidelines are recommended.

First, the issue of *grid independence* is explored. To address this issue, we can begin by solving the flow problem with reasonable mesh sizes of Δx, Δy, and Δz (and a time step of Δt for a transient problem) based on acquired experience. The computational results may look qualitatively good, but let us assume that the problem is repeated with twice as many grid points, thus halving the mesh sizes in each direction by $\Delta x/2$, $\Delta y/2$, and $\Delta z/2$. If the results obtained do not differ significantly from the results obtained in the original grid layout, we can conclude that the discretization error is at an acceptable level. But if the values for the transport variables are quite different for the second calculation, then the solution is a function of the number of grid points. In all practical cases, the grid needs to be refined by increasing the number of grid points until a solution is achieved where no significant changes in the results occur. This indicates that the discretization error is reduced to an acceptable error and grid independence is reached. This issue is investigated in a test case described in Section 5.7.1.

Second, the majority of CFD calculations are usually performed in single precision to avoid overburdening the computational resources. Nevertheless, if the round-off error is found to be significant, the flow calculations can be repeated using double precision while holding the mesh sizes (and the size of the time step in transient problems) constant. If the results do not change considerably, we conclude that the round-off error is not a problem in the CFD solution. However, if the changes are larger than expected, we may attempt to reduce the total number of calculations by either increasing the mesh sizes or changing the order of computations, such as adopting a higher approximation to evaluate the first-order spatial derivatives of the convective terms and/or the first-order time derivatives in the conservation equations. As seen in Figure 5.2, discretization error increases with increasing mesh size; the user should therefore acknowledge this important trend and seek some reasonable compromise.

Third, selection of appropriate turbulence models or other approximate models can be a daunting task, especially in attempting to minimize the physical modeling errors. The desired level of simplification that can be accepted to adequately model the physical flow problem is not straightforward because it depends on the choice of models that govern and characterize the particular flow physics. Also, carelessness in setting up a feasible geometrical model and an improper choice of boundary conditions are some of the human errors that frequently arise in CFD. Succinct practical guidelines for eliminating such pitfalls are discussed in the next chapter.

5.5.3 Verification and Validation

In addition to errors, uncertainties can also arise while performing a numerical simulation. The uncertainties can be due to the improper modeling of physics, such as a misunderstanding of the phenomena leading to falsifying assumptions, or an incorrect computational design, such as making wrong approximations and simplifications about the parameters that govern the fluid dynamics. For a CFD solution to be credible, it requires performance of a detailed analysis to quantify the modeling and numerical uncertainties in the simulation. Verification and validation procedures are the means by which a CFD solution can be properly assessed through quantitative estimation of the inherent errors and uncertainties.

Verification and validation have very distinct definitions. Although there is no universal agreement about the details of their definitions, there is fairly standard and consistent agreement on their usage. In this book, we adopt definitions that are focused on numerical errors and uncertainties where they cannot be considered negligible or overlooked.

An important aspect to remember about verification and validation is that they are applied in two very distinct ways. We can differentiate verification and validation in the following. *Verification* can be defined as *a process for assessing the numerical simulation uncertainty and when conditions permit, estimating the sign and magnitude of the numerical simulation error and the uncertainty in that estimated error.* This procedure concerns primarily the input parameters used for geometry, initial conditions, and boundary conditions. They are required to be carefully checked and systematically documented. It is also important that mesh and time-step sensitivity studies are extensively performed to bound the errors (whether they are insufficient spatial discretizations, too large temporal advancement, lack of iterative convergence, or computer programming errors) that are associated with the discrete approximations employed for the partial differential equations. On the other hand, *validation* can be defined as *a process for assessing simulation model uncertainty by using benchmark experimental data and when, conditions permit, estimating the sign and magnitude of the simulation modeling error itself.* This procedure simply means validating the calculations by establishing a range of physical conditions obtained from the calculations and by performing comparisons of the results from the CFD code with experiments that span the range of conditions. This represents the final phase of the credibility process of the models applied and is interpreted as the stage to determine the degree to which a model corresponds to an accurate representation of the real physical flow problem that is solved.

We demonstrate the use of the verification and validation procedures in the case studies described in Section 5.7.

5.6   EFFICIENCY

The rapid advancement in, and reduced costs of, computer hardware and resources have revolutionized the use of CFD. Nowadays, users have the luxury of accessing many commercial or shareware computer codes. More complex engineering systems and applications can be solved because of the feasibility of constructing high-quality grids to resolve the physical flow structures. Such flow problems usually require a substantial amount of grid points to envelope the whole physical domain and to achieve adequate resolution. If the mesh requires further refinement, almost all iterative solution methods suffer from slower convergence on the finer grids. The rate of convergence depends on the particular numerical method; the number of iterations for many methods is linearly proportional to the number of grid nodal points in one coordinate direction. This behavior is related to the fact that, for iterative procedures, information has to travel back and forth across the domain several times.

To overcome the convergence problem, applications of conjugate gradient and multi-grid methods have received unprecedented attention in enhancing the *efficiency* of a CFD procedure. The former is essentially a method that seeks the minimum of a function that belongs to the class of *steepest-descent methods*. The basic method in itself converges rather slowly and is not very useful, but when it is used in conjunction with some *preconditioning* of the original matrix, major enhancements in its speed of convergence have been recorded. The preconditioning is achieved either by applying the *incomplete Cholesky* factorization for symmetric matrices or by applying *biconjugate gradients* for asymmetric matrices. The latter, which is discussed in Chapter 4, employs a hierarchy of meshes. In the simplest case, the coarse ones are just subsets of the fine ones. Efficiency may be improved by the decision to switch from one grid to another through the V-cycle, W-cycle, combination of V-W cycles, or other possible strategies. (Additional details of the multi-grid method are found in section 8.2.4.) The conjugate gradient and multi-grid methods are ideal for solving the Poisson-like pressure or pressure-correction equation, such as the SIMPLE method.

Another approach to achieve computational efficiency is parallel computing. The increase in the capability of single-processor computers has almost reached its peak. It now appears that further increases in speed will require multiple processors—parallel computers. One of the advantages of parallel computers over classical vector supercomputers is scalability. Parallel computers employ standard processing chips and are therefore cheaper to produce and to obtain.

Commercially available parallel computers may have thousands of processors, gigabytes of memory, and computing power measured in gigaflops. The basic idea of parallel computing involves the sub-division of the solution domain into sub-domains and the assignment of each sub-domain to one processor. In such a case, the same computer code runs all processors. Since each processor needs data that reside in other sub-domians, exchange of data between processors and storage overlap are necessary. This demonstrates that parallel computing environments require redesign of algorithms. Good parallelization therefore needs modification of the solution algorithm.

Explicit marching methods are relatively easy to parallelize, since all operators are performed on data from preceding time steps. It is only necessary to ensure that data are exchanged at the interface regions between neighboring sub-domains after each step is completed. The sequence of operations and the results are identical on one and many processors. Of course, the problem remains handling the solution of the Poisson-like pressure equation. Implicit methods are more difficult to parallelize because the iterative solvers that are efficient in serial computations are not suitable to performance in parallel. Some solvers can be parallelized and can perform the same sequence of operations on multi-processors as on a single one. However, these arithmetic operations are inefficient and the communication overhead is extremely large.

One possible way to achieve parallelization for implicit methods is by data parallelism or domain decomposition. For steady flow problems, the concept of spatial domain decomposition is to divide the solution domain into a number of sub-domains, with the objective of maximizing efficiency and thus distributing the same amount of work on each processor. The usual approach is to split the global matrix coefficients that make up the central coefficient a_P and neighboring coefficients a_{nb} into a system of diagonal blocks. Each of these blocks is assigned to one processor and data are then transferred on two levels of communications. Local communication takes place between processors operating on neighboring blocks and global communication gathers information from all blocks in a "master" processor while broadcasting some information back to the other processors. For transient flow problems, domain decomposition in space can also be equally applied in time. More details of parallel computations (see section 8.2.5 and references given within) are left to the interested reader.

5.7   CASE STUDIES

A selection of case studies is presented to demonstrate the relevance of the computational solutions in *consistency*, *stability*, *convergence*, and *accuracy*. These cases individually show two or three of the following effects: discretization error (*truncation error*), iteration or convergence error (*grid convergence*), and physical modeling error (application uncertainty of boundary conditions, geometry, and CFD models, for example, in a turbulence model). They also include comparisons with an analytic solution (*verification*) and with experimental data (*validation*). It is important to note that the use of a test case from a particular code is not intended to provide any endorsement or the acceptance of the code for this particular purpose. Likewise, the absence of a test case from any particular code does not provide a statement on the unsuitability of the code for this particular application.

5.7.1 Test Case A: Channel Flow

This test case was calculated using an in-house finite-volume CFD computer code.

Model description: The geometry of the test problem is a two-dimensional laminar flow between two parallel plates, as used in previous worked examples. For this test problem, the channel has the dimensions height $H = 0.1$ m and length $L = 1.0$ m, with air taken as the working fluid.

Grid: The governing equations are discretized on a co-located grid arrangement. Velocity and pressure are co-located (cell-centered) as described in Figure 4.8. A uniform mesh is generated spanning the height and length of the channel. *Grid convergence (independence)* is performed with meshes of 5 × 10, 10 × 20, and 20 × 40 control volumes.

Features of the simulation: The numerical technique is based on the finite-volume discretization. The algorithm for the solution of the Navier–Stokes equations relies on the implicit segregated velocity–pressure formulation, such as the SIMPLE scheme. This leads to a Poisson equation for the pressure correction that is solved through a default iterative solver within the in-house CFD code. To avoid non-physical oscillations of the pressure field and the associated difficulties in obtaining a converged solution, the Rhie and Chow (1983) interpolation scheme is employed.

The fluid is incompressible and its density and viscosity have values of 1.2 kg/m^3 and 4×10^{-5} kg/m·s, respectively. With the inlet velocity specified at 0.01 m/s, the corresponding Reynolds number (see Eq. (3.26) in Chapter 3) based on this inlet velocity and the height of the channel is 30.

Results: As demonstrated in Example 5.7, a stable and converged solution is evidenced by diminishing residuals as the number of iterations increases. The decreasing trends of the residuals for a coarse grid distribution of 5 × 10, medium grid distribution of 10 × 20, and fine grid distribution of 20 × 40 control volumes also exhibit the desirable aspects of stability and convergence. Since the diminishing nature of the residuals is represented in all three respective grid distributions, only indicative residuals obtained from the fine grid distribution are illustrated, as shown in Figure 5.3. The numerical calculations are terminated for all three different meshes when all the velocities, pressure correction, and mass residuals fall below the convergence tolerance of $\varepsilon = 1 \times 10^{-7}$. It is noted that the pressure-correction residual is the same as the mass residual; only this mass residual is presented in Figure 5.3.

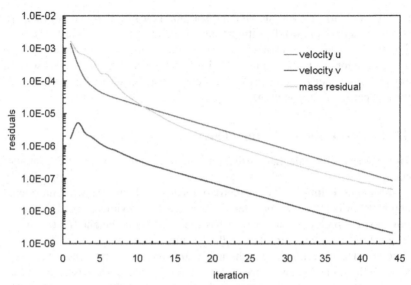

FIGURE 5.3 Convergence histories of the horizontal velocity u, vertical velocity v, and mass residual for the fine grid distribution.

The numerical solution is verified against the analytical solution developed in Example 3.4. Recalling Eq. (3.4-A) in Chapter 3, which was developed for $-H/2 \leq y \leq H/2$, the horizontal velocity u can be determined as

$$u(y) = \frac{3}{2} U_m \left[1 - \frac{y^2}{(H/2)^2} \right]$$

where the average velocity U_m in the test problem corresponds to the uniform inlet velocity $u_{in} = 0.01$ m/s. Figure 5.4 compares the numerical solutions for the three meshes against the analytical solution in the fully developed flow region. Controlling the solution error due to spatial discretization is fundamentally addressed through this parametric study. As the mesh is refined, the approximate solution approaches the exact solution. Herein, *consistency* is achieved, since the truncation error diminishes with an increasing mesh resolution. Table 5.1 presents the solution error evaluated for the maximum magnitude of the horizontal velocity u between the analytical value and predicted results for four different meshes. The results demonstrate that *accuracy* is attained for the numerical solution on grid distributions of 20×40 and 40×80 control volumes. With the additional refinement made to the channel geometry consisting of 40×80 control volumes, *grid independence* is thus achieved, since this approximate solution is the same as the results obtained for the grid distribution of 20×40 control volumes.

As shown in Table 5.1, the solution error appears to be leveling off, even though the mesh is further refined from a grid distribution of 20×40

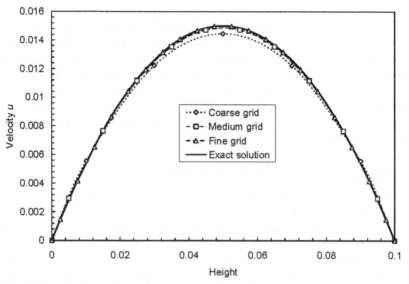

FIGURE 5.4 Velocity profiles in the fully developed region for the computational results of three different grid distributions and analytical solution.

TABLE 5.1 Solution errors between the predicted and analytical maximum velocities for the three different grid distributions

Mesh	Predicted Max Velocity u	Analytical Max Velocity u	Error $= (u_{analytical} - u_{predicted})/ u_{analytical} \times 100\%$
5×10	0.01444	0.015	3.73
10×20	0.01472	0.015	1.87
20×40	0.01496	0.015	0.27
40×80	0.01496	0.015	0.27

to 40×80 control volumes. Here, the solution error registers below 1%, which could not be totally eliminated due to the presence of round-off error during the numerical computations. This round-off error is due to *single precision* during the numerical calculations. As shown in Figure 5.2, further refinement of the mesh may lead to an increase in the round-off error. One possible way to significantly reduce this round-off error is to extend the number of significant digits during the iterative procedure to *double precision*. Bear in mind that *double precision* tends to incur additional computational burden. In practice, some trade-off of the solution *accuracy* to achieve a quicker turnaround of the numerical computations is generally required, especially for fluid flows that are complex.

Conclusion: In this test case, *consistency*, *stability*, *convergence*, and *accuracy* are succinctly illustrated for a two-dimensional channel flow problem. The error contribution to the numerical solution is investigated on the basis of evaluating the discretization or truncation error. The increasing mesh resolution demonstrates two important outcomes. First, as the spatial discretization error becomes smaller, a grid-independent solution is achieved. Second, the excellent agreement between the approximate and analytical solutions verifies the numerical algorithm that has been adopted and provides credibility to the computational solution.

5.7.2 Test Case B: Flow over a 90° Bend

This test case was calculated using a commercial finite-volume CFD computer code, ANSYS-FLUENT, Version 6.1.

Model description: The geometry of the test problem is a three-dimensional turbulent flow over a 90° bend. Figure 5.5 shows a schematic view of the experimental setup, which comprises an open-circuit suction wind-tunnel system for the 90° duct bend, made up of a 3.5-m-long horizontal duct, a 90° bend with a radius ratio of 1.5, and a 1.8-m-long vertical straight duct. Air flows through a 10-mm-thick Perspex square test section, with the bulk gas velocity U_b adjusted

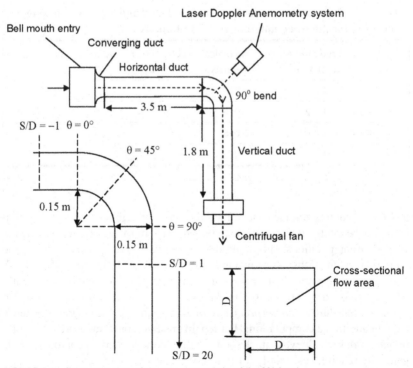

FIGURE 5.5 Schematic view of the experimental rig of the 90° bend.

with the aid of a variable-frequency controller. Experimental data are obtained on this setup using flow visualization and a laser Doppler anemometry (LDA) system.

Grid: For the 90° square-section bend, the computational domain begins at a distance of 2D upstream from the bend entrance and extends to 20D downstream from the bend exit. A structured mesh of $325 \times 43 \times 41$ control volumes in the respective directions along the streamwise, width, and height is generated for the whole computational domain.

Features of the simulation: This test case illustrates the importance of evaluating the choice of the turbulence models, standard $k-\varepsilon$ (Launder and Spalding, 1974) and Reynolds stress (Launder et al., 1975; Launder, 1989), for computing the flow separation around the 90° bend and validating the computational solutions against experimental data.

The algorithm for the solution of the Navier–Stokes equations relies on the implicit segregated velocity–pressure formulation, such as the SIMPLE scheme. This leads to a Poisson equation for the pressure correction that is solved through the default iterative solver, normally the multi-grid solver in the ANSYS-FLUENT computer code. Finite-volume discretization is employed to approximate the governing equations. To avoid non-physical oscillations of the pressure field and the associated difficulties in obtaining a converged solution on a co-located grid arrangement, the Rhie and Chow (1983) interpolation scheme is employed.

The working fluid, air, is taken to be incompressible and the default initial conditions implemented in the computer code are used for the simulations.

Boundary conditions: At the inlet, Dirichlet conditions are used for all variables. The bulk velocity U_b was taken as constant, with a value of 10 m/s. With the density and viscosity of air having values of 1.2 kg/m^3 and 2×10^{-5} kg/m · s, respectively, the corresponding Reynolds number based on this inlet velocity and height of the channel is 90,000. The turbulent kinetic energy k and dissipation ε are determined from the measured turbulence intensity I of about 1% at the center of the duct cross-section. For the Reynolds stresses, the diagonal components are taken to be equal to 2/3 k, whereas the extra-diagonal components are set to zero (assuming isotropic turbulence). At the outlet, Neumann boundary conditions are applied for all the transported variables. The non-equilibrium wall function (described in the next chapter) is employed for the air flow at solid walls because of its capability for better handling complex flows where the mean flow and turbulence are subjected to severe pressure gradients and rapid change, such as flow separation, reattachment, and impingement.

Results: The comparison between the measured and calculated longitudinal mean velocities normalized by the bulk velocity U_b at the bend exit ($\theta = 90°$) and 0.5D after the bend exit is illustrated in Figure 5.6. The prediction of the streamwise velocities using the Reynolds stress model is observed to yield better agreement than the standard k-ε model, which is due to the capability of the Reynolds stress model for capturing the anisotropic behavior of the flow-separation region around the 90° bend.

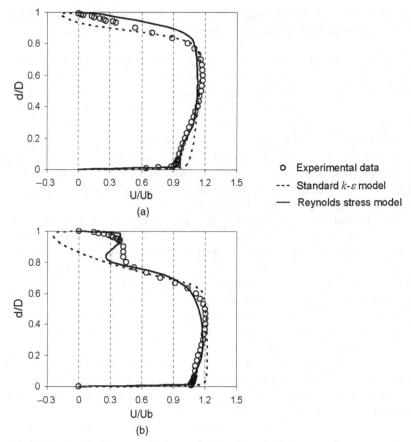

FIGURE 5.6   Comparison between measured and calculated longitudinal mean velocities normalized by the bulk velocity U_b: (a) streamwise velocity at bend exit ($\theta = 90^\circ$) and (b) streamwise velocity at 0.5D after the bend exit (see Figure 5.6 for more description).

The predicted longitudinal mean velocity normalized by the bulk velocity U_b using the Reynolds stress model is further compared against the measure data at different locations in the duct center plane, as represented in Figure 5.7. At the bend entrance ($\theta = 0^\circ$), the turbulence model successfully predicts the acceleration of the air flow near the inner wall. The fluid deceleration caused by the unfavorable pressure gradient is also captured near the outer wall. More important, the turbulence model adequately reproduces the distorted longitudinal velocity profiles at the angles of $\theta = 30^\circ$, $\theta = 45^\circ$, and $\theta = 60^\circ$ after the bend entrance.

To better understand the flow characteristics around the 90° bend, Figure 5.8 presents the calculated velocity vectors and pressure distribution of the air flow obtained through the Reynolds stress model at different cross-sections of the duct flow. Figure 5.8(a) and Figure 5.8(b) show the calculated air velocity vector and pressure distribution at the cross-sectional middle plane of the duct flow. Favorable (positive) and unfavorable (negative) longitudinal pressure gradients

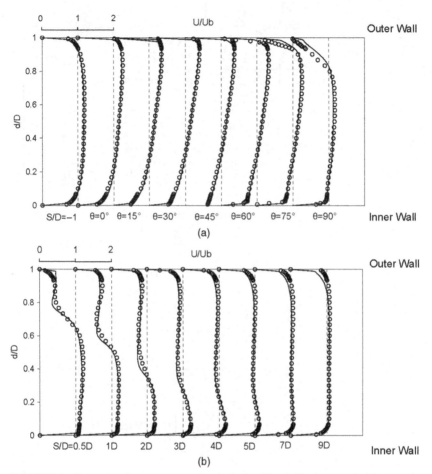

FIGURE 5.7 Comparison between measured and calculated longitudinal mean velocities normalized by the bulk velocity U_b at different locations in the duct center plane: (a) between bend entrance and exit and (b) downstream of bend exit (see Figure 5.6 for more description).

persist near the inner and outer walls of the bend entrance. The favorable and unfavorable pressure gradients are caused by the balance of centrifugal force and radial pressure gradient in the bend (Humphrey et al., 1981). This physical phenomenon is typical of curved-duct flows. Secondary flows are developed as the direct consequence of the cross-stream pressure gradient. The predicted secondary flow vectors are clearly depicted by the three square cross-sectional flow areas at angles of $\theta = 45°$ and $\theta = 90°$ and distance of 3D after the bend exit (S/D = 3) in Figure 5.8 (c), (d), and (e), respectively.

Conclusion: This test case focuses on the use of approximate models, such as the turbulence models, to predict the physical characteristics of the turbulent flow around a 90° bend. Owing to the absence of analytical solutions, *validation* of the computational solutions is performed by comparing the predictions against the experimental data in order to address the simulation model uncertainty

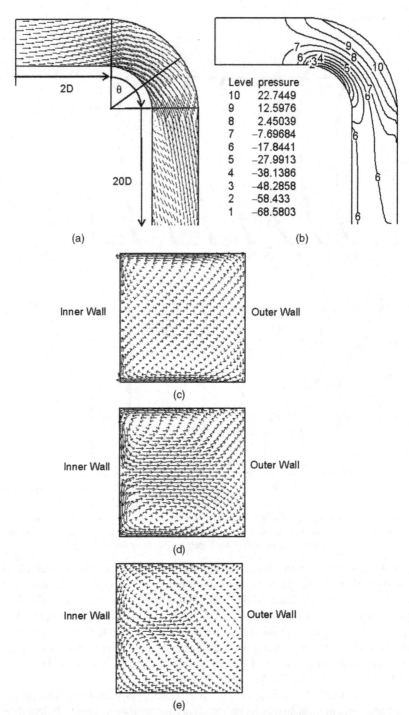

FIGURE 5.8 Calculated flow field: (a) air flow in the duct middle plane; (b) static pressure distribution in the duct middle plane; (c) secondary flow at a cross-sectional flow area of location $\theta = 45°$; (d) secondary flow at a cross-sectional flow area of location $\theta = 90°$ (bend exit); and (e) secondary flow at cross-sectional flow area of location $S/D = 3$ after the bend exit.

and the degree to which models correspond to an accurate representation of the real physical flow. The flow around a 90° bend, though geometrically simple, exhibits complex flow structures due to the existence of secondary flows in the vicinity of the bend region, which are generally anisotropic in nature. The results from the more sophisticated Reynolds stress model are shown to better capture the anisotropy behavior of the flow, in contrast to the standard k-ε model, which assumes isotropy in its original model formulation.

5.8 SUMMARY

The credibility of a computational solution was analyzed and assessed in this chapter through the consideration of *consistency*, *stability*, *convergence*, and *accuracy*. It is usually possible to demonstrate whether a discretized form of the governing fluid-flow equations is consistent and also if the algebraic form of the equations is stable. Convergence requires, however, implications from consistency and stability, i.e., *consistency* + *stability* = *convergence*. In any numerical calculations, errors and uncertainties affect the accuracy of the computational solution. It is imperative that the errors and uncertainties be systematically reduced so that the computational solution better represents the real physical flow problem that is being solved. The conceptual framework linking the various aspects of consistency, stability, convergence, and accuracy, beginning from the governing partial differential equations considered in Chapter 3 and arriving at the approximate solution of the algebraic equations described in Chapter 4, can be seen in Figure 5.9.

The application of a Taylor series expansion to the discrete approximation of the governing equations results in an explicit expression for the truncation error (Section 5.2). This error measures the accuracy of the finite-difference or finite-volume approximation and determines the rate at which the error decreases as the finite quantities, time step, and/or mesh spacing, diminish. It is noted that because of the close correlation between truncation error and solution error (Section 5.5.1), reducing the truncation error has beneficial consequences in also reducing the solution error.

Numerous worked examples and test cases presented throughout this chapter have aimed to illustrate the conceptual properties of consistency, stability, convergence, and accuracy while solving the discretized form of the partial differential equations. In addition, it is important that the computational solution be subjected to the rigorous processes of *verification* and *validation*, which are demonstrated through two test cases. For the simple two-dimensional channel flow problem in Test Case A, it is feasible to verify the computational solution against an analytical relationship. However, for a more complicated flow problem that employs a computational solution in a three-dimensional domain, the absence of an analytical solution requires the dependency of benchmark and/or experimental data to validate the computational solution. This is evidenced by Test Case B, three-dimensional air flow around a 90° bend. In this case, we demonstrated the use of appropriate turbulence models to better capture the physical flow behavior in a 90° bend. This aspect highlights the uncertainty that

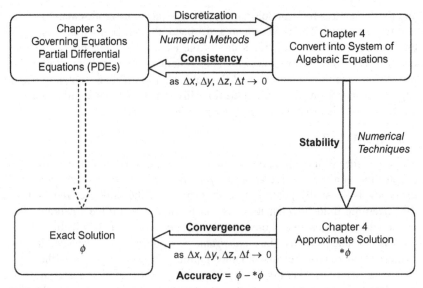

FIGURE 5.9   A conceptual framework linking the various aspects of consistency, stability, convergence, and accuracy in arriving at a solution for the transport equations.

arises through the use of approximate models for solving turbulent flow, which is investigated further in the next chapter. More practical guidelines for handling and solving a range of CFD problems are described in Chapter 6.

REVIEW QUESTIONS

5.1   Why do the results obtained through numerical methods differ from the exact solutions solved analytically? What are some of the causes for this difference?

5.2   In the analysis of CFD results, what does consistency imply?

5.3   What are the key aspects of consistency?

5.4   If a system of algebraic equations is equivalent to the partial differential equation as the grid spacing tends to zero, does this also mean that the solution of the system of algebraic equations will approach the exact solution of the partial differential equation? Why?

5.5   Explain why the following Taylor series expansion of the one-dimensional transient diffusion equation $\left[\frac{\partial \phi}{\partial t} - \alpha \frac{\partial^2 \phi}{\partial x^2} + \alpha \left(\frac{\Delta t}{\Delta x}\right)^2 \frac{\partial^2 \phi}{\partial t^2}\right]_i = 0$ does not show consistent properties.

5.6   Describe the concept of stability.

5.7   What are the stability criteria produced by the von Neumann analysis?

5.8   What is the Courant number and what is its function?

5.9   Consider the following discretized equation and locate the Courant number and discuss the *Courant-Friedrichs-Lewy* condition for stability in this case: $\phi_i^{n+1} = \phi_i^n + \alpha \frac{\Delta t}{\Delta x^2} \left(\phi_{i+1}^n - 2\phi_i^n + \phi_{i-1}^n\right)$.

5.10 Provide a definition of the concept of convergence.

5.11 State *Lax's equivalence theorem* for convergence. Does it apply to non-linear problems?

5.12 What are the three important rules when considering *iterative convergence*?

5.13 How is the concept of residuals applied to describe the discretized equation of the system of transport equations?

5.14 Differentiate between a local residual and a global residual.

5.15 What is implied when the residuals become negligible with increasing iterations?

5.16 What is the usual recommended residual tolerance level?

5.17 Define the under-relaxation factor. State its advantages and disadvantages when using a small value.

5.18 Discuss ways in which convergence can be accelerated.

5.19 Is a converged solution also an accurate solution? Why?

5.20 Discuss some types of errors that can cause a solution to be inaccurate.

5.21 What are discretization errors? What is the difference between a global error and a local error?

5.22 What methods can be used to minimize discretization errors?

5.23 What are round-off errors and what calculations are most affected by them?

5.24 Which method can be used to minimize round-off errors?

5.25 What does it mean to perform a grid convergence (independence) test?

5.26 What is the difference between verification and validation? Why are these two steps important in analyzing results?

5.27 Discuss briefly how multi-grid methods are employed to increase the computational efficiency of solving CFD problems.

5.28 Discuss briefly how parallel computing is used to achieve computational efficiency.

5.29 Consider the following algebraic equation: $a_p\Phi_p = \Sigma a_{nb}\Phi_{nb} + b$ (described in Chapter 4). In matrix form, a_p represents the diagonal element while a_{nb} is the neighboring element. The condition for convergence stipulates that $\sum |a_{nb}|/a_P \leq 1$, which simply means that the sum of the neighboring elements divided by the diagonal elements must be less than unity, at all grid nodal points. Analyze the condition for convergence given the system of equations below.

Case 1 : $\quad \Phi_1 = 0.5\Phi_2 + 1.5 \quad (1)$
$\qquad\quad \Phi_2 = \quad \Phi_1 + 2 \quad\ (2)$

By re-arranging the above equations, we have

Case 2 : $\quad \Phi_1 = \quad \Phi_2 - 2 \quad (2)$
$\qquad\quad \Phi_2 = 2\Phi_1 - 3 \quad (1)$

Chapter 6

Practical Guidelines for CFD Simulation and Analysis

6.1 INTRODUCTION

The need to understand the physics of the flow problem at hand, the basis of the numerical methods employed to solve the governing equations, and the means to obtain the most accurate and consistent results given available computing resources are some of the common challenges faced by the CFD user. This chapter particularly addresses these common aspects of CFD, with the aim of providing some practical guidelines for carrying out a CFD simulation, solution assessment, and analysis.

In Chapter 2, grid generation is viewed as a key consideration in the preprocess stage, following the definition of computational domain geometry. The generation of a mesh is an important numerical issue, where the type of mesh chosen for a given flow problem can determine *success* or *failure* in attaining a computational solution. Because of this, grid generation has become an entity by itself in CFD. In grid generation, the mesh must be sufficiently fine to provide an adequate resolution of the important flow features and geometrical structures. For flows with bounded walls, it is recommended that recirculation vortices or steep flow gradients within the viscous boundary layers be properly resolved through locally refining or clustering the mesh in the vicinity of wall boundaries. Mesh concentration may also be required for fluid flows having high-shear and/or high-temperature gradients. Furthermore, the quality of the mesh has significant implications for the convergence and stability of the numerical simulation and the accuracy of the computational result obtained. All of these grid-generation issues are examined in the next section.

Also in Chapter 2, special attention was given to specifying a range of boundary conditions commonly applied for a given CFD problem. The physical meanings of the various boundary conditions employed to close the fluid-flow system were illustrated in Chapter 3. Boundary conditions have serious implications for the CFD solution. This further reinforces the requirement to define suitable boundary conditions that appropriately mimic the real physical representation of the fluid flow. In many real applications, there is great difficulty in defining in detail some of the boundary conditions at the inlet and outlet of a flow domain that are required for an accurate solution. A typical example is the specification of turbulence properties (turbulent intensity and length scale) at the inlet flow boundary, as these are arbitrary in many CFD problems. Nevertheless, by carrying out an uncertainty analysis, the user can develop a good feel for the appropriateness and inappropriateness of the boundary conditions imposed within the physical context of the CFD problem being solved. Some useful guidelines for the specification of inlet, outlet, wall, and other types of boundary conditions for different classes of problems are examined and discussed in this chapter.

Since most flows of engineering significance are turbulent, the classical two-equation modeling approach of handling turbulence was briefly introduced in Chapter 3, where the basic formulation of the standard k-ε model was described. Nonetheless, we demonstrated in the test case of flow over a $90°$ bend in Chapter 5 that the use of a more sophisticated turbulence model, the Reynolds stress model, is necessary to better capture the anisotropic behavior of the flow separation around the bend region. A turbulence model is a computational procedure to close the system of mean flow governing equations. Nowadays, the two-equation and Reynolds stress models form the basis of turbulence calculations in numerous commercial, shareware, and in-house CFD codes. On the other hand, due to some inherent limitations of the standard k-ε model, development of other dedicated models for limited categories of flows has led to the formulation of many variants of the standard model. The CFD user is therefore confronted with the pressing choice of a suitable turbulence model. The provision of appropriate guidelines is therefore important. Although many industrial problems fall within the limited class of flow that can be resolved by the standard k-ε model, the only probable way to validate more specific flow problems is to conduct a case-by-case examination to determine the optimum turbulence model. Pertinent turbulence-modeling issues are explored later in this chapter.

As we continue to unravel the mysteries of CFD, this chapter by and large assembles all the essential knowledge gathered in the previous chapters, thereby bringing to fruition the realization of CFD simulation and analysis. We begin with practical guidelines for handling grid generation.

6.2   GUIDELINES ON GRID GENERATION

Grid generation presents an important consideration in computing numerical solutions to the governing partial differential equations of the CFD problem. A well-constructed mesh wields great influence on the numerical computation. It not only removes problems that can lead to an apparent instability or lack of convergence but also increases the likelihood of attaining the eventual solution of a CFD problem.

CFD requires the sub-division of the computational domain into a number of smaller mesh or grid cells overlying the whole domain geometry. It is therefore generally expected that the discretized domain needs to adequately resolve the important physics and capture all the geometrical details of the domain within the flow region. Designing a suitable grid is by no means trivial. The quest to yield a well-constructed mesh deserves as much attention as prescribing the necessary physics to the flow problem.

It is not the intention of this book to dwell on or elaborate on the various methods of grid generation. There are many books that focus on this subject alone, such as Thompson et al. (1985), Arcilla et al. (1991), and Liseikin (1999). We also note that many existing commercial codes have their own in-built powerful mesh generators and that there is also an available selection of independent grid-generation packages, such as GRIDGEN by Pointwise (ww.pointwise.com) and GRIDPRO by Program Development Company (www.gridpro.com). Although these independent grid-generation packages as well as the built-in mesh generators in the commercial codes have been designed to be very user-friendly, proficient management of these software packages still relies on the reader's aptitude.

Here, we concentrate on an overview of the types of grids that are common in many of today's CFD problems. More significantly, we draw the reader's attention to the important properties that a grid should have, such as grid quality, and the need for local grid refinement. These topics are discussed further in Sections 6.2.5 and 6.2.6.

6.2.1   Structured Mesh

Of all the grid-generation techniques to be discussed, the simplest is the application of a structured mesh. For the purpose of illustration, let us consider the fluid flow within a rectangular conduit. For convenience, a uniformly distributed Cartesian mesh can be generated where the spacing of the grid points along both the x direction and the y direction is uniform, as illustrated in Figure 6.1. An exploded view of a section of the discrete mesh in the x-y plane is also drawn to depict the arrangement of the discrete points within the domain. Note the regularly shaped four-nodal grid points of the rectangular element in Figure 6.1.

In a two-dimensional structured mesh, the grid points are normally addressed by the indices (i, j), where the index i represents points that run in the x direction while the index j represents points that run in the y direction. If (i, j) are the indices for the point P, the neighboring points immediately to the right, to the left, directly above, and directly below are defined by increasing or reducing one of the indices by unity. By allocating appropriate discrete values for Δx_i and Δy_j, the coordinates in the x direction and y direction inside the physical space can be incrementally determined, resulting in a rectangular mesh covering the whole domain.

For uniformly distributed grid points, the spacing of Δx_i or Δy_j is essentially a single representative value in the x direction or y direction. For non-uniformly distributed grid points, the spacing of Δx_i or Δy_j can effectively take a number of discrete values; hence, we could have easily dealt with totally unequal spacing in the x direction or y direction. For example, a finely spaced mesh in the

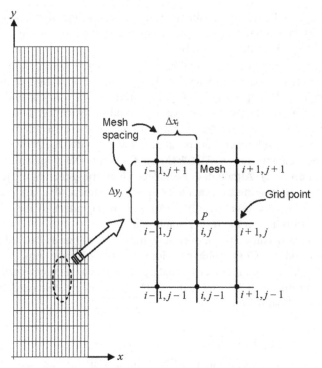

FIGURE 6.1   A uniform rectangular mesh.

x direction can be generated to adequately resolve the viscous boundary layer of the multi-phase flow in the vicinity of the wall geometry, while a uniformly spaced mesh is retained in the y direction, as shown in Figure 6.2. This particular arrangement is usually regarded as a "stretched" mesh, where the grid points are considered as being biased toward the wall boundaries.

In three dimensions, any grid point in space must be addressed by the indices (i, j, k), where the index k represents points that run in the z direction. The consideration of an additional dimension requires knowledge of the spacing of Δz_k, in addition to the spacing of Δx_i and the spacing of Δy_j, to construct the appropriate mesh covering the three-dimensional geometry.

6.2.2   Body-Fitted Mesh

Consider the problem of a flow inside a 90° bend, such as illustrated in Figure 6.3. In order to apply an orthogonal mesh to the geometry, compromises are required to be made, especially on the curved section, through characterizing the boundaries by staircase-like steps. Nevertheless, this approach raises two problems, First, such an approximate boundary description is tedious and time-consuming to set up. Second, the steps at the boundary introduce

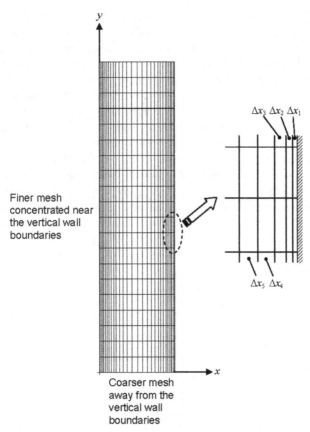

FIGURE 6.2   A non-uniform rectangular mesh.

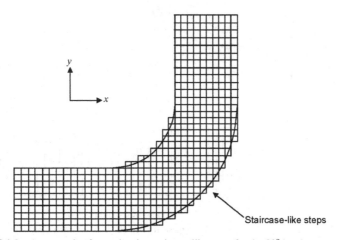

FIGURE 6.3   An example of a mesh using staircase-like steps for the 90° bend geometry.

errors in computations of the wall stresses, heat fluxes, boundary-layer effects, etc. The treatment of the boundary conditions at stepwise walls generally requires a fine Cartesian mesh to cover the wall regions, but the requirement for the highly regular structure of gridlines may cause further wastage of computer storage due to unnecessary refinement in interior regions that are of minimal interest. This example clearly shows that CFD methods that are based on Cartesian coordinate systems have limitations in irregular geometries. Therefore, it would be more advantageous to work with meshes that can handle curvature and geometric complexity more naturally.

In applying the body-fitted mesh to the 90° bend geometry, it would be appropriate to make the walls coincide with lines of constant η (see Figure 6.4). Location along the geometry, say from A to B or D to C, subsequently corresponds to specific values of ξ in the computational domain. Corresponding points on AB and CD connected by a particular η line will have the same value of ξ_i but different η values. At a particular point (i, j) along this η line, $\xi = \xi_i$ and $\eta = \eta_j$. A point $x = x(\xi_i, \eta_j)$ and $y = y(\xi_i, \eta_j)$ in the computational domain has a corresponding point in the physical domain.

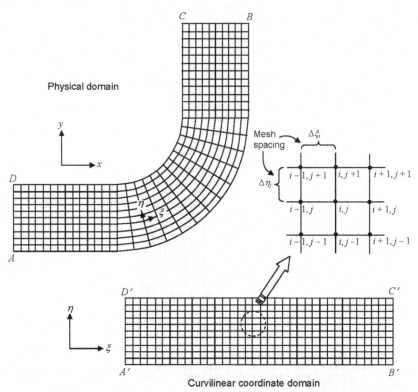

FIGURE 6.4 An example of a body-fitted mesh for the 90° bend geometry and corresponding computational geometry.

In Figure 6.4, the transformation must be defined so that there is a one-to-one correspondence between the rectangular mesh in the computational domain and the curvilinear mesh in the physical domain. The algebraic forms of the governing equations for the multi-phase problems are carried out in the computational domain, which has uniform spacing of $\Delta\xi$ and uniform spacing of $\Delta\eta$. Computed information is then fed directly back to the physical domain via the one-to-one correspondence of grid points. Because of the need to solve the equations in the computational domain, they have to be expressed in terms of curvilinear coordinates, rather than Cartesian coordinates, which means that they must be transformed from (x, y) to (ξ, η) as the new independent variables.

The mesh construction of the internal region of the physical domain can normally be achieved via two approaches. On the one hand, the Cartesian coordinates may be algebraically determined through interpolation from the boundary values. This methodology requires no iterative procedure and is computationally inexpensive. On the other hand, a system of partial differential equations of the respective Cartesian coordinates may be solved numerically with the set of boundary values as boundary conditions in order to yield a highly smooth mesh in the physical domain. The former is commonly known as the *transfinite interpolation method* and the latter is the *elliptic grid-generation method* (Smith, 1982; Thompson, 1982). More details are left to interested readers.

6.2.3   Unstructured Mesh

As an alternative to a body-fitted mesh, an unstructured mesh could be constructed to fill the interior region of the 90° bend geometry (Figure 6.5). It can be observed in Figure 6.5 that there is no regularity to the arrangement of the cells in the overlay mesh. There are no coordinate lines that correspond to the curvilinear directions

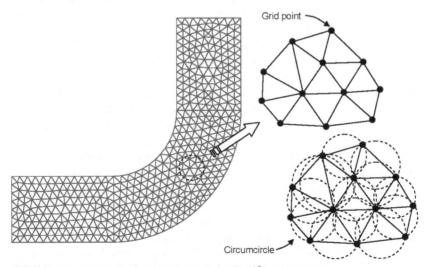

FIGURE 6.5   An example of a triangular mesh for the 90° bend geometry.

ζ and η (as in a body-fitted mesh), and the cells are totally unstructured. Maximal flexibility is therefore allowed in matching the cells, especially with highly curved boundaries, and in inserting the required cells to resolve the flow regions where they matter most, that is, in areas of high gradients. Triangle meshing and tetrahedral meshing are by far the most common forms of unstructured-mesh generation.

In Delaunay meshing, the initial set of boundary nodes of the geometry are triangulated according to the Delaunay triangulation criterion. The most important property of a Delaunay triangulation is that it has the *empty circumcircle* (*circumscribing circle*) *property* (Shewchuk, 2002). By definition, the circumcircle of a triangle is the unique circle that passes through the triangle's three vertices. The Delaunay triangulation of a set of vertices can therefore be regarded as the triangulation (usually, but not always, unique) in which every triangle has an empty circumcircle—meaning that the circle encloses no vertex of the triangulation. As depicted in Figure 6.5, it can be shown that the circumcircle of every Delaunay triangle of the mesh generated for the 90° bend geometry is empty. All algorithms for computing Delaunay triangulations rely on fast operations for detecting when a grid point is within a triangle's circumcircle and an efficient data structure for storing triangles and edges. The most straightforward way of computing the Delaunay triangulation is to repeatedly add one vertex at a time and then retriangulate the affected parts thereafter. When a vertex is added, a search is done for all triangles' circumcircles containing the vertex. Then those triangles are removed whose circumcircles contain the newly inserted point. All new triangulation is then formed by joining the new point to all boundary vertices of the cavity created by the previous removal of intersected triangles. Delaunay triangulation techniques based on point insertion extend naturally to three dimensions by considering the *circumsphere* (*circumscribing sphere*) associated with a tetrahedron. In hindsight, the Delaunay method is normally more efficient than the advancing-front method owing to the simplicity of the algorithm. The reader is strongly encouraged to refer to the excellent review paper by Mavriplis (1997) and the book by de Berg et al. (2000) for a more thorough understanding on the basic concepts of Delaunay triangulation and meshing.

However, relying on the Delaunay triangulation alone does not solve the many associated problems of unstructured meshing. In particular, the Delaunay method tends to maximize the minimum angle of the triangle but the angle may be too small and it might not conform to the domain boundaries. The obvious solution is to add more vertices, but the question remains, "Where should they be inserted?" This has brought about the concerted development of a number of Delaunay refinement algorithms. For the particular refinement algorithm proposed by Chew (1989), the order of insertion, which is based on the minimum angle of any triangle, continues until the minimum angle is greater than a predefined minimum ($\approx 30°$). Such an approach has been shown to generate high-quality meshes, albeit without any guarantees of grading or size optimality. The problem was rectified by Ruppert (1993), who introduced an algorithm to produce a mesh with good grading as well as size optimality.

Besides the Delaunay method, other meshing algorithms in unstructured-grid generation include the advancing-front method (Lo, 1985; Gumbert et al., 1989; Marcum and Weatherill, 1995) and the quadtree/octree method (Yerry and Shepard, 1984; Shepard and Georges, 1991).

The advancing-front method centers on the idea that the unstructured mesh is generated by adding individual elements one at a time to an existing front of generated elements. Once the boundary nodes are generated, these edges form the initial front that is to be advanced out into the field. Triangular cells are formed on each line segment, and, in turn, these cells create more line segments on the front. The front thus constitutes a stack, and edges are continuously added to or removed from the stack. The process terminates when the stack is empty, which is when all fronts have merged upon each other and the domain is entirely covered. For three-dimensional grid generation, a surface grid is first constructed by generating a two-dimensional triangular mesh on the surface boundaries of the domain. This mesh forms the initial front, which is then advanced into the physical space by placing new points ahead of the front and forming tetrahedral elements. The required intersection checking involves triangular front faces, rather than edges as in the two-dimensional case.

The quadtree/octree method performs grid generation through a recursive subdivision of the physical space down to a prescribed (spatially varying) resolution. The vertices of the resulting quadtree or octree structure are used as grid points, and the tree quadrants or octants are divided up into triangular or tetrahedral elements, in two or three dimensions, respectively. It should be noted, though, that the quadtree/octree cells intersecting the boundary surfaces and the vertices at boundaries must somehow be required to be displaced or wrapped in order to coincide with the boundary. The method is relatively simple and inexpensive and produces good-quality meshes in interior regions of the domain. One drawback of the method is that it has a tendency to generate an irregular cell distribution near boundaries.

6.2.4 Comments on Mesh Topology

Application of a *structured* mesh in any aspect of grid generation has certain advantages and disadvantages. The advantage of such a mesh is that the points of an elemental cell can be easily addressed by double indices (i,j) in two dimensions or triple indices (i,j,k) in three dimensions. The connectivity is straightforward because cells adjacent to a given elemental face are identified by the indices, and the cell edges form continuous mesh lines that begin and end on opposite elemental faces, as illustrated in Figure 6.6. In two dimensions, the central cell is connected by four neighboring cells. In three dimensions, the central cell is connected by six neighboring cells. Structured mesh also allows easy data management and connectivity occurs in a regular fashion, which makes programming easy. Nevertheless, the disadvantage of adopting such a mesh, particularly for more complex geometries, is the increase in grid

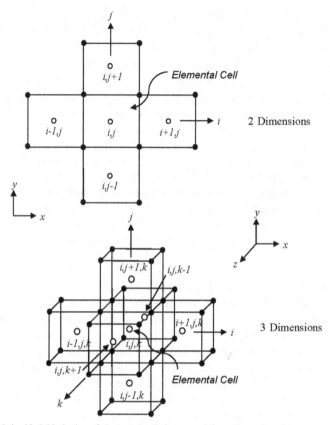

FIGURE 6.6 Nodal indexing of elemental cells in two and three dimensions for a structured mesh.

non-orthogonality or skewness that can cause unphysical solutions due to the transformation of the governing equations. The transformed equations that accommodate the non-orthogonality act as the link between the structured coordinate system (such as Cartesian coordinates) and the body-fitted coordinate system, but contain additional terms, thereby augmenting the cost of numerical calculations and difficulties in programming. Because of this, such a mesh may also affect the accuracy and efficiency of the numerical algorithm that is being applied.

The use of an *unstructured* mesh has become more prevalent in CFD applications. The majority of commercial codes nowadays are based on the unstructured-mesh mesh approach. Here, the cells are allowed to be assembled freely within the computational domain. The connectivity information for each face thus requires appropriate storage in some form of a table. The most typical shape of an unstructured element is a triangle in two dimensions or a tetrahedron in three dimensions. Nevertheless, any other elemental shape, including quadrilateral or hexahedral cells, is also possible.

In general, *structured* meshes do not necessarily have to be constrained to have matching cell faces. Figure 6.7 illustrates a structured Cartesian mesh without matching cell faces near the bottom boundary for the rectangular-type geometry. We can view this type of mesh as a special case of a local grid-refinement strategy. It is noted that the locally refined region in Figure 6.7 could also have been achieved through the use of other types of elements, such as triangular elements or a combination of both triangular and quadrilateral elements. Unstructured meshes are certainly well suited for handling arbitrary shape geometries, especially for domains having high curvature boundaries. Figure 6.8 illustrates the interior of a circular cylinder that has been filled through the *structured* and *unstructured* grids. For such a geometrical feature, the structured body-fitted non-orthogonal grid has a tendency to generate highly skewed cells at the four vertices, as indicated in Figure 6.8, since the interior of the domain must be built to satisfy the geometrical constraints imposed by the domain boundary. This type of mesh generally leads to numerical instabilities and deterioration of the computational results. It might have been preferable to re-mesh the geometry with an unstructured triangular mesh (see Figure 6.8).

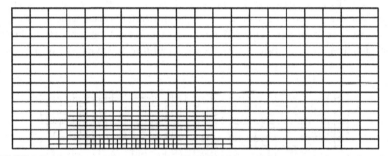

FIGURE 6.7   A structured Cartesian mesh without matching cell faces near the bottom boundary.

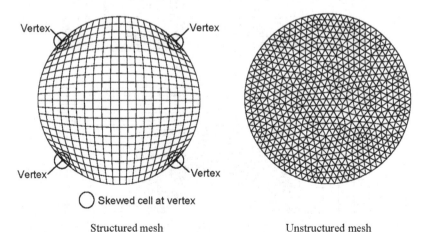

Structured mesh   Unstructured mesh

FIGURE 6.8   A structured and an unstructured mesh for a circular cylinder.

Despite its many advantages, the reader should also be aware of the disadvantages of employing an unstructured mesh for CFD simulations. In comparison to a structured mesh, the points of an elemental cell for an unstructured mesh generally cannot be simply treated or addressed by double indices (i,j) in two dimensions or triple indices (i,j,k) in three dimensions. An elemental cell may have an arbitrary number of neighboring cells attaching to it, making the data treatment and connection arduously complicated. Triangular (two-dimensional) or tetrahedral (three-dimensional) cells, in comparison to quadrilateral (two-dimensional) or hexahedral (three-dimensional) cells, are usually ineffective for resolving wall boundary layers. In most cases, the grid yields very long, thin, triangular or tetrahedral cells adjacent to the wall boundaries, thereby creating major problems in the approximation of the diffusive fluxes. Another disadvantage in connection with data treatment and connectivity of elemental cells is the requirement for more complex solution algorithms to solve the flow-field variables. This may result in increased computational times in obtaining a solution and may erode the gains in computational efficiency.

Block-structured or multi-block mesh is another special case of a structured mesh. For the sake of simplicity, the mesh is assembled from a number of structured blocks attached to each other. Here, the attachments of each face of adjacent blocks may be regular (i.e., having matching cell faces), or arbitrary (i.e., having non-matching cell interfaces), as shown in Figure 6.9. Generation of grids,

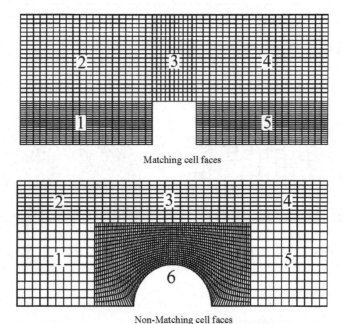

Matching cell faces

Non-Matching cell faces

FIGURE 6.9 Multi-block structured mesh with matching and non-matching cell faces.

especially with non-matching cell interfaces, is certainly much simpler than the creation of a single-block mesh fitted to the whole domain. This approach provides the flexibility to choose the best grid topology for each of the subdivided blocks. The user may select an appropriate grid topology based on a structured H-, O-, or C-grid, or an unstructured mesh of tetrahedral or hexahedral elements to fill each block. Figure 6.10 presents an example of an H-grid designed for the flow calculation in a symmetry segment of a staggered tube bank. It is also possible to remove the highly skewed cells for the circular cylinder in Figure 6.8 through the use of an O-grid, as exemplified in Figure 6.11.

Instead of a block-structured mesh, where the attachment of a number of adjacent blocks is realized at block boundaries, the use of overlapping grids to cover the irregular flow domains is another grid-generation approach for handling complex geometries. Here, rectangular, cylindrical, spherical, or non-orthogonal grids can be combined with the parent Cartesian grids in the solution domain. An example of an overlapping grid for a cylinder in a channel with inlet–outlet mappings is shown in Figure 6.12. This approach is attractive because the structured-mesh blocks can be placed freely in the domain to fit any geometrical boundary while satisfying the essential resolution requirements. Information between the different grids is achieved through the interpolation process. Block-structured grids with overlapping blocks are sometimes referred to as *chimera* grids. The advantages of employing such grids are that complex domains are treated with ease and they can especially be employed to follow moving bodies in stagnant surroundings. Some examples can be found in Tu and Fuchs (1992) and Hubbard and Chen (1994, 1995). The disadvantages of

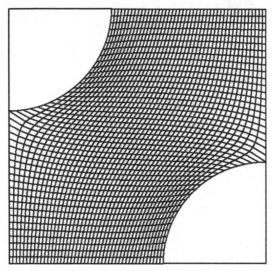

FIGURE 6.10 The generation of a structured H-grid for the flow calculation in a symmetry segment of a staggered tube bank.

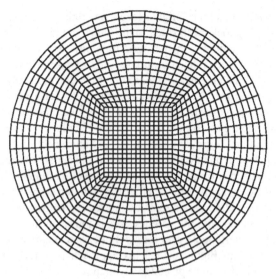

FIGURE 6.11 The generation of a structured O-grid for a circular cylinder.

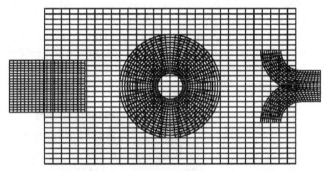

FIGURE 6.12 A structured overlapping grid for a cylinder in a channel with inlet–outlet mappings.

these grids are that conservation is usually not maintained or enforced at block boundaries, and the interpolation process may introduce errors or convergence problems if the solution exhibits strong variation near the interface.

The use of hybrid grids that combine different element types, such as triangular and quadrilateral elements in two dimensions or tetrahedra, hexahedra, prisms, and pyramids in three dimensions, can provide maximal flexibility in matching mesh cells with the boundary surfaces and in allocating cells of various element types in other parts of the complex flow regions. As a common practice, grid quality is usually enhanced through the placement of quadrilateral or hexahedral elements in resolving boundary layers near solid walls, while triangular or tetrahedral elements are generated for the rest of the flow domain.

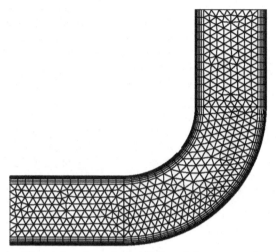

FIGURE 6.13   A grid consisting of structured quadrilateral elements near the walls and unstructured triangular elements in the remaining part of the 90° bend geometry.

This generally leads to both accurate solutions and better convergence for the numerical solution methods.

Figure 6.13 illustrates an example of a mesh consisting of quadrilateral elements near the walls and triangular elements for the rest of the flow domain. Note that a "stretched" mesh (as previously considered for the rectangular conduit in Figure 6.2) has been constructed in the vicinity of the wall boundaries for the 90° bend geometry. Currently, there is increasing interest in the development of a mesh containing *polyhedral* cells in resolving a range of practical flow problems. A polyhedral mesh can be created by combining tetrahedral cells into polyhedral cells. Considering the tetrahedral mesh that has been generated for the 90° bend geometry in Figure 6.5, a polyhedral mesh, such as that shown in Figure 6.14, can be created via cell agglomeration, which results in a considerable reduction of the overall cell count. More important, cell agglomeration has the capacity to improve the original mesh by converting particular regions with highly skewed tetrahedral cells to polyhedral cells, thereby improving mesh quality. The use of a polyhedral mesh also leads to quicker convergence of the numerical solution. A clear potential benefit of applying polyhedral mesh is that it allows the flexibility of an unstructured mesh to be applied to a complex geometry without the computational overheads associated with a large tetrahedral mesh. Although the application of polyhedral mesh is still very much in its infancy, it is gaining significant traction in the CFD community. Polyhedral mesh has been shown thus far to have considerable advantages over tetrahedral mesh with regard to the attained accuracy and efficiency of numerical computations. Practical guidelines for grid quality and grid design are discussed next.

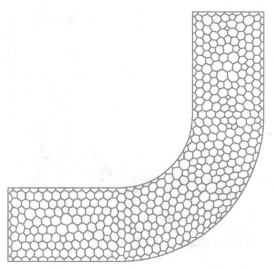

FIGURE 6.14   A grid consisting of polyhedral elements of the 90° bend geometry.

6.2.5   Guidelines for Grid Quality and Grid Design

As mentioned, grid generation is by no means a trivial exercise. The mesh for each class of problem is usually unique to the specific CFD geometry, thereby demanding a more measured consideration during the grid-generation process.

First, adopting the practice of applying a coarse mesh topology as the initial step for a particular CFD problem is one of the recommended strategies for grid design because it offers the opportunity to evaluate the specific computer code's storage and running time. More important, the utilization of a suitable coarse mesh allows a number of "test runs" to be carried out, with the aim of assessing the convergence or divergence behavior of the numerical calculations. When a solution is found to be converging, mesh refinement in the flow domain can then be undertaken to achieve the eventual CFD solution. For a diverging solution, the user needs to negotiate and investigate the problems arising during the numerical calculations. We note that "test runs" performed on such a mesh also provide the means of rectifying possible sources of solution errors, such as *physical modeling* and *human errors*, that may be present during the course of the numerical simulations.

Evidently, "test runs" are not recommended for a fine mesh, since it could take hours or days to examine the numerical solutions. Care should also be taken against hastily applying a fine mesh at the first instance because of the diverging tendency that may occur during the iterative procedure. Let us elucidate this point by examining the convective (first-order gradients) and diffusive (second-order gradients) terms along the Cartesian x and y directions in the governing transport equation, which can be approximated according to the following discretizations:

$$u\frac{\partial \phi}{\partial x} \approx u\frac{\Delta \phi}{\Delta x}, v\frac{\partial \phi}{\partial y} \approx v\frac{\Delta \phi}{\Delta y} \quad \text{and} \quad \frac{\partial^2 \phi}{\partial x^2} \approx \frac{\Delta^2 \phi}{\Delta x^2}, \frac{\partial^2 \phi}{\partial y^2} \approx \frac{\Delta^2 \phi}{\Delta y^2} \quad (6.1)$$

On a fine mesh, the mesh spacing of Δx and Δy may be very small, resulting in

$$\frac{\Delta^2 \phi}{\Delta x^2}, \frac{\Delta^2 \phi}{\Delta y^2} \gg u\frac{\Delta \phi}{\Delta x}, v\frac{\Delta \phi}{\Delta y} \quad (6.2)$$

When Eq. (6.2) is coupled with poor initial conditions for the transport variable ϕ during the first iteration, we observe that the values of the second-order gradients can become extremely large because of the step change of ϕ being small. This can cause the flow solver to misbehave and the iterative procedure to subsequently diverge. One practical way to overcome the poor initial guesses or unresolved steep gradients in the flow field is through the use of *under-relaxation factors*, as previously discussed in Section 5.4.3, to curtail the iterative advancement of the numerical calculations. The other strategy that is worthwhile noting and is available in many commercial codes is to initially solve the problem in a coarse mesh. The solution of this mesh is later interpolated onto a fine mesh in the subsequent calculations to aid convergence and promote numerical stability and efficiency.

Second, we address the grid quality of a generated mesh that depends on the consideration of the cell shape: aspect ratio, skewness, warp angle, or included angle of adjacent faces.

Figure 6.15 illustrates a quadrilateral cell having mesh spacing of Δx and Δy and an angle of θ between the gridlines of the cell. Accordingly, we can define the grid *aspect ratio* of the cell as $AR = \Delta y/\Delta x$. One pertinent guideline to bear in mind during the course of grid generation is that large aspect ratios should always be avoided in important regions inside the interior flow domain as they can degrade the solution accuracy and may result in possible poor iterative

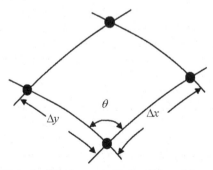

FIGURE 6.15 A quadrilateral cell having mesh spacing of Δx and Δy and an angle of θ between the gridlines of the cell.

convergence (or divergence), depending on the flow solver, during the numerical computations. Whenever possible, it is recommended that AR be maintained within the range of $0.2 < AR < 5$ in the interior region. For near wall boundaries, the condition for AR can, however, be relaxed. If the fluid flow is in the y direction, the need to appropriately choose small Δx mesh spacing in the x direction will generally yield $AR > 5$. In such a case, the approximated first- and second-order gradients are now only biased in the y direction, mimicking a one-dimensional flow behavior along this direction, since

$$\frac{\Delta^2 \phi}{\Delta x^2} \ll \frac{\Delta^2 \phi}{\Delta y^2} \quad \text{and} \quad \frac{\Delta \phi}{\Delta x} \ll \frac{\Delta \phi}{\Delta y} \tag{6.3}$$

This behavior is also exemplified where AR is < 0.2 if the fluid flow is in the x direction. Eq. (6.3) can assist in alleviating convergence difficulties and in enhancing the solution accuracy, especially in appropriately resolving the wall boundary layers where rapid solution change exists along the perpendicular direction of the fluid flow.

The next aspect concerning cell shape is grid *distortion* or *skewness*, which relates to the angle θ between the gridlines, as indicated in Figure 6.15. It is normally desirable that the gridlines be optimized in such a way that the angle θ is approximately 90° (orthogonal). If the angle θ is $< 45°$ or $> 135°$, the mesh contains highly skewed cells and often exhibits a deterioration of the computational results or leads to numerical instabilities. A typical example of highly skewed cells can be seen in Figure 6.8 for the structured non-orthogonal body-fitted grid filling the interior circular cylinder. For some complicated geometries, there is a high probability that the generated mesh may contain cells that are just bordering the skewness angle limits. The convergence behavior of such a mesh may be hampered due to the significant influence of additional terms in the discretized form of the transformed equations. Such a case may be remedied through the use of under-relaxation factors to increase the diagonal dominance in the matrix solver, thereby gradually improving the solution between iterations. It is also necessary to avoid non-orthogonal cells near the geometry walls. The angle between the gridlines and the boundary of the computational domain (especially the wall, inlet, or outlet boundaries) should be maintained as close as possible to 90°. The reader should pay special attention to this requirement, and it is stronger than the requirement given for the gridlines in the flow field far away from the domain boundaries.

If an unstructured mesh is adopted, special care needs to be taken to ensure that the *warp angles* measured between the surfaces normal to the triangular parts of the faces are not greater than 75°, as indicated by the angle β in Figure 6.16. Cells with large deviations from the co-planar faces can lead to serious convergence problems and deterioration in the computational results. In many grid-generation packages, the problem can be overcome by a grid-smoothing algorithm to improve the element warp angles. Whenever possible,

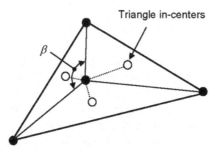

FIGURE 6.16 A triangular cell having an angle of β between the surfaces normal to the triangular parts of the faces connected to two adjacent triangles.

the use of tetrahedral elements should be avoided in wall boundary layers. Prismatic or hexahedral cells are preferred because of their regular shape and their ability to adjust in accordance with the near-wall turbulence model requirements, which are discussed in a later section.

Third, special grid-design features, such as the H-grid, O-grid, or C-grid, introduced in Section 6.2.4, and careful consideration in locating block interfaces in a sensible manner can assist en masse in improving the overall quality of a block-structured mesh. The presence of arbitrary mesh coupling, non-matching cell faces, or extended changes of element types at block interfaces should always be avoided in critical regions of high-flow gradients or high shear. Wherever possible, finer and more regular mesh should be employed in these critical regions, which may also include significant changes in geometry, or where suggested by error estimates. In all cases, it is recommended that the CFD user check the assumptions made when setting up the grid with regard to the critical regions of high-flow gradients and large changes agreeing with the result of the computation, and proceed to rearrange the grid nodal points if necessary.

Finally, it is desirable that a grid independence study be performed to analyze the suitability of the mesh and to yield an estimate of the numerical errors in the simulation for each class of problem. Ideally, at least three different and significant grid resolutions will be employed. Strictly speaking, grid independence could be examined by doubling the grid twice in each direction and then comparing the fine grid numerical solution with the results obtained by using the Richardson extrapolation of the original coarse grid solution (Roache, 1997). If this is not feasible, selective local refinement of the grid in critical flow regions of the domain, a subject discussed in the next section, can be applied. Otherwise, the user may attempt to compare different orders of spatial discretizations on the same mesh, which is further expounded upon in Section 6.5.

6.2.6 Local Refinement and Solution Adaptation

Adequate mesh resolution within critical flow regions has an enormous effect on the stability and convergence of the numerical procedure. It also significantly affects the accuracy of the eventual computational solution. This section places particular emphasis on the use of *local refinement* and *solution adaptation* to capture important flow features.

One *local refinement* technique that is widely used in many CFD applications is the concept of a stretched grid near domain walls. For a viscous flow bounded with solid boundaries, the motivation to cluster a large number of small cells within the physical boundary layer is more than just attempting to minimize the truncation error with the closely spaced grid points. Rather, it is a matter of utmost importance that the actual flow physics be appropriately encapsulated. Let us revisit the case of fluid flowing between two stationary parallel plates, as investigated in previous chapters. In a real physical flow, there will be a developing boundary layer that will grow in thickness as the fluid enters the left boundary and migrates downstream along the bottom wall of the domain, as illustrated in Figure 6.17. By denoting the local thickness of the boundary layer as δ, where $\delta = \delta(x)$, it is clear that a coarse uniform mesh, in essence, misses the physical boundary layer. The predicted *viscous-like* velocity profile shown at some point downstream in the fully developed region is simply due to the application of the no-slip boundary condition at the bottom wall. In contrast, a coarse stretched grid at the very least catches some of the essential features of the actual physical boundary layer. It is therefore not surprising that the accuracy of the computational solution is greatly influenced by the grid distribution inside the boundary-layer region. We can further investigate this by invoking the concept of *consistency* (as discussed in Chapter 5) for incompressible mass conservation in two dimensions, where the truncation error yields the following:

$$\frac{\Delta x^2}{6}\frac{\partial^3 u}{\partial x^3} + \frac{\Delta y^3}{6}\frac{\partial^3 v}{\partial y^3} \qquad (6.4)$$

If the solution error is expected to follow the truncation error, it is imperative that the grid be appropriately refined to determine the steep gradient of the velocity profile that exists within this region. This will help to minimize the solution error associated with the truncation error. Local refinement is also important to resolving specific fluid-dynamics problems like upward stagnation flow and backward-facing step geometry. The latter is one of the basic geometries commonly used in many engineering applications to better understand the phenomena of flow separation, flow re-attachment, and free shear jet. The placement of stretched grids along both the horizontal and vertical directions to capture the essential feature of the recirculation vortex is illustrated in Figure 6.18.

In applying the stretched grid described above, care must be exercised to avoid sudden changes in the grid size. The mesh spacing should be continuous and grid-size discontinuities should be removed as much as possible in regions of large flow changes, particularly when dealing with multi-block meshing of arbitrary mesh coupling, non-matching cell faces, or extended changes

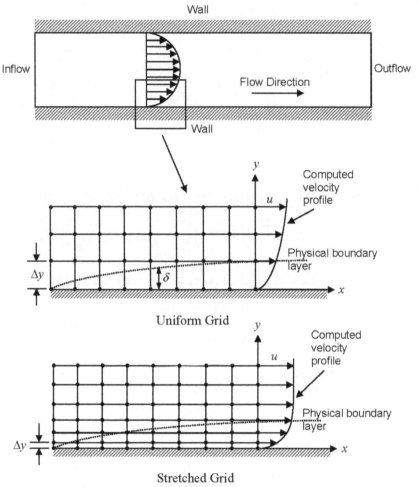

FIGURE 6.17   Two schematic illustrations demonstrating the need for local refinement in the near vicinity of the bottom wall to resolve the physical boundary layer.

of element types. Discontinuity in the grid size destabilizes the numerical procedure due to the accumulation of truncation errors in the critical flow regions. These errors usually contain the diffusive terms (second-order derivatives), as expressed by Eq. (6.1), where the discretization imposed on these derivatives requires very smooth grid changes. Making sure that the grid changes slowly and smoothly away from the domain boundary as well as within the domain interior will assist in overcoming divergence tendencies of the numerical calculations. It is also worthwhile to note that most built-in mesh generators in commercial codes and independent grid-generation packages have the means of prescribing suitable mesh stretching or expansion ratios (rates of change of cell size for adjacent cells). The specification

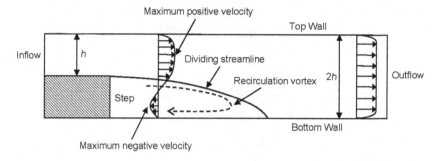

Schematic drawing

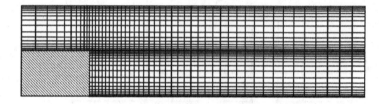

Computational grid

FIGURE 6.18 A schematic drawing for the backward-facing step geometry and the computational grid to capture the essential feature of the recirculation vortex.

of these ratios should always be negotiated within the codes' requirements while generating the appropriate stretched grid.

Just by the construction of a stretched grid, the local refinement technique provides possibilities for allocating additional grid nodal points to resolve the important fluid-flow action and for reducing or removing the grid nodal points from other regions where there is little or no action. Nevertheless, it should be noted that a stretched grid is an algebraically generated grid prescribed *prior* to the solution of the flow field being calculated. The question that needs to be carefully addressed is whether the new generated stretched grid sufficiently captures the major fluid-flow action or, the opposite, whether the real flow action is far away from the area which is targeted to be resolved by the generated stretched-grid region. This cannot be known a priori. This leads to the next subject of this section, solution adaptation.

Solution adaptation, usually through the use of an adaptive grid, is a grid network that automatically or dynamically clusters the grid nodal points in regions where large gradients exist in the flow field. It therefore employs the solution of the flow properties to locate the grid nodal points in the physical flow domain. During the course of the solution, the grid nodal points in the physical flow domain *migrate* in such a manner as to *adapt* to the evolution of the large flow gradients. Therefore, the actual grid nodal points are constantly in motion during the solution of the flow field and become stationary when the flow solution approaches some quasi-steady-state condition. An adaptive grid is therefore intimately linked to the flow-field solution and alters as the flow field develops, unlike the stretched grid described above, where the grid generation is completely separate from the flow-field solution. Unstructured meshes are well suited for this purpose, especially in automating the generation of elements such as triangular or tetrahedral meshes of various sizes to solve the critical flow regions. A sample illustration of solution adaptation through triangular meshes for the fluid flowing over two cylinders (investigated in Chapter 2) is given in Figure 6.19. For this particular flow problem, the wake region has been further resolved to capture the essential formation and shedding of vortices behind the two cylinders.

Moreover, if a solution method can be applied on an unstructured mesh with cells of varying topology, the adaptive grid is subjected to fewer constraints. Further subdividing the cells into smaller ones makes possible local grid refinement, whereby a non-refined neighboring cell, although it retains its original shape, becomes a *polyhedron* since its cell face is now replaced by a set of sub-faces. An adaptive grid with polyhedral elements certainly offers many challenges and potential prospects in CFD. The advantages for an adaptive grid are twofold: *increased accuracy for a fixed number of grid points* and *fewer grid points required for a given accuracy*. Most commercial codes nowadays offer some form of automated grid adaptation. However, the reader should bear in mind that the grid quality (skewness and/or aspect ratio) might not improve as a consequence of activating the feature in these codes. Adaptive grids are still progressively being enhanced in CFD. The concept of grid adaptation described in this section is by no means exhaustive. We strongly encourage the reader to consult the literature for more in-depth understanding of this extensive subject (see section 8.2.2.2 and reference within).

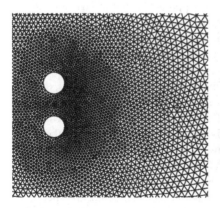 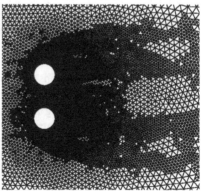

Before   After

FIGURE 6.19   A demonstration of solution adaptation through the use of triangular meshes for the fluid flowing over two cylinders.

6.3   GUIDELINES FOR BOUNDARY CONDITIONS

6.3.1   Overview of Setting Boundary Conditions

Setting proper physical boundary conditions also governs the computational stability as well as the numerical convergence of the CFD problem. In many real applications, there is great difficulty in prescribing some of the boundary conditions at the inlet and outlet of the computational domain. Boundary conditions for fluid flow are generally more complex due to the coupling of velocity fields with the pressure distribution. In the context of CFD, defining suitable boundary conditions generally encompasses the specification of two types of boundary conditions: the *Dirichlet* and *Neumann* boundary conditions. The physical meanings of these boundary conditions are described in section 3.7. For brevity, the Dirichlet boundary condition can be simply defined for the transport property ϕ as the requirement for specifying the physical quantity over the boundary, such as

$$\phi = f(\text{analytic}) \tag{6.5}$$

The Neumann boundary condition involves, however, the prescription of its derivative at the boundary, given by

$$\frac{\partial \phi}{\partial n} = 0 \tag{6.6}$$

A practice that is widely adopted for inflow boundaries is to set the transported quantities of either a uniform or some predetermined profile over the boundary surface (Dirichlet). For outflow boundaries, the convective derivative normal to the boundary face is set equal to zero; the transported quantities at the boundaries are extrapolated along the streamwise direction of the fluid flow (Neumann). However, the use of such an approach is not straightforward in some selected applications and some difficulties may arise during the implementation of such boundary conditions. For example, *non-physical* reflection of outgoing information back into the calculation domain (Giles, 1990), such as the fluid that may inadvertently re-enter the domain through the outflow boundaries as well as in regions of possible high swirl, large curvatures, or pressure gradients, may significantly affect the convergence behavior of the iterative procedure. In addressing some of these difficulties, the specification of radial equilibrium of a pressure field is deemed to be preferable to the usual constant static pressure for swirling flows at an outlet. Also, when strong pressure gradients are present, special non-reflecting boundary conditions are sometimes required for the inflow and outflow boundaries (Giles, 1990).

Some other difficulties that are also important to note include the specification of suitable inlet turbulence intensity, length scales, turbulence kinetic energy and dissipation for a turbulent flow entering the flow domain, the correct description of the boundary layer velocity profile on the walls and at the inlet, and the precise distribution of scalar or species concentrations at the inflow boundary. It is therefore not surprising that a CFD user must be fully aware of these problems and needs to develop a good feel for the certainty or uncertainty of the boundary conditions that are being imposed.

To ensure consistency of the boundary conditions imposed with the CFD models, it is vital that the boundary conditions strongly reflect the application that is being calculated. Some general guidelines are provided. Whenever possible, it is useful to examine the potential for modifying the computational domain by moving the domain boundaries to a position where the boundary conditions are more readily identifiable and where they can be more precisely specified. It is also important to be mindful of upstream or downstream obstacles, such as bends, contractions, diffusers, blade rows, etc., outside of the flow domain, which may significantly affect the flow distribution. Often, information on these components upstream and downstream of the domain is lacking; it may be necessary to incorporate these pertinent components in the calculations so that appropriate flow predictions are obtained. In handling a wide range of different CFD applications, it may be beneficial to carry out a sensitivity analysis in which the boundary conditions are systematically altered within certain limits to detect any significant variation of the computational results. Should any of these variations prove to greatly influence the simulated results and lead to large changes in the simulation, it is then necessary to obtain more accurate data on the boundary conditions that are being specified. When commercial codes are used, default settings of the boundary conditions for the domain boundaries are given and the user is usually not required to specify any boundary conditions. Nonetheless, the user still needs to ensure that appropriate boundary conditions are specified for the application that is being solved. Next we concentrate on specific guidelines for the inlet, outlet, wall, and symmetry and periodicity boundary conditions.

6.3.2   Guidelines for Inlet Boundary Conditions

Since flows inside a CFD domain are driven by boundary conditions, carrying out a sensitivity analysis by systematically changing the inlet boundary conditions within allowable limits is highly recommended. The aim of this practice is to ensure that appropriate boundary conditions are applied at the inflow boundaries. Depending on the flow problem, certain key parameters can be scrutinized to ascertain the suitability of the inlet boundary conditions that are specified. Some useful parameters that can be examined are

- Inlet flow direction and the magnitude of the flow velocity
- A uniform distribution of a parameter or a profile specification, for example, either a uniform inlet velocity or some predetermined inlet velocity profile
- Variation of physical properties
- Variation of turbulence properties at inlet

The reader should also note that an inlet boundary represents a potential *mass source* for the fluid flow. It is therefore imperative that an outlet boundary be imposed within the flow domain to characterize the *mass sink* of the fluid flow. Without an outlet boundary, the mass is not conserved. Not surprisingly, CFD calculations in this circumstance will tend to "blow up" swiftly.

By revisiting the case for the fluid flowing between two stationary parallel plates, the various combinations of the inlet and outlet boundary conditions can be appropriately demonstrated. The configurations that are typically adopted are illustrated in Figure 6.20. The first configuration, which involves the specification of a prescribed velocity profile at the inlet and a constant pressure at the outlet, is commonly employed in many CFD calculations. For rapid convergence and computational robustness, most commercial codes generally recommend the implementation of this type of configuration. Nonetheless, other highlighted configurations may be equally applicable if proper initial conditions are accommodated prior to the CFD calculations. At steady state, the CFD solution for all the configurations should yield identical velocity profiles, especially at the fully developed region situated near the channel exit.

For certain types of geometries, the specification of inlet boundary conditions for the upstream flow may require special attention. At the pre-designated inlet location, the exact distribution may be unknown. The possibility of moving the inlet boundary to a position where the fluid flow is allowed to develop

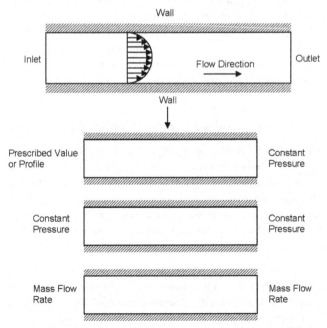

FIGURE 6.20 Configurations for a simple flow between two stationary parallel plates.

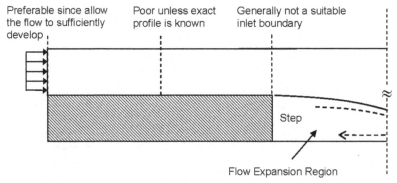

FIGURE 6.21   Inlet locations for the backward-facing step flow problem.

through some distance inside the domain should therefore always be examined. Let us consider a typical flow problem of the backward-facing step geometry described in Section 6.2.6. For high accuracy, it is necessary to demonstrate that the interior solution is unaffected by the choice of location of the inlet by carrying out a sensitivity analysis of the effect of different upstream distances from the flow expansion region, as indicated in Figure 6.21. If the inlet is too close, the velocity profile will be changing in the flow direction before the flow expansion region, which significantly affects the recirculation vortex and subsequently the re-attachment length. Of course, such a solution cannot be fully trusted and a re-examination of the inlet location is required.

6.3.3   Guidelines for Outlet Boundary Conditions

It is vital that the boundary condition imposed at the outlet always be selected on the basis that it has a weak influence on the upstream flow. The most suitable outflow conditions, as demonstrated in Figure 6.16, are weak formulations involving the specification of a constant pressure, such as static pressure, at the outlet plane. Whenever possible, the outlet boundary must be placed as far away as possible from the region of interest and should be avoided in regions of strong geometrical changes or in wake regions with recirculation. Let us illustrate the potential hazards of the latter by revisiting the backward-facing step geometry flow problem.

Figure 6.18 shows typical velocity profiles downstream from the step and the appropriate choice for the location of the outlet boundary. As mentioned, a common practice that is adopted for outflow boundaries in many CFD calculations is to apply the Neumann boundary condition, where the convective derivative normal to the boundary face is enforced to zero and the boundary value is later evaluated based on the streamwise extrapolation of the transported quantities. If the outlet is located at *Position 1*, a plane in this position cuts across the wake region of recirculation (Figure 6.22). Not only is the assumed

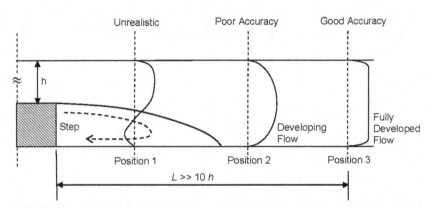

FIGURE 6.22   Outlet locations for the backward-facing step flow problem.

boundary condition not valid, but the problem is also exacerbated by the apparent area of reverse flow where the fluid enters the domain. Further downstream, at *Position 2*, there may not be any reverse flow but the zero-gradient condition still does not hold, since the velocity profile is changing in the flow direction. As a practical guideline, the outlet boundary should be placed much further downstream, by at least 10 step heights ($>10\,h$) to give accurate results—*Position 3*. It is also necessary to ensure and to demonstrate that the interior solution is not affected by the location of the outlet boundary. It is imperative that the downstream distance between the step and outlet boundary be extended until no significant change to the interior solution is observed.

For some CFD problems where the outlet boundaries may not be feasibly relocated in the domain, care should be exercised that the possibility of fluid flow entering through the outflow boundaries does not lead to progressive instability of the numerical procedure or to the extent of attaining an incorrect solution. In such an event, some reconfiguration of the model geometry may be required and the flow area at these outlets may need to be restricted, provided that the outlet boundaries are situated at a location substantially far away from the region of interest. Nonetheless, if the outflow boundary condition permits flow to re-enter the domain, then appropriate Dirichlet boundary conditions should be imposed for all transported quantities.

6.3.4   Guidelines for Wall Boundary Conditions

Wall boundary conditions are generally employed for solid walls bounding the flow domain. Here, care should always be taken to ensure that the boundary conditions imposed on these walls are consistent with both the numerical model used and the actual physical features of the flow geometry. For fluid flowing between two stationary parallel plates (Figure 6.17), wall boundary conditions are enforced for the channel walls. Nonetheless, these boundary conditions are also often used to bound fluid and solid regions, as was applied for the presence of the step within the flow domain in Figure 6.18 or the two cylinders inside the open surrounding flow environment in Figure 6.19.

For stationary walls, the default consideration is to assume that the no-slip condition applies, which simply means that the velocities are taken to be zero at the solid boundaries. This condition implies that the fluid flow comes to rest at the solid walls. We may also explore the possibility of modeling the boundary conditions on the solid walls as a free-slip condition, which assumes that the flow is parallel to the wall at this point. This condition corresponds to the absence of viscous effects in the continuum equations and is applied to the problem where the continuum approach breaks down as the fluid approaches the wall in viscous flow. For a general fluid-flow case, where the transport of heat takes precedence within the domain, great care must be taken to specify boundary conditions (for example, adiabatic walls or local heat fluxes) on the solid boundaries of the numerical model that properly represent the heat-transfer characteristics of the solid walls in the actual physical model.

In flow cases with moving or rotating walls, it is important that the boundary conditions that need to be specified are consistent with the motion of the solid walls. A lid-driven cavity and a rotating cylinder in a fluid environment, as shown in Figure 6.23, are some typical examples of the moving-boundary problems in CFD. These problems allow a positive or negative tangential velocity to be imposed at the top boundary of the lid-driven cavity or a clockwise or anti-clockwise rotational speed to be specified on the circumferential surface of the rotating cylinder. Other, more complex CFD problems may require the use of sliding or moving meshes to better emulate the motion of a rotating impeller stirring the fluid in a tank or ocean waves hitting a ship's hull. The use of sliding or moving meshes is beyond the scope of this book. Interested readers are advised to consult the literature for more information (see for example, Ferziger and Perić, 1999).

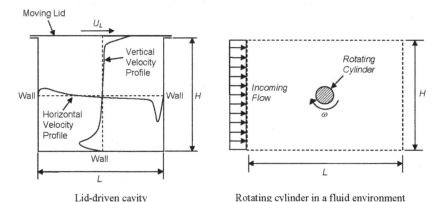

Lid-driven cavity   Rotating cylinder in a fluid environment

FIGURE 6.23   Illustrative fluid-flow examples with moving or rotating walls.

6.3.5   Guidelines for Symmetry and Periodic Boundary Conditions

These boundary conditions are generally used when the flow geometry possesses some symmetry or periodic properties that allow the flow problem to be simplified by solving only a fraction of the domain.

Symmetry boundary conditions are applied when the physical geometry and the flow field have mirror symmetry. At the symmetry plane, the following requirements must be satisfied:

- The normal velocity is zero.
- The normal gradients for all transported properties are zero.

Consider the flow problem where air is flowing through a square duct, as described in Figure 6.24. The application of symmetry planes can significantly reduce the computational effort, since only a quarter of the flow domain needs to be modeled. Care should always be exercised, especially in prescribing the mass flow rates at the inlet and outlet boundaries. For this simplified flow geometry, the correct mass flow rates to be imposed at these boundaries are only a quarter of the actual mass flow rates.

To apply the periodic boundary conditions, it is imperative that the periodic planes come in pairs. The key requirement in employing these boundary conditions is not only that they be physically identical but also that the mesh distribution in the respective planes be identical. In CFD applications, this condition implies that the flow leaving one of the periodic planes is equal to the flow entering the other. For wall-bounded flows, such as the fluid flowing between two stationary parallel plates (as in Figure 6.17) or air flowing through a square duct (as in Figure 6.24), the inlet and outlet boundaries can otherwise be prescribed as periodic planes where the flow field leaving the outlet boundary is taken to be the same as that entering the inlet boundary. Figure 6.25 further illustrates the application of periodic boundary conditions for the fluid flow in a mixing tank and across an in-line arrangement of heat-exchanger tubes. For the former, the flow field is known as rotationally or cyclically periodic, where the flow field leaving "Boundary 2" is enforced as an inlet flow condition at

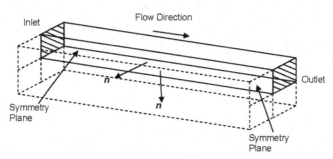

FIGURE 6.24   Application of symmetry boundary conditions for air flow through a square duct.

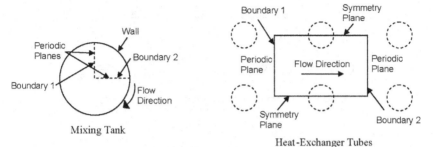

FIGURE 6.25   Examples of the application of periodic boundary conditions.

"Boundary 1." For the latter, the flow field is considered translationally periodic. As before, the flow field leaving "Boundary 2" is used as an inlet flow field at "Boundary 1." It is worthwhile noting that this flow case is particularly similar to the aforementioned wall-bounded flows.

6.4   GUIDELINES FOR TURBULENCE MODELING

6.4.1   Overview of Turbulence-Modeling Approaches

Some pertinent properties of turbulent flows were discussed in Chapter 3. The reader may wish to revisit section 3.5.1 and review the various aspects detailed therein. In brief, turbulent flows can be classified as being highly unsteady and random. They are also known to contain large coherent structures responsible for the mixing or stirring processes. Nonetheless, the fluctuating property across a broad range of length and time scales, particularly from the modeling standpoint, makes the direct numerical simulation of turbulent flows very difficult. Turbulent flows are constantly encountered in engineering systems, and they tend to demand more computational resources than laminar flows. For research and design purposes, CFD analysts and engineers are expected to understand and to predict the effects of turbulence. In this section, we review state-of-the-art turbulence-modeling approaches that have been developed and applied to date.

   Advanced techniques: The most accurate approach to turbulence simulation is to solve directly the governing transport equations without undertaking any averaging or approximation other than the numerical discretizations performed on them. Through such simulations, all of the fluid motions contained in the flow are considered to be resolved. This approach is commonly known as direct numerical simulation or by its well-established acronym, DNS. Since DNS requires all significant turbulent structures to be adequately captured (i.e., the domain for which the computation is carried out needs to accommodate the smallest and the largest turbulent eddy), it can be very expensive to use. Alternatively, we can consider an approach where the structure of turbulent flow can be viewed as the distinct transport of large- and small-scale motions.

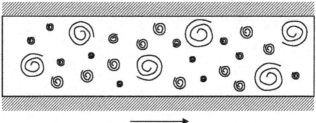

Flow Direction

FIGURE 6.26   Schematic representation of a turbulent motion.

Figure 6.26 is a schematic illustration of such flow. Since the large-scale motions are generally much more energetic and by far the more effective transporters of the conserved properties, a simulation that treats the *large eddies* exactly but approximates the *small eddies* makes sense. Such an approach is recognized as large eddy simulation (LES). It is still expensive, but it is much less costly than DNS. In general, though, DNS is the preferred method because it is more accurate. LES is, however, the preferred method for flows in which the Reynolds number is too high or the geometry is too complex to allow application of DNS.

The results of a DNS or LES simulation contain very detailed information about the flow, producing an accurate realization of the flow while encapsulating the broad range of length and time scales. For design purposes, this is far more information than any engineer needs. DNS and LES approaches usually require high usage of computational resources and often cannot be used as a viable *design tool* because of the enormity of the numerical calculations and the large number of grid nodal points. Therefore, questions arise about their usefulness and the role they can play in CFD. Because of the wealth of information they yield, DNS and LES taken as *research tools* can provide a qualitative understanding of the flow physics and also can construct a quantitative model allowing other, similar, flows to be computed. More important, in some cases they can assist in improving the performance of currently applied turbulence models in practice. (We demonstrate the use of DNS as a research tool in some sample aerodynamic investigations in section 7.4.5.)

Practical techniques: For most engineering purposes, it is unnecessary to resolve the details of turbulent fluctuations. In general, the effects of turbulence on the mean flow are usually sufficient to quantify the turbulent flow characteristics. For a turbulence model to be practical and useful in a general-purpose CFD code, it must have wide applicability and it must be simple, accurate, robust, and economical.

In Chapter 3, the Reynolds-averaged approach to turbulence resulted in the formulation of the two-equation turbulence model—the *standard k-ε model* proposed by Launder and Spalding (1974). This model is well established and widely validated and gives sensible solutions to most industrially relevant flows. Nonetheless, numerous limitations are also identified, especially for flows with large, rapid, extra strains (for example, highly curved boundary layers and diverging passages), since the model generally fails to fully describe the subtle effects of the streamline curvature on turbulence. This weakness was illustrated in Chapter 5 in section 5.7.2. The test case results revealed that the *Reynolds stress model* works better because of the ability of the model to accommodate the *anisotropic* turbulent stresses occurring around the 90° bend. The standard k-ε model, which is a consequence of the eddy viscosity concept, assumes that the turbulent stresses are linearly related to the rate of strain by a scalar turbulent viscosity, and that the principal strain directions are aligned to the principal stress directions—*isotropic* stresses. Owing to the deficiencies in the treatment of the normal stresses, secondary flows that exist in the 90° bend, which are driven by *anisotropic* normal Reynolds stresses, could not be properly predicted.

The Reynolds stress model, also called the second-moment closure model, dispenses with the notion of turbulent viscosity and determines the turbulent stresses directly by solving a transport equation for each stress component. This requires the solution of six additional equations ($\overline{u_1'^2}$, $\overline{u_2'^2}$, $\overline{u_3'^2}$, $\overline{u_1'u_2'}$, $\overline{u_1'u_3'}$, and $\overline{u_2'u_3'}$), which are solved accordingly in the form of Eq. (3.53) in order to account for the directional effects of the Reynolds stress field. An additional equation for ε is also solved to provide a length-scale–determining quantity. For a comprehensive explanation of this particular type of turbulence model, interested readers can consult relevant reference texts by Launder (1989) and Rodi (1993). In a similar manner, the turbulent heat fluxes can be determined directly by solving three additional equations, one for each flux component, thereby removing the notion of a turbulent Prandtl number. There is no doubt that the Reynolds stress model has greater potential to represent the turbulent flow phenomena than the standard k-ε model. This model can handle complex strain and, in principle, can cope with non-equilibrium flows. The shortcoming of this model is the very large computing costs that may be incurred because of the extra governing equations. Also, the model's success thus far has been moderate, specifically for axisymmetric and unconfined recirculating flows, where it has been shown to perform as poorly as the standard k-ε model.

A lot of research is still being performed in this field, and new models are constantly being proposed. The turbulent states that can be encountered across the whole range of industrially relevant flows are rich, complex, and varied. It is now accepted that no single turbulence model can span these states, since none is expected to be universally valid for all flows. Accordingly, Bradshaw (1994) refers to turbulence as *the invention of the Devil on the seventh day of creation, when the Good Lord wasn't looking*. Because of the difficult nature of turbulence, in the next section we propose some useful strategies for selecting appropriate turbulence models in handling turbulent-fluid–engineering problems.

6.4.2   Strategy for Selecting Turbulence Models

In CFD, different types of turbulent flows require different applications of turbulence models. In the event that insufficient knowledge precludes the selection of an appropriate model, we strongly encourage the use of the two-equation model, such as the standard k-ε model, as a starting point for turbulence analysis. This model offers the simplest level of closure since it has no dependence on the geometry or flow-regime input. As a first step to turbulence model selection, the standard k-ε model is robust and stable, and it is as good as any more sophisticated turbulence models in some applications. The majority of in-house and commercial codes generally set this model as the default option for handling turbulent flows. This is not entirely surprising because the model has been a de-facto standard in industrial applications and still remains the workhorse of industrial computations.

Nevertheless, the standard k-ε model is not without its weaknesses. It is therefore imperative that the major weaknesses associated with this model be catalogued, so as to instigate palliative actions that might be fruitfully considered for improving the numerical predictions. These advisory actions, which are given below, should be viewed not as definitive cures but as recommendations whereby possible alternatives to the standard k-ε model can be systematically investigated. There is also no guarantee that the specific advice will yield significantly improved results. The necessity for carrying out careful *validation* and *verification* remains a defining step in justifying the application of turbulence models for the particular CFD problem being solved.

Guideline for a particular weakness of the standard *k-ε* model: Historically, the five adjustable constants C_μ, σ_k, σ_ε, $C_{1\varepsilon}$, and $C_{2\varepsilon}$ in the standard k-ε model have been calibrated against comprehensive data for a wide range of turbulent flows that are nonetheless of simple geometrical flow origins. This has been evidenced by the model's remarkable successes in handling thin shear layer, recirculating, and confined flows. Deviation from such flow behaviors has, however, resulted in poor performance of the standard k-ε model, especially in important flow cases having

- Flow separation (Baldwin and Lomax, 1978)
- Flow re-attachment (Kato and Launder, 1993)
- Flow recovery (Ince and Launder, 1995)
- Some unconfined flows (e.g., free shear jet) (Apsley et al., 1997)
- Secondary flows in complex geometrical configurations (e.g., flow around a poppet valve) (Ferziger and Perić, 1999)

Let us demonstrate a specific palliative technique for the weakness of the standard k-ε model under consideration in a worked example of the backward-facing step flow problem described earlier. It is well known that the re-attachment length l_r (Figure 6.27) is generally poorly predicted by this model under turbulent flow because of the over-prediction of the turbulent kinetic

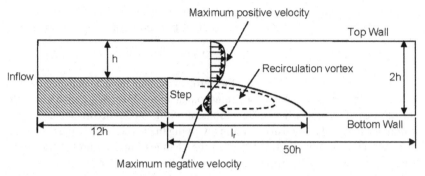

FIGURE 6.27 A schematic illustration of the re-attachment length location and respective dimensions for the regions before and after the step for the backward-facing step geometry.

energy. The high turbulence levels predicted upstream following the flow expansion at the step are transported downstream, and the real boundary layer development is subsequently swamped by this effect. For illustration purposes, we investigate the use of other, more sophisticated, turbulence models, such as the *RNG k-ε model* and *realizable k-ε model* proposed by Yakhot et al. (1992) and Shih et al. (1995), respectively, to exemplify the improvements that can be achieved in the numerical predictions.

For this flow problem, the CFD commercial code ANSYS-FLUENT, Version 6.1, is utilized to predict the continuum gas phase under steady-state conditions through solutions of the conservation of mass and momentum in two dimensions. A computational domain with a size of 12 *h* (length) × 1 *h* (height) before the step and 50 *h* (length) × 2 *h* (width) after the step is considered. The Reynolds number based on the free-stream velocity u_∞ of 40 m/s and step height *h* for this investigation is evaluated as 64,000.

Some pertinent differences are worthwhile mentioning with regard to each of the turbulence models before proceeding to discuss the numerical results. The RNG k-ε model includes a modification to the transport ε-equation where the source term is solved as

$$S_\varepsilon = \frac{\varepsilon}{k}(C_{\varepsilon 1}P - C_{\varepsilon 2}D) - R \tag{6.7}$$

In the standard k-ε model, the rate-of-strain term R in the above equation is absent (compare Eq. (3.52) in Chapter 3). The presence of this R term is formulated in the form of

$$R = \frac{C_\mu \eta^3 (1 - \eta/\eta_o)}{1 + \beta\eta^3} \frac{\varepsilon^2}{k} \tag{6.8}$$

where β and η_o are constants with values of 0.015 and 4.38. The significance of the inclusion of this term is its responsiveness to the effects of rapid rate of strain and streamline curvature, which cannot be properly represented by the standard k-ε model. According to renormalization group theory (Yakhot and Orszag, 1986), the constants in the turbulent transport equations are given by

$$C_\mu = 0.0845, \ \sigma_k = 0.718, \ \sigma_\varepsilon = 0.718, \ C_{\varepsilon 1} = 1.42, \ C_{\varepsilon 2} = 1.68$$

For the realizable k-ε model, the term *realizable* means that the model satisfies certain mathematical constraints on the normal stresses, consistent with the physics of turbulent flows. Development of this model involved the formulation of a new eddy-viscosity formula involving the variable C_μ in the turbulent viscosity relationship. It also differs in the changes imposed on the transport ε-equation (based on the dynamic equation of the mean-square vorticity fluctuation) where the source term is now solved according to

$$S_\varepsilon = C_1 \rho \left(2S_{ij}^2 \right)^{1/2} \varepsilon - C_2 \rho \frac{\varepsilon^2}{k + \sqrt{\nu_T \varepsilon}}, \qquad S_{ij} = \frac{1}{2} \left(\frac{\partial u_i}{\partial x_j} + \frac{\partial u_j}{\partial x_i} \right) \qquad (6.9)$$

and the variable constant C_1 is expressed as

$$C_1 = \max \left[\frac{\eta}{\eta + 5} \right], \qquad \eta = \frac{k}{\varepsilon} \left(2S_{ij}^2 \right)^{1/2}$$

The variable C_μ, no longer a constant, is evaluated from

$$C_\mu = \frac{1}{A_o + A_s \frac{kU^*}{\varepsilon}} \qquad (6.10)$$

Consequently, model constants A_o and A_s are determined as

$$A_o = 4.04, \quad A_s = \sqrt{6} \cos\varphi, \quad \varphi = \frac{1}{3}\cos^{-1}\left(\sqrt{6}W \right), W = \frac{S_{ij}S_{jk}S_{ki}^3}{\tilde{S}^3}, \ \tilde{S} = \sqrt{S_{ij}^2}$$

while the parameter U^* is given by

$$U^* \equiv \sqrt{S_{ij}^2 + \tilde{\Omega}_{ij}^2}, \quad \tilde{\Omega}_{ij} = \Omega_{ij} - 2\varepsilon_{ijk}\,\omega_k, \quad \Omega_{ij} = \tilde{\Omega}_{ij} - \varepsilon_{ijk}\omega_k$$

Other constants in the turbulent transport equations for this model are $C_2 = 1.9$, $\sigma_\kappa = 1.0$, and $\sigma_\varepsilon = 1.2$. The k-equation in both the RNG k-ε model and the realizable k-ε model is the same as that in the standard k-ε model except for the model constants.

The computed velocity profiles normalized by the free-stream velocity u_∞ at locations of $x/H = 0$, 1, 3, 5, 7, and 9 behind the step, predicted through each turbulence model for the backward-facing step geometry, are illustrated in Figure 6.28. In order to validate the numerical predictions, the results are compared against measurements by Ruck et al. (1988). We can observe that the flow velocities are better predicted by the RNG k-ε model and the realizable k-ε

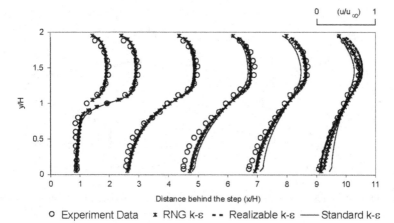

FIGURE 6.28   Measured and predicted normalized velocity profiles at locations of $x/H = 0, 1, 3, 5, 7,$ and 9 behind the step for the backward-facing step geometry.

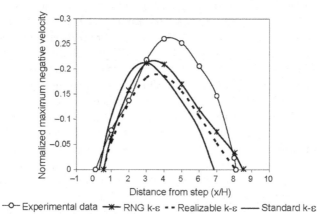

FIGURE 6.29   Maximum measured and predicted negative velocity profiles of the flow in the recirculation zone.

model and that significant deviations from the experimental data are evident for the standard k-ε model at downstream locations of $x/H = 5$, 7, and 9.

Under the same flow condition, the maximum negative velocity profiles of the flow in the recirculation zone are shown in Figure 6.29. Much lower values of the maximum negative velocities are predicted by all the three turbulence models in comparison to the experimental value. Among these three turbulence models, the re-attachment length of the realizable k-ε model comes closest to the measurement (a predicted value of $x/H = 8.2$, compared with the measured value of $x/H = 8.1$). Note the re-attachment length l_r as described in Figure 6.27. The RNG k-ε model marginally over-predicts the re-attachment length, with a value of $x/H = 8.5$. The standard k-ε model claims the last spot, as it severely

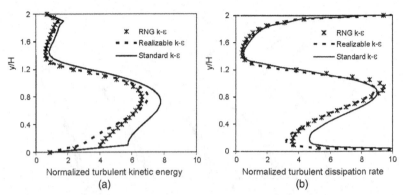

FIGURE 6.30   Normalized profiles at location $x/H = 5$ (recirculation region): (a) turbulent kinetic energy and (b) turbulent dissipation rate.

under-predicts the re-attachment length by a substantial margin, yielding a value of $x/H = 6.9$.

As mentioned, the standard k-ε model generally over-predicts the gas turbulence kinetic energy k in the recirculation region, which leads to the evaluation of a higher turbulent viscosity $v_T \; (= C_\mu k^2/\varepsilon)$. Figure 6.30(a) demonstrates the normalized profiles of the turbulent kinetic energy k at location $x/H = 5$, while Figure 6.30(b) provides the normalized profiles of the turbulent dissipation rate ε at the same location determined by the three respective turbulence models. It is evident that the standard k-ε model predicts higher turbulent kinetic energy in the recirculation zone. This therefore results in the production of excessive mixing in the standard k-ε model, which significantly reduces the intensity of this zone (as also confirmed by Murakami, 1993). Another possible cause is the ε-transport equation of the standard k-ε model. The modifications in the ε-transport equation of the RNG k-ε model and realizable k-ε model accommodate the possibility of handling large rates of flow deformation. Both the rate of production of k and rate of destruction of ε can be reduced to yield smaller eddy viscosity. We note that, even though the turbulence dissipation rate ε is predicted lower by the RNG k-ε model and reliazable k-ε model in Figure 6.30(b), the turbulent viscosities predicted through these two models are still much lower because of the apparent k^2 evaluation in the turbulent viscosity relationship, which is dominated by the higher turbulent kinetic energy k predicted by the standard k-ε model.

Other useful guidelines: It is noted that, for wall-attached boundary layers, such as that found within a simple flow between two stationary parallel plates or the backward-facing step geometry, turbulent fluctuations are suppressed adjacent to the wall and the viscous effects become prominent in this region, known as the *viscous sub-layer*. This modified turbulent structure generally precludes the application of the two-equation models, such as standard k-ε model, RNG k-ε model, and realizable k-ε model, or even the Reynolds stress model, at the near- wall region, which thereby requires special near-wall modeling procedures. Selecting an appropriate near-wall model is another important strategy in turbulence modeling. Here, the user has to decide whether to adopt the so-called wall-function method, in which the near-wall region is bridged with *wall functions*, or a *low-Reynolds-number (LRN) turbulence model*, in which the flow structure in the viscous sub-layer is totally resolved. This decision will certainly depend on the availability of computational resources and the accuracy requirements for resolution of the boundary layer. Some useful guidance on the application of relevant LRN models that can be employed all the way through the wall is provided herein. More discussion and practical guidelines are given for near-wall treatments using the wall-function method in the next section.

The *k-ω model* developed by Wilcox (1998), where ω is a frequency of the large eddies, has been shown to perform splendidly close to walls in boundary-layer flows. Such a model is common in the majority of commercial codes and it works exceptionally well, particularly under strong adverse pressure gradients, which explains its popularity in aerospace applications. As in the *standard k-ε model*, a modeled transport equation is solved for ω to determine its local distribution within the fluid flow. Nonetheless, the model is very sensitive to the free-stream value of ω and unless great care is taken in prescribing this value, spurious results are obtained in both boundary-layer flows and free-shear flows. In general, the standard k-ε model is less sensitive to the free-stream values but is often inadequate under adverse pressure gradients. To overcome such problems, Menter (1994a, 1994b) proposed to combine both the standard k-ε model and the k-ω model, which retains the properties of *k-ω* close to the wall and gradually blends into the standard k-ε model away from the wall. *Menter's model* has been shown to eliminate the free-stream sensitivity problem without sacrificing the *k-ω* near-wall performance.

To account for strong non-equilibrium effects, the SST (shear stress transport) variation of Menter's model (1993,1996) leads to a significant improvement in handling non-equilibrium boundary layer regions, such as those found behind shocks and close to separation. It is therefore highly recommended for flow separation, since the real flow is more likely to be much closer to separation (or more separated) than the calculations from the standard k-ε model suggest. Bear in mind that SST should not be viewed as a universal cure for turbulence modeling because it still inherits noticeable weaknesses. SST, for example, is less able to cope with flow recovery following flow re-attachment. For this, a promising possibility is the use of a length-scale–limiting device, as proposed by Ince and Launder (1995). Interested readers may also wish to refer to Patel et al. (1985) for the applications of other LRN versions of the standard k-ε model and the Reynolds stress model, where modifications to the governing transport equations are used to deal with near-wall effects, allowing these models to be deployed directly through to the wall. Alternatively, the standard k-ε model and the Reynolds stress model can be employed in the interior of the flow and can be coupled to the one-equation *k-L model* (Wolfshtein, 1969), which is dedicated to resolving mainly the wall region (see the review by Rodi, 1991), a so-called two-layer model. It is imperative that whatever LRN models are adopted, a sufficient number of grid nodal points be placed into a very narrow region adjacent to the wall to adequately capture the rapid variation in the flow variables.

6.4.3 Near-Wall Treatments

It is inevitable that appropriate near-wall models are required for handling wall-bounded turbulent flow problems. In addition to turbulent models that can be applied all the way through the wall, another modeling procedure commonly adopted is *wall functions*. The use of wall functions is prevalent in industrial practice and can be found in practically every CFD commercial and in-house computer code. In this approach, the difficult near-wall region is not explicitly resolved within the numerical model, but is bridged using such functions (Launder and Spalding, 1974; Wilcox, 1998). To illustrate the modeling procedure, attention is primarily directed to consideration of the flow domain having smooth walls; the treatment of rough walls is subsequently considered as a special case.

To construct these functions, the region close to the wall is usually characterized in terms of dimensionless variables with respect to the local conditions at the wall. If we let y be the normal distance from the wall and U be the time-averaged velocity parallel to the wall, then the dimensionless velocity U^+ and wall distance y^+ can be appropriately described as U/u_τ and $y\rho u_\tau/\mu$, respectively. Within these dimensionless parameters, the wall friction velocity u_τ is defined with respect to the wall shear stress τ_w as $\sqrt{\tau_w/\rho}$. If the flow close to the wall is solely determined by the conditions at the wall, then for some limiting value of the dimensionless wall distance y^+, the dimensionless velocity U^+ can be expected to be a *universal* (wall) function, as

$$U^+ = f(y^+) \tag{6.11}$$

For wall distance $y^+ < 5$, the layer is dominated by viscous forces that produce the no-slip condition and is consequently called the *viscous sub-layer*. We may assume that the shear stress is approximately *constant* and equivalent to the wall shear stress τ_w. A linear relationship between the time-averaged velocity and the distance from the wall can thus be obtained, and making use of the definitions of U^+ and y^+ leads to

$$U^+ = y^+ \tag{6.12}$$

Outside the viscous sub-layer, turbulent diffusion effects are felt and a logarithmic relationship is usually employed to account for this. The profile is

$$U^+ = \frac{1}{\kappa}\ln(Ey^+) \tag{6.13}$$

The above relationship is often called the *log-law* and the layer where the wall distance y^+ lies between the range of $30 < y^+ < 500$ is known as the *log-law layer*.

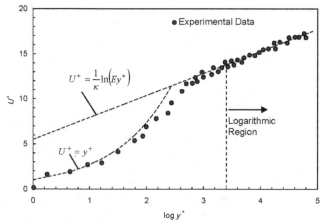

FIGURE 6.31   The turbulent boundary layer: dimensionless velocity profile as a function of the wall distance in comparison to experimental data.

Figure 6.31 illustrates the validity of Eqs. (6.14) and (6.15) inside the turbulent boundary layer. The values for κ (~ 0.4) and E (~ 9.8) are universal constants valid for all turbulent flows past smooth walls at high Reynolds numbers. For rough surfaces, the constant E in Eq. (6.15) is usually reduced. Additionally, the formula of the wall law must also be modified by scaling the normal wall distance y on the equivalent roughness height, h_o (i.e., y^+ is replaced by y/h_o), and appropriate values must be selected from data or literature. A similar universal, non-dimensional function can also be constructed for the heat and scalar fluxes. This can be used to bridge the near-wall region when solving the energy and scalar equations.

The reader should be aware that the universal profiles presented above have been derived based on an attached two-dimensional Couette flow configuration, with the assumptions of *small pressure gradients, local equilibrium of turbulence* (production rate of k equals its dissipation rate), and *a constant near-wall stress layer*. Therefore, it is imperative that care always be exercised to check the validity of wall functions in the CFD problem that is being solved. The calculated flow must be consistent, or nearly consistent, with the specific assumptions made, in arriving at the wall-function relationships. Applying the wall functions outside this application range will lead to significant inaccuracies in the CFD solution.

To remove some of the limitations imposed by the above standard wall functions, a two-layer-based, non-equilibrium wall function is also available. A more in-depth discussion regarding the formulation of such a wall function can be found in Kim and Choudbury (1995). Briefly, the key elements are (1) the log-law, which is now sensitized to pressure-gradient effects, and (2) the two-layer-based concept, which is adopted to calculate the cell-averaged turbulence kinetic energy production and dissipation in wall-adjacent cells. Based on the latter, the turbulence kinetic energy budget for the wall-adjacent cells is sensitized to the proportions of the viscous sub-layer and the fully turbulent layer, which can significantly vary from cell to cell in highly non-equilibrium flows. This effectively relaxes the *local equilibrium of turbulence* that is adopted by the standard wall functions. Because of their capability to partly account for the effects of pressure gradients and departure from equilibrium, the non-equilibrium wall functions are recommended for complex flows that may involve flow separation, flow re-attachment, and flow impingement. In such flows, improvements are obtained in the CFD solution, particularly in the prediction of wall shear and heat transfer. We note that the non-equilibrium wall functions were employed for the results obtained in the worked example of the backward-facing step turbulent flow problem in Section 6.4.2.

Another useful near-wall modeling method worth considering is the enhanced wall treatment that combines a two-layer model with enhanced wall functions. The main thrust of this approach is to achieve the goal of implementing the standard two-layer approach for fine meshes while not significantly reducing the accuracy for coarse meshes. By formulating the law of the wall as a single wall law for the entire wall region, the enhanced wall function extends its applicability throughout the near-wall region. This can be achieved by blending the linear and logarithmic laws of the wall as

$$U^+ = e^\Gamma U_{lam}^+ + e^{1/\Gamma} U_{turb}^+ \tag{6.14}$$

where Γ is a blending function. Similarly, the general equation for $\partial U^+/\partial y^+$ can also be expressed by

$$\frac{\partial U^+}{\partial y^+} = e^\Gamma \frac{\partial U_{lam}^+}{\partial y^+} + e^{1/\Gamma} \frac{\partial U_{turb}^+}{\partial y^+} \tag{6.15}$$

The above equation allows the turbulent law to be easily modified and extended to account for effects like pressure gradients or variable properties. It also guarantees the correct asymptotic behavior for large and small values of the wall distance y^+ and reasonable representation of the velocity profiles in cases where y^+ lies inside the wall buffer region ($3 < y^+ < 10$). More details of this approach can be found in Kader (1993).

Near-wall meshing guidelines for wall functions: Since the purpose of wall functions is to relate the flow variables to the first computational mesh point, thereby removing the requirement to resolve the structure in between, the lower limit of y^+ at this point must be carefully placed so that it does not fall into the viscous sub-layer. In such a case, the meshing should be arranged so that the values of y^+ at all the wall-adjacent integration points are considered only slightly above the recommended limit, typically between 20 and 30. This procedure offers the best opportunity to resolve the turbulent portion of the boundary layer. Besides checking the lower limit of y^+, it is important that the upper limit of y^+ also be investigated during the computational calculation. For example, a flow with a moderate Reynolds number has a boundary layer that extends up to y^+ between 300 and 500. If the first integration point is placed at a value of $y^+ = 100$, then this will certainly yield an impaired solution due to insufficient resolution for the region. Adequate boundary-layer resolution generally requires at least eight to ten grid nodal points in the layer, and it is recommended that a post-analysis of the CFD solution be undertaken to determine whether the degree of resolution is achieved or the flow calculation is subsequently performed with a finer mesh. This can be achieved by plotting the ratio of the turbulent diffusion and the molecular diffusion (due to molecular viscosity), which is generally high inside the boundary layer.

Near-wall meshing guidelines with LRN turbulence models: A universal near-wall behavior over a practical range of y^+ may not be realizable everywhere in a flow, such as that found for LRN flows. Under such circumstances, the wall function concept breaks down and its use will lead to significant errors. The alternative is to fully resolve the flow through the wall, which can be achieved by LRN turbulence models, as already mentioned, but it should be noted that the cost of the solution is around an order of magnitude greater than when wall functions are used, because of the additional grid nodal points involved. With the intention of resolving the viscous sub-layer inside the turbulent boundary layer, y^+ at the first node adjacent to the wall should be set preferably close to unity (i.e., $y^+ = 1$). Nevertheless, a higher y^+ is acceptable so long as it is still well within the *viscous sub-layer* ($y^+ = 4$ or 5). Depending on the Reynolds number, the user should ensure that there are between five and ten grid nodal points between the wall and the location where y^+ equals 20, which is within the viscosity-affected near-wall region, in order to resolve the mean velocity and turbulent quantities. This most likely will require 30 to 60 grid nodal points inside the boundary layer to achieve adequate boundary-layer resolution.

6.4.4 Setting Boundary Conditions

Specifying appropriate boundary conditions is particularly important in turbulence modeling. In this section, we survey the various approaches to handling the various types of boundaries within the flow domain and provide some useful guidelines for setting proper boundary conditions for turbulence.

In many practical CFD problems, specification of the turbulence quantities at the inlet can be difficult, and some sensible engineering judgment usually needs to be exercised. This is because the magnitude of turbulent kinetic energy k and dissipation ε can have a significant influence on the CFD solution. In most cases, readily accessible measurements of k and ε are rare in practice. In exploratory design computations, the problem is compounded by the nonexistence of any boundary condition information to operate the turbulence models. Preferably, experimentally verified quantities are always applied as inlet boundary conditions for k and ε. Nonetheless, if they are not available, then the values need to be prescribed using sensible engineering assumptions, and the influence of the choice taken must be examined against sensitivity tests with different simulations.

For the specification of the turbulent kinetic energy k, appropriate values can be specified through turbulence intensity I, which is defined by the ratio of the fluctuating component of the velocity to the mean velocity. In general, the inlet turbulence is a function of the upstream flow conditions. Approximate values for k can be determined according to the following relationship:

$$k_{inlet} = \frac{3}{2}(U_{inlet}I)^2 \qquad (6.16)$$

In external aerodynamic flows over airfoils, the turbulence intensity level is typically 0.3%. For atmospheric boundary-layer flows, the level can be as high as two orders of magnitude, i.e., 30%. In internal flows, a turbulence level between 5% and 10% is deemed appropriate. Similarly, the specification of the dissipation ε can be approximated by the following form:

$$\varepsilon_{inlet} = C_\mu^{3/4}\frac{k^{3/2}}{L} \qquad (6.17)$$

where L is the characteristic length scale. If the k-ω model is employed, ω can be approximated by

$$\omega_{inlet} = \frac{k^{1/2}}{C_\mu^{1/4}L} \qquad (6.18)$$

For external flows remote from the boundary layers, a value determined from the assumption of a ratio of turbulent and molecular viscosity between 1 and 10 is a reasonable guess. For internal flows, a constant length scale derived from a characteristic geometrical feature can be employed, such as 1% to 10% of the inlet hydraulic diameter. If the Reynolds stress model is applied, each stress component ($\overline{u_1'^2}$, $\overline{u_2'^2}$, $\overline{u_3'^2}$, $\overline{u_1'u_2'}$, $\overline{u_1'u_3'}$, and $\overline{u_2'u_3'}$) is required to be properly specified. If they are unavailable, as is often the case, the diagonal components ($\overline{u_1'^2}$, $\overline{u_2'^2}$, and $\overline{u_3'^2}$) are taken to be equal to 2/3 k, whereas the extra-diagonal components ($\overline{u_1'u_2'}$, $\overline{u_1'u_3'}$, and $\overline{u_2'u_3'}$) are set to zero (assuming isotropic turbulence). In cases where problems arise in specifying appropriate turbulence quantities, the inflow boundary should be moved sufficiently far away from the region of interest so that the inlet boundary layer and subsequently the turbulence are allowed to develop naturally.

For solid walls, boundary conditions for k and ε or ω are substantially different, depending on whether LRN turbulence models or the wall-function method is employed. For the former, it is appropriate to set $k=0$ at the wall, but the dissipation ε is determined through either

$$\frac{\partial \varepsilon}{\partial n} = 0 \quad \text{or} \quad \varepsilon = \left(\frac{\partial v_t}{\partial n}\right)^2 \qquad (6.19)$$

where v_t is the velocity component tangential to the wall. For the k-ω *model*, the rough-wall method of Wilcox (1993) can be adopted. The surface value for ω can be written as

$$\omega_{wall} = \frac{u_\tau}{v_{wall}} S_R \tag{6.20}$$

The variable v_{wall} is the kinematic viscosity on the wall, while S_R is a non-dimensional function determining the degree of surface roughness of the wall. However, when law-of-the-wall-type boundary conditions are employed instead, the diffusive flux of k through the wall is usually taken to be zero, yielding

$$\frac{\partial k}{\partial n} = 0 \tag{6.21}$$

The dissipation ε is, however, derived from the assumption of *local equilibrium of turbulence*. It is noted that ε is not applied at the wall but is calculated at the first computational mesh point. For the finite-volume method, ε_P is evaluated at the control volume center and is given by

$$\varepsilon_P = \frac{C_\mu^{3/4} k_P^{3/2}}{\kappa\, n_P} \tag{6.22}$$

where n_P denotes the first computational mesh point normal to the wall. In cases where the non-equilibrium wall function is used instead, ε_P is calculated according to

$$\varepsilon_P = \frac{1}{2n_P}\left[\frac{2\mu}{\rho y_v} + \frac{C_\mu^{3/4} k_P}{\kappa} \ln\left(\frac{2n_p}{y_v}\right)\right] k_P \tag{6.23}$$

where y_v is the physical viscous sub-layer thickness computed from $y^*\mu/\rho C_\mu^{1/4} k_P^{1/2}$ and y^* is set at 11.225. More details concerning the formulation of Eq. (6.25) can be found in Kim and Choudbury (1995).

At the outlet or symmetry boundaries, the Neumann boundary conditions are applicable, namely,

$$\frac{\partial k}{\partial n} = 0; \quad \frac{\partial \varepsilon}{\partial n} = 0; \quad \frac{\partial \overline{u_i' u_j'}}{\partial n} = 0 \tag{6.24}$$

In the free-stream flow where the computational boundaries are far away from the region of interest, the following boundary conditions can be used:

$$k \approx 0; \quad \varepsilon \approx 0; \quad \overline{u_i' u_j'} \approx 0 \tag{6.25}$$

which results in the turbulent viscosity $\mu_T \approx 0$.

6.4.5   Test Case: Assessment of Two-Equation Turbulence Modeling for Hydrofoil Flows

This test case was calculated using a commercial finite-volume CFD computer code, ANSYS-FLUENT, Version 6.1.

Model description: The geometry is a two-dimensional hydrofoil, spanning the test section with a distance of 3.05 m and a chord length (C) of 2.134 m. The cross-section profile is represented by a generic naval propeller of moderate thickness (t) and camber (f), utilizing a NACA-16 airfoil profile ($t/C = 8\%$ and $f/C = 3.2\%$) with two modifications. A detailed diagram of the hydrofoil geometry is shown in Figure 6.32, while the anti-singing trailing-edge geometry is detailed in Bourgoyne et al. (2000). Experiments performed on this test hydrofoil were conducted in the world's largest water tunnel, the William B. Morgan Large Cavitation Channel (LCC) in Memphis, Tennessee.

Grid: For the hydrofoil geometry, the computational domain extends $1.5 \times C$ upstream of the leading edge, $1.5 \times C$ above and below the pressure surface, and $3 \times C$ downstream from the trailing edge. A mesh overlay of quadrilateral elements is constructed for the flow domain. In the mesh generation particular attention is directed toward an offset *inner region* encompassing the hydrofoil. Within this region, a considerably fine O-type mesh is applied to sufficiently resolve the hydrofoil surface and the boundary-layer region. For the *wake region* (downstream from the trailing edge of the inner region), a considerably fine H-type mesh is applied to accurately resolve the near- and far-wake flow behavior. The remaining *outer region* of the domain is subsequently filled with a coarser H-type mesh. A total number of 208,416 grid nodal points are generated for the whole computational domain, with an average distribution of y^+ at 2.31 and a minimum and maximum y^+ at 0.09 and 4.06, respectively. The computational grid is shown in Figure 6.33.

Features of the simulation: This test case illustrates the importance not only of evaluating the choice of various turbulence models described above but also, more important, of assessing the wall functions employed to model the near-wall region of the hydrofoil at high Reynolds numbers. Three wall

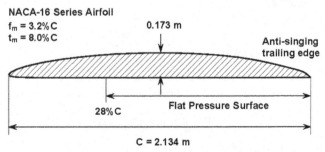

FIGURE 6.32   Schematic view of the two-dimensional hydrofoil geometry.

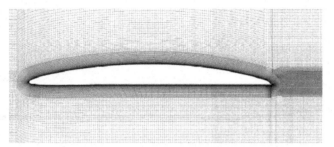

FIGURE 6.33 Close-up view of the computational grid around the foil.

treatments—standard logarithmic wall function, non-equilibrium wall function, and enhanced wall treatment—are investigated.

The algorithm for the solution of the Navier–Stokes equations utilized an implicit segregated velocity–pressure formulation, such as the SIMPLE scheme. This led to a Poisson equation for the pressure correction, which was solved through the default iterative solver, normally the multi-grid solver, of the ANSSYS-FLUENT computer code. Finite-volume discretization was employed to approximate the governing equations. To avoid non-physical oscillations of the pressure field and the associated difficulties in obtaining a converged solution on a co-located grid arrangement, the Rhie and Chow (1983) interpolation scheme was employed. A second-order upwind scheme was used for the convection, while the central-differencing scheme was used for the diffusion terms.

The working fluid, water, was taken to be incompressible and default initial conditions implemented in the computer code were used for the simulations.

Boundary conditions: At the inlet, Dirichlet conditions were used for all variables. The inlet velocity based on the free-stream velocity U_{ref} was taken as constant, with values of 3 m/s and 6 m/s. With the density and viscosity of water having values of 995.1 kg/m^3 and 7.69 $\times$ 10^{-4} kg/m $\cdot$ s, the corresponding Reynolds numbers based on the inlet velocities and the chord length are 8.284 $\times$ 10^6 and 1.657 $\times$ 10^7, respectively. At the free-stream field, the inlet velocity was also applied to the computational domain walls above and below the hydrofoil. The turbulent kinetic energy k and dissipation ε were determined from the measured turbulence intensity I of about 0.1%. At the outlet, zero-gradient conditions were applied for all the transported variables. No-slip wall boundary conditions were applied to the pressure and suction surfaces of the hydrofoil.

Results: The comparison between the measured and calculated coefficients of pressure distribution at the surface of the hydrofoil using different wall treatments is illustrated in Figure 6.34. In this figure, the realizable k-ε model is adopted in conjunction with the three wall treatments to predict the coefficients of pressure distribution. It is apparent that the use of different wall-treatment approaches affects the solution behavior. For strong adverse pressure gradients and boundary-layer separation flows, the enhanced wall treatment produces the most accurate distribution. Prediction by the standard logarithmic wall function

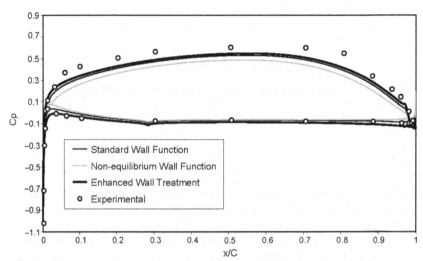

FIGURE 6.34   Wall-treatment analysis: pressure coefficient (C_p) distribution at the surface of the hydrofoil ($U_b = 3$ m/s).

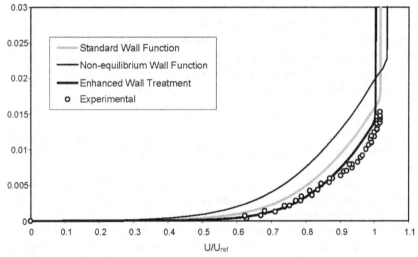

FIGURE 6.35   Wall-treatment analysis: pressure-surface boundary-layer normalized mean velocity profile at 93% C ($U_{ref} = 3$ m/s).

is slightly less accurate but still performs well at the leading and trailing edges. Nevertheless, the use of non-equilibrium wall functions produces questionable results, particularly at the leading edge, where the coefficients of pressure distributions at the suction and pressure surfaces cross each other.

Figure 6.35 represents the experimental and predicted pressure-surface boundary-layer velocity profiles at 93% C at a free-stream velocity of 3 m/s.

The realizable k-ε model is also employed here in conjunction with the three wall treatments to predict the pressure-surface boundary-layer velocity profiles. As expected in the near-wall region, the use of different wall-treatment procedures results in different predictions of the velocity profiles. Among these three different wall treatments, good agreement is achieved between the measurements and the enhanced wall-treatment approach. The use of the standard logarithmic wall function produces a boundary-layer velocity profile similar to that produced by the enhanced wall treatment, except it has a larger boundary-layer thickness. An even greater boundary-layer thickness is predicted for the nonequilibrium wall function.

Figures 6.36 and 6.37 demonstrate the measured and predicted pressure-surface boundary-layer velocity profiles with different turbulence models applied at 93% C for free-stream velocities of 3 m/s and 6 m/s, respectively. It is observed that the three turbulence models, standard k-ε, standard k-ω (Wilcox's), and SST k-ω (Menter's), generally over-predict the boundary-layer thickness when compared with the experimental result. Realizable k-ε appears to be the only turbulence model to accurately predict the boundary-layer velocity profile. Close to the surface ($y/C < 0.5\%$), this model predicts the boundary layer exceptionally well and continues to maintain a high degree of correlation further from the surface. Unlike the other models, the realizable k-ε model produces a definable gradient change where the velocity profile becomes blunt.

Conclusion: This test case focuses on the evaluation of turbulence model applications—the standard k-ε, standard k-ω (Wilcox's), SST k-ω (Menter's), and realizable k-ε—in conjunction with three wall-treatment approaches—standard logarithmic wall function, non-equilibrium wall function, and enhanced wall treatment—for a turbulent boundary-layer flow over a hydrofoil at high

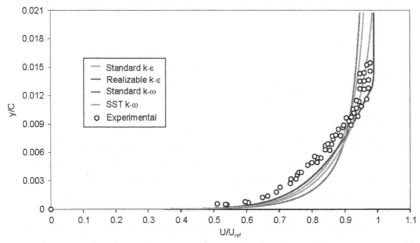

FIGURE 6.36 Turbulence model performance: pressure-surface boundary-layer normalized mean velocity profile at 93% C ($U_{ref} = 3$ m/s).

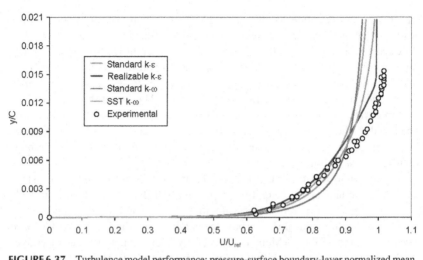

FIGURE 6.37 Turbulence model performance: pressure-surface boundary-layer normalized mean velocity profile at 93% C (U_{ref}=6 m/s).

Reynolds numbers. As for the test case presented in Chapter 5, *validation* of the computational solutions is performed by comparing the predictions against experimental data to address the simulation model uncertainty and the degree to which models correspond to an accurate representation of the real physical flow in the absence of analytical solutions. The realizable k-ε model is found to accurately predict the pressure coefficient distribution at the surface of the hydrofoil, leading to good overall predictions of the pressure-derived lift and drag coefficients. It also resolves the velocity profile and correctly predicts the thinning of the boundary layer, the commencement of boundary-layer separation, and the full separation point of the turbulent boundary layer moving rearward with increasing Reynolds number. For flows with adverse pressure gradients and boundary-layer separation, the enhanced wall treatment is considered to be a good candidate for handling such near-wall flow complexities. Nevertheless, it is noted that the numerical results obtained pertain only to the test case, and they may well be different for other types of flow problems considered.

6.5 SUMMARY

Practical guidelines for grid generation, specification of appropriate boundary conditions within the CFD context, and turbulence modeling were discussed in this chapter. The guidelines provided thus far are by no means exhaustive, but they should put the reader in an advantageous position for understanding how an appropriate CFD solution can be attained that is physically valid and meaningful. It is also shown in this chapter that CFD simulation and analysis are more than a mechanically driven exercise. They actually encapsulate, on the one hand, an understanding of the application of the essential fundamental theory of conservation equations and numerical methods to solve the CFD problem and, on the other hand, an appreciation of setting a viable mesh and defining suitable boundary conditions and models to arrive at a proper CFD solution.

Although they are not expounded upon within this chapter, the reader may well benefit by considering the use of either the steady-state or the transient approach to achieve a steady-state solution. Generally, a steady-state calculation yields a solution in a shorter execution time than a transient calculation does. Nonetheless, there may be some underlying circumstances where convergence and/or stability cannot be guaranteed by the steady-state approach and a transient calculation may be required to march the numerical procedure toward the steady-state condition. Other issues concerning *computational accuracy* and *efficiency* can also have a strong influence on the CFD solution. Subject to the availability of computational resources, it is nearly always inevitable that some compromise has to be reached for solving complex CFD problems. By increasing the number of cells, i.e., with decreasing mesh spacing, in the computational domain geometry, the *accuracy* of the computational solution is usually enhanced. There is, however, a trade off with increased computer storage and running time. One possible way that comparable *accuracies* can be obtained on a coarser mesh while maintaining *computational efficiency* is to employ higher-order discretization schemes to solve such problems. More sophisticated CFD approaches, such as DNS and LES techniques for pertinent applications, can yield solutions of high *computational accuracy* due to the application of fine meshes but are still subject to stringent *computational efficiency* due to computer hardware requirements. In practice, DNS and LES techniques should be employed only as a last resort, when nothing else succeeds, or to check the validity of a particular turbulence model being applied. Hence, it is still beneficial to explore other practical turbulence models to solve a number of real engineering flows. As demonstrated in the above test case, a more advanced turbulence model based on the realizable k-ε with enhanced wall treatment can be applied to successfully resolve flows with adverse pressure gradients and boundary-layer separation, especially those occurring with hydrofoil or even airfoil geometries.

In the next chapter, we apply CFD techniques to a variety of practical problems of varing degrees of complexity. The culmination of theory and practice is demonstrated through these worked examples.

REVIEW QUESTIONS

6.1 What are some of the benefits of a well-designed grid?

6.2 What are some of the advantages of a structured mesh?

6.3 Why is it difficult to write CFD programs that involve a structured mesh for complex geometries?

6.4 Discuss some of the advantages of an unstructured grid.

6.5 What are some of the difficulties that arise regarding programming of CFD problems for an unstructured mesh?

6.6 What conditions and constraints apply if you have to use a structured mesh for the geometry below? What about for an unstructured mesh? Discuss the advantages and disadvantages for this case.

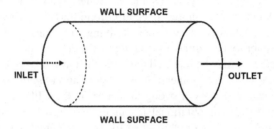

6.7 For the geometry below, discuss how using a block-structured mesh has advantages over a single-structured or unstructured mesh.

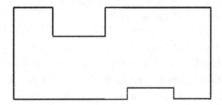

6.8 Why is it more favorable to start off with a coarse mesh when solving a CFD problem?

6.9 What is the aspect ratio of a mesh element? Why should a large aspect ratio be avoided, especially in important regions within the flow field? (See diagram below.)

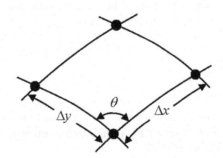

6.10 What is the skewness of a mesh element? Why is it best to avoid highly skewed elements?

6.11 Discuss why tetrahedral elements are a poor choice for meshing near walls and boundaries.

6.12 What techniques can be used in the solver when highly skewed and high-aspect-ratio elements exist within the mesh?

6.13 A computational domain with different boundary types for the flow around a hydrofoil is shown below. Show where a fine mesh should be appropriately located.

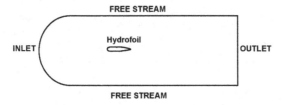

6.14 The stretched-grid technique is an example of a local refinement technique. What is one problem that is associated with the application of this technique?

6.15 Describe the way in which solution adaptation uses an adaptive grid for mesh refinement.

6.16 What are some problems/difficulties in setting up correct boundary conditions?

6.17 What types of conditions may be applied for an inlet boundary and why do you need a corresponding outlet condition?

6.18 What is the Neumann boundary condition? Explain how it is used as an outlet boundary condition.

6.19 The geometry for an air-conditioning problem in two rooms separated by a partitioned wall is shown below. Label the boundaries that have to be defined and discuss what types of conditions may be applied.

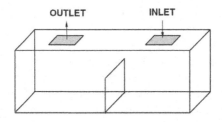

6.20 What requirements must be satisfied for a symmetry boundary condition, and what are the benefits of using this condition?

6.21 Discuss the main difference between a symmetry boundary condition and a periodic boundary condition.

6.22 The geometry of a staggered tube-bank heat exchanger in two dimensions is shown below. Make a sketch showing how you would define a computational domain for this geometry, using the periodicity or the symmetry boundary condition.

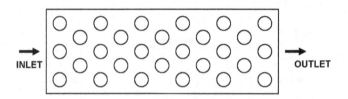

6.23 In general, describe how LES (large eddy simulation) solves a turbulent flow. How does LES differ from DNS (direct number simulation)?

6.24 Why do engineers prefer the Reynolds-averaged turbulence models, such as the k-ε model, over the complex LES model?

6.25 What are some of the major problems experienced with the k-ε model?

6.26 The standard k-ε turbulence model over-predicts the turbulent kinetic energy k when compared with the *realizable* and *RNG* turbulence models in the recirculation zones. For the backward-facing step case given below, provide an explanation of your final predictions of the re-attachment length L_R for these turbulence models in comparison with the experimental result (L_E is the actual length). In modeling turbulent flow, we effectively solve the following equation:

$$u\frac{\partial \phi}{\partial x} + v\frac{\partial \phi}{\partial y} = \mu_t\left[\frac{\partial^2 \phi}{\partial x^2} + \frac{\partial^2 \phi}{\partial y^2}\right] + S_\phi$$

(a) Use the relationship between $\mu_t = \rho C_\mu k^2/\epsilon$ and k to discuss the flow prediction.

(b) Which area within the domain requires fine mesh and where can coarse mesh be used?

(c) What kinds of boundary conditions do we impose for the inlet and outlet, and at the wall?

(d) How should the flow pattern at the outlet be described?

(e) What are the values of the horizontal u velocity (positive or negative) before, after, and at the re-attachment point?

(f) Make a sketch to show the development of the velocity profile throughout the flow from the inlet to the outlet at X_1, X_2, and X_3.

(g) Keeping the inlet velocity the same, what would the L_E be if the working fluid were changed from air to water?

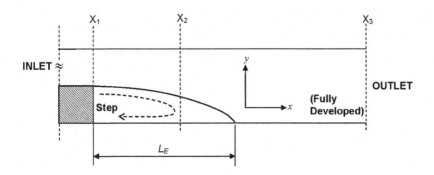

6.27 What is y^+ and why is it important in the context of near-wall turbulence modeling?

6.28 What should a typically recommended y^+ value be for the first adjacent node next to the wall in order to account for the viscous sub-layer?

6.29 Without experimental data for turbulent inlet profiles, what is the recommended method for considering turbulence effects?

6.30 (a) For the modeling of flow around a simplified car model given below, state and justify the values of y^+ to be used around the car if an "enhanced wall function" model is employed.

(b) If C_d (coefficient of drag) is used for testing mesh independence, on what basis would you think of achieving it? Substantiate your answer with an x-y plot.

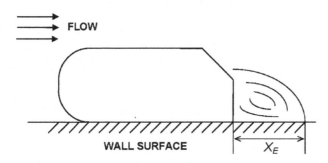

Chapter 7

Some Applications of CFD with Examples

7.1 INTRODUCTION

The demand for CFD in resolving numerous fluid-flow problems is growing within the scientific community. Needless to say, this situation could only have arisen because of CFD's meticulous development by persistent researchers and, more recently, CFD's extension to industrial usage by dedicated code developers who have made available a number of commercial CFD packages. The cornerstones of CFD analysis are the transport equations and efficient numerical techniques. Throughout this book, the authors stress the importance of grasping the essential conservation equations (as described in Chapter 3) and the basic understanding of numerical approximations (considered in Chapters 4 and 5). In Chapter 6, the authors further provide useful guidelines for handling practical flow problems, such as the requirement for suitable turbulence models to resolve real fluid-flow processes, which incidentally exemplifies the range of computed results presented in Chapter 1.

We should also be aware that CFD is not confined to predicting only fluid-flow behavior. Consideration of adequate physical models that may appropriately handle chemical reactions (e.g., combustion), multi-phase flows (e.g., transport of gas–liquid, gas-solid, liquid–solid, or even gas–liquid–solid mixtures), or phase changes (e.g., solidification or boiling) is increasingly being incorporated into the CFD framework to tackle some of these complex and challenging industrial processes. Obtaining real engineering solutions through CFD is now a very realizable prospect, due not only to the evolution of computer hardware but also to the mature development of numerical methods.

After the many important aspects of CFD presented in previous chapters, this chapter culminates the knowledge gathered by aptly describing the application of theory in practice. From a practical viewpoint, the examples that are illustrated in the subsequent sections, examples that range from rudimentary to complex flow physics and simple to complicated geometrical domains, will assist in establishing a basis for undertaking any CFD problem in a wide range of engineering disciplines. Therefore, the important aims of this chapter are

- To illustrate the basic steps required for the user to solve a practical flow problem
- To guide the reader on how to appropriately apply the knowledge gained within this book and any additional knowledge that may be required to solve other complex flow problems
- To demonstrate how CFD can be adopted as a research tool in better comprehending particular flow behaviors
- To demonstrate how CFD can be employed as a design tool in enhancing performance through better understanding of flow systems

7.2 TO ASSIST IN THE DESIGN PROCESS—AS A DESIGN TOOL

As described in Chapter 1, CFD is progressively being adopted to optimize existing equipment and/or to predict the performance of new designs even while they are in the conceptual stage of development. To illustrate how CFD can function as a design tool, a specific flow system is solved in the subsequent section—for example, a three-dimensional air flow in an office room layout. In practice, this flow problem is prevalent in the design of ventilation systems.

The flow process within air-flow systems is generally turbulent, and which turbulence models (Chapters 3 and 6) are suitable to be applied to flows characterized by low-Reynolds-number (LRN) turbulence are assessed and discussed in the first half of this example. CFD is also increasingly considered to be the preferred approach in the ventilation design process since scale-up experimental room measurements can be difficult to perform due to the expensive costs of instrumentation. The use of CFD as a design tool in enhancing the diffuser outlet design of a room ventilation system is demonstrated in the second half of this example.

7.2.1 Indoor Air-Flow Distribution

Understanding the air-flow distribution in enclosed environments is integral to indoor air quality control. It is well known that air flow is usually a function of the building's ventilation systems. With the availability of CFD computer codes, building engineers are increasingly embracing CFD methodology as an attractive alternative tool for predicting air-flow distribution, instead of employing scale-model methods.

Despite encouraging success, some uncertainties still remain, particularly in the application of turbulence models for ventilation design. An important aspect of modeling indoor air flows is the characterization of LRN turbulence. The improper handling of LRN turbulence can contribute to inaccurate calculations, since the air flow is strongly affected by the air phase velocity and turbulent fluctuations. Before CFD can be confidently applied to assist engineers in ventilation-system design, it is imperative to evaluate and to validate the range of available turbulence models.

In this specific example (Tian et al., 2006), the application of three turbulence models—standard k-ε, RNG k-ε, and RNG-based LES models—is investigated for an indoor air-flow environment. The calculated air-flow velocities are evaluated and validated against experimental data obtained by Posner et al. (2003) in a ventilated model room.

Model description: The geometrical structure of the model room used in this study is illustrated in Figure 7.1. The width, depth, and height of the room are, respectively, 91.4 cm, 45.7 cm, and 30.5 cm. Within the model room, a partition with a height of 15 cm is located in the middle of the room. Air is allowed to enter the room through one ceiling vent and to leave through the other, as indicated by the outlet vent shown in Figure 7.1.

Grid: For the room geometry, a structured mesh with rectilinear elements, distributed uniformly, is allocated for the whole physical domain, yielding an elemental volume size of 0.8 cm × 0.8 cm × 0.8 cm. The computational mesh, shown in Figure 7.2, results in a total number of 246,924 finite volumes

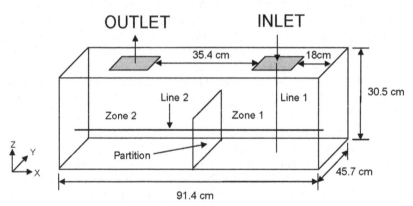

FIGURE 7.1   Schematic view of the ventilation inside the model room geometry.

FIGURE 7.2   Computational mesh for the model room geometry.

generated for the whole computational domain. In any CFD calculation, it is recommended that grid-independence analysis (Chapter 5) be performed to assess the numerical errors generated by the computed results. For this flow problem, the mesh is further refined to an elemental volume size of 0.5 cm × 0.5 cm × 0.5 cm. The difference in the air phase velocity predicted by the turbulence RNG k-ε model for the original mesh and the refined mesh is found to be less than 1%. For computational efficiency, the results presented below are obtained with a mesh with an elemental volume size of 0.8 cm × 0.8 cm × 0.8 cm.

Features of the simulation: The vertical inlet velocity (U_{inlet}) of 0.235 m/s and a characteristic length of 0.1 m provides a flow Reynolds number of 1500. All computations are performed in transient state, in which the time-dependent terms appearing in the transport equations are handled through an implicit second-order backward differencing in time. A non-dimensional time step of 0.035 is used. This time step has been defined by $t' = U_{inlet}\, t/H$, where U_{inlet} is the inlet air velocity as given above, t is the physical time step with a value of 0.05 seconds, and H is the room height. The transport equations are discretized using the finite-volume method (Chapter 4). For the Reynolds-Avergaed Navier-Stokes (RANS) approach employing the standard k-ε and RNG k-ε models (see Chapters 3 and 6 for more detailed description), a third-order interpolation scheme, such as the *QUICK* scheme (see Appendix B), is used to approximate the convective terms at the faces of the control volumes. In LES, energetic eddies that exist near the cut-off wave number can significantly influence the spatial discretization errors (Park et al., 2004). The contribution of the LES sub-grid scale force may be overwhelmed by the use of *upwind* and *upwind-biased* schemes (Mittal and Moin, 1997) and therefore a central-difference scheme is adopted for the LES calculations. For the pressure–velocity coupling, the SIMPLE algorithm is employed (Chapter 4). As the air flow in near-wall meshes can be at a very low Reynolds number ($y^+ \approx 1$), this study employs an enhanced wall treatment, a near-wall modeling method that combines a two-layer model with enhanced wall functions, for the k-ε models (Chapter 6). For the LES model, a very fine mesh is required to resolve the wall layer, which is very computationally expensive, especially for engineering applications. Therefore, a wall model, like the RANS approach, is used to bridge the wall with the adjacent turbulent air flow. Convergence for the air-flow governing variables (velocities, pressure, k, and ε) is assumed to have been reached when the iteration residuals are reduced by five orders of magnitude (e.g., 1×10^{-5}).

Model validation against experimental data: The case of the ventilated model room, as investigated by Posner et al. (2003), is used to evaluate turbulent indoor air flow using the standard k-ε, RNG k-ε, and RNG-based LES models. The initial condition of the flow field in the room is assumed to have a randomly perturbed velocity about the magnitude of the mean velocity U_{inlet}. To ensure that the solution for the LES computations achieves sufficient statistical independence from the initial state, time-averaged results are obtained from the instantaneous transient values after the air-flow simulation is marched for 2000 non-dimensional time steps, which represents 100 seconds in physical time. After this time, the instantaneous values, such as the air-flow velocities, are averaged over 10,000 non-dimensional time steps (500 seconds in physical time). The simulated vertical air velocity component along the vertical inlet jet axis (line 1 in Figure 7.1) and the vertical air velocity component along the horizontal line at mid-partition height (line 2 in Figure 7.1) are validated against the measured results.

Figure 7.3 shows the comparison between the predicted and measured vertical air velocity component along the horizontal line at mid-partition height. Between the location of $x=0$ m and the partition position, all three turbulence models yield almost similar results. Here again, the RNG-based LES model provides a slightly better prediction in the region between $x=0.2$ m and the partition position. In the near-wall regions about the locations $x=0.46$ m and $x=0.9$ m, the RNG-based LES model successfully captures the highest positive vertical velocities, while the two k-ε models significantly under-predict the velocities. From these results, it is apparent that better prediction is achieved through the RNG-based LES model, as demonstrated by the excellent agreement with the experimental results in the region from the partition position to the horizontal location of $x=0.6$ m. Considerable under-prediction of the negative vertical velocity by the k-ε models, found in the region immediately beneath the inlet, may well be attributed to the excessive diffusion caused by the eddy-viscosity modeling. Marginal discrepancy between the measured data and the simulation results found in the region about the location $x=0.85$ m shows that the k-ε model's results are slightly better than the RNG-based LES model's results. Over all, all three turbulence models perform well; good agreement is achieved between the predictions and measured data and the flow

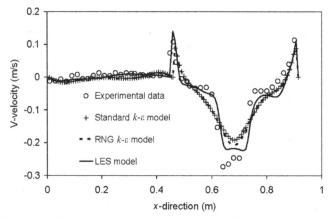

FIGURE 7.3 Comparison between predicted and measured results of the vertical velocity component along the horizontal line at mid-partition height.

trends are successfully captured. One significant aspect of this study is that the RNG-based LES model has been shown to provide significantly better results, especially in zone 2, because of the model's inherent ability to better capture the fluid-flow characteristic within a confined space. Despite this important discovery, we observe that the velocities predicted by the two k-ε models are still within reasonable limits of the measurements. Practically, these models can still be applied with some degree of confidence and are generally sufficient for the majority of engineering applications.

Design of a room ventilation system: The practical use of CFD is illustrated here. As a design tool, CFD can be systematically employed to parametrically investigate a number of ventilation-system design features. Let us investigate the scenario of the ventilated compartment depicted in Figure 7.4. In this example, cold air is supplied through the top plenum duct. A portion of the main air supply diverts into the diffuser grills at the ceiling and consequently exhausts at the bottom of the diffuser outlet into an open space inside the enclosure. The diffuser grill is connected to the top plenum duct by an adjoining duct at a length H. Two design features—Design A and Design B—in the vicinity of the diffuser outlet are shown in Figure 7.5. From a ventilation perspective, Design A generally results in a distortion of the air distribution as it exits through the diffuser grill. In order to achieve a more symmetrical diffusion pattern, guiding vanes near the adjoining duct are introduced as part of an improved design to better distribute the air flow before it reaches the diffuser outlet.

CFD simulations of Design A and Design B with the adjoining duct of length H are presented in Figure 7.6. The color spectrum (shown in shades of gray) of the velocity vectors exemplifies the velocity magnitude: the lowest speed and the highest speed are both indicated. As observed, the proposed design change

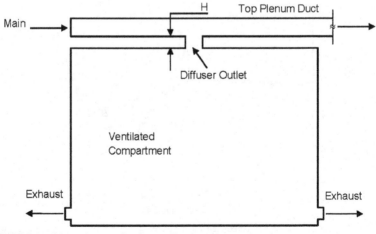

FIGURE 7.4 Schematic scenario of a ventilated compartment.

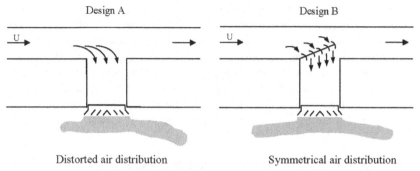

FIGURE 7.5 Distorted and symmetrical air diffusion through the diffuser grill.

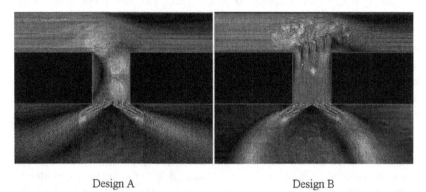

FIGURE 7.6 Comparison between Design A and Design B for the predicted air distribution in the vicinity of the diffuser outlet with an adjoining duct of length H.

with the guiding vanes being positioned upstream of the adjoining duct in Design B has a dramatic impact on the air distribution at the diffuser exit. The air flow appears to be more uniformly distributed before reaching the diffuser grill. A symmetrical flow pattern is obtained. However, this is not the case seen in Design A, where a stagnant flow region persists and the air flow beyond the diffuser grill behaves more like a jet stream being injected into the ventilated compartment.

Conclusion: Two key points are demonstrated in this example. First, the mature development achieved in turbulence modeling demonstrates the feasibility of employing such models for engineering applications with some degree of confidence. Second, it exemplifies the possible use of CFD in assisting the design process, because multiple parametric investigations can be performed at low cost and hence engineers can be provided with the opportunity to better assess numerous design options before the eventual selection of a viable system.

Before we proceed, the authors should point out an important aspect of application of turbulence models. Despite the significant advances achieved in the area of turbulence, even to the point where LES models nowadays can be feasibly applied with current computational resources, the standard k-ε model still remains the workhorse model that provides reliable predictions for many engineering flow problems. In practice, the standard k-ε model should be promoted for carrying out initial exploratory design calculations. The results obtained may well be within the acceptable limits of engineering solutions, but in the event that the model fails to capture the essential flow physics or to crystallize any viable solutions, more advanced turbulence models can be applied as the next step of the design process to improve the CFD results.

7.3   TO ENHANCE UNDERSTANDING—AS A RESEARCH TOOL

In retrospect, CFD was established as a research tool for comprehending the many important physical aspects of a flow field. As we briefly observed in Chapter 1, CFD can be applied nowadays to elucidate some interesting flow structures over cylindrical obstacles. In addition to the many significant advances and the maturity achieved in CFD research on single-phase flow problems, CFD is currently being used in two-phase flow investigations. To further demonstrate the versatility of the tool, we concentrate here on another challenging area for CFD, the numerical study of gas-particle flow.

Gas-particle flows are encountered in numerous important processes, such as in the minerals, petroleum, chemical, metallurgical, and energy industries, as well as in environmental engineering. Scaled-up experiments involving such flows are usually difficult to perform, mainly due to the inherent complexity of the flow phenomena (e.g., clusters appearing within the gas-particle flows) that require high-precision instrumentation. Figure 7.7 illustrates an experimental flow visualization of solid particles (glass beads with a mean diameter of 66 μm) suspended in the gas flow through a 90° bend.

During the last decade, significant efforts were made in both academic and industrial research to better understand dispersed two-phase flows using CFD techniques. Note that this CFD example should be construed as providing only

FIGURE 7.7   Flow visualization of solid particles suspended in a gas flow.

a brief overview of the additional modeling effort that is required in modeling dispersed two-phase flows. Handling such flows generally requires the consideration of additional conservation equations and appropriate models, where the gas phase is usually regarded as a carrier and the particle phase is taken to be dispersed within the gas flow. First, the Lagrangian approach can be adopted by treating the gas phase as a continuum fluid with discrete particles in the fluid space. Second, both the gas and particle phases can alternatively be assumed to be continuum fluids, which is the essence of the Eulerian approach. Transport equations (mathematical models) describing the gas-particle flow according to the two different approaches are discussed in the next section.

7.3.1   Gas-Particle Flow in a 90° Bend

Two different CFD computer codes, an in-house research code (Tu, 1997) and a commercial CFD code, ANSYS-FLUENT, Version 6.1, are employed to simulate the gas-particle flow. The discrete approach described by the Lagrangian model is handled by the ANSYS-FLUENT code, while the continuum approach that is essentially the Eulerian model is solved through the in-house research code. The simulated results are validated against the experimental data of Kliafas and Holt (1987).

Model description: The geometry of this example is a three-dimensional turbulent air flow with particles traveling around a 90° duct bend, as exemplified in Figure 7.10. The schematic view of the experimental setup for this geometry is similar to Figure 5.6 for Test Case B in Chapter 5, and is comprised of a 1.2 m long horizontal duct, a 90° bend with a radius ratio of 1.76 (bend radius to bend hydraulic diameter), and a 1 m long vertical straight duct. Air enters the square test section with a bulk gas velocity U_b adjusted using a variable-frequency controller. Experimental data for both the gas and particle fields are obtained using laser Doppler anemometry (LDA).

Grid: For the 90° square-section bend, the computational domain begins at 1 m (10 D) upstream from the bend entrance and extends to 1.2 m (12 D) downstream from the bend exit. A structured mesh with hexahedral elements is generated for the in-house computer code. On the other hand, an unstructured mesh comprised of tetrahedral elements fills the whole 90° bend geometry, as illustrated in Figure 7.8, for the ANSYS-FLUENT computer code.

Physics and mathematical models: For both approaches, the governing equations for the gas phase are the same, as described by Eq. 3.53 (Chapter 3), which can be re-written as

$$\frac{\partial \phi}{\partial t} + \frac{\partial (u\phi)}{\partial x} + \frac{\partial (v\phi)}{\partial y} + \frac{\partial (w\phi)}{\partial z} = \frac{\partial}{\partial x}\left[\Gamma_\phi \frac{\partial \phi}{\partial x}\right] + \frac{\partial}{\partial y}\left[\Gamma_\varphi \frac{\partial \phi}{\partial y}\right]$$
$$+ \frac{\partial}{\partial z}\left[\Gamma_\phi \frac{\partial \phi}{\partial z}\right] + S_\phi \qquad (7.1)$$

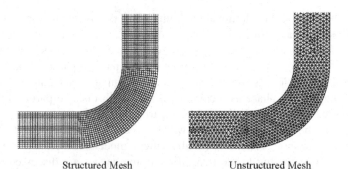

Structured Mesh   Unstructured Mesh

FIGURE 7.8   Computational meshes for the 90° square-section bend.

By setting the transport property ϕ equivalent to 1, u_g, v_g, w_g, k_g, ε_g, and selecting appropriate values for the diffusion coefficient Γ_ϕ and source terms S_ϕ, we obtain the set of transport equations presented in Table 3.2 (see Chapter 3) for each partial differential equation describing the conservation of mass and momentum and the turbulent quantities of the gas phase.

Governing equations of particle phase using the Lagrangian approach: This approach calculates the trajectory of each individual discrete particle by integrating Newton's second law, written in a Lagrangian reference frame (Chiesa et al., 2005). Appropriate forces, such as the drag and gravitational forces, can be incorporated into the equation of motion. The equation can thus be written as

$$\frac{du_p}{dt} = F_D(u_g - u_p) + \frac{g(\rho_p - \rho_g)}{\rho_p} \tag{7.2}$$

where $F_D(u_g - u_p)$ is the drag force per unit particle mass, and F_D is given by

$$F_D = \frac{18\mu_g}{\rho_p d_p^2} \frac{C_D Re_p}{24} \tag{7.3}$$

The relative Reynolds number Re_p is defined as

$$Re_p = \frac{\rho_p d_p |u_p - u_g|}{\mu_g} \tag{7.4}$$

while the drag coefficient C_D is determined from

$$C_D = a_1 + \frac{a_2}{Re_p} + \frac{a_3}{Re_p^2} \tag{7.5}$$

In Eq. (7.5), the coefficients denoted by a_1, a_2, and a_3 are empirical constants for smooth spherical particles over several ranges of particle Reynolds number (Morsi and Alexander, 1972). The *eddy lifetime* model is used to account for the effect of gas-phase turbulence on the particle phase. More details regarding this model can be found in Tian et al. (2005).

Governing equations of particle phase using the Eulerian approach: For this approach, the governing equations for the particle phase can also be expressed in the form of Eq. (7.1) described above. The set of transport equations can be similarly obtained by setting the transport property ϕ_p equivalent to 1, u_p, v_p, w_p, k_{gp} k_p, ε_{gp}, and selecting appropriate values for the diffusion coefficient $\Gamma_{\phi p}$ and source terms $S_{\phi p}$; the special forms tabulated in Table 7.1 describe the partial differential equations for the conservation of mass and momentum and the turbulent quantities of the particle phase. Details on the derivations of these equations can be found in Tu (1997).

Features of the simulation: The governing transport equations are discretized using a finite-volume approach with a non-staggered grid system (see Chapter 4 for more explanation). A third-order *QUICK* scheme (Appendix B) is used to approximate the convective terms, while a second-order accurate central-difference scheme is adopted for the diffusion terms. The velocity–pressure linkage is realized through the SIMPLE algorithm (Chapter 4). In order to mimic the experimental conditions, a uniform velocity ($U_b = 52.19$ m/s) for both gas and particle phases is imposed at the top inlet 1 m away from the bend entrance, which corresponds to a Reynolds number of 3.47×10^5. The inlet turbulence intensity is prescribed at 1%, whereas the particles are taken to be glass spheres of density $\rho_s = 2990$ kg/m^3 and size 50 μm. The corresponding particulate loading and volumetric ratios are 1.5×10^{-4} and 6×10^{-8}, respectively, for which the particle suspension is considered to be very dilute. At the outflow, the normal gradient for all dependent variables is set to zero. A no-slip boundary condition is employed along the wall for the gas phase as well as the particulate phase at the wall. Appropriate boundary conditions also need to be specified to represent the particle–wall momentum transfer for the Eulerian approach, of which more details can be found in Tu and Fletcher (1995).

TABLE 7.1 General form of governing equations for the particle flow in a Eulerian reference frame

ϕ_p	$\Gamma_{\phi p}$	$S_{\phi p}$
ρ_p	0	0
u	v_{pT}	$F_D^u + F_G^u + F_{WM}^u$
v	v_{pT}	$F_D^v + F_G^v + F_{WM}^v$
w	v_{pT}	$F_D^w + F_G^w + F_{WM}^w$
k_{gp}	$\frac{v_{pT}}{Pr_T}$	$P_{kgp} - \overline{\rho}_p \varepsilon_{gp} - \Pi_{gp}$
k_p	$\frac{v_{pT}}{\sigma_k}$	$P_{kp} - I_{gp}$
ε_{gp}	$\frac{v_{pT}}{\sigma_\varepsilon}$	$\frac{\varepsilon_g}{k_g}\left(C_{\varepsilon 1}P_g - C_{\varepsilon 2}D_g\right)$

For the Lagrangian model, particle transport using a discrete random walk (DRW) model is computed from the converged solution of the gas flow. For the DRW model, 20,000 individually tracked particles are released from 10 uniformly distributed points across the inlet. The independence of statistical particle phase predictions is tested using 10,000, 20,000, and 50,000 particles. The difference in the maximum positive velocities of 20,000 and 50,000 particles is found to be less than 1%.

In the Eulerian approach, the numerical method for the solution of governing equations of the particle phase is similar to the method for scalar transport variables, such as temperature, for the gas phase. This is due to the fact that there is no pressure term in the particle momentum equations based on an assumption of no collisions among particles in a diluted gas-particle flow. All the governing equations for both gas and particle phases are solved sequentially at each iteration step. They are iteratively solved via the strongly implicit procedure (SIP) solver. The above solution process is marched toward steady state until convergence is attained.

The CPU time for the Lagrangian approach is generally much greater than that for the Eulerian approach. This is not surprising since significantly more computations are required to determine each individual particle trajectory in the Lagrangian approach when compared with the much reduced computational effort needed to calculate the particle phase as one entity in the Eulerian approach.

Model validation against experimental data: The mean quantities of both gas and particle phases employing the two different numerical approaches are compared against the well-established experimental results of Kliafas and Holt (1987). In this example, an important dimensionless parameter, particularly for gas-particle flow, is the Stokes number, which is defined as the ratio between the particle relaxation time and a characteristic time of the fluid motion, in other words, $St = t_p/t_s$. This dimensionless number determines the kinetic equilibrium of the particles with the surrounding gas. The system relaxation time t_s in the Stokes number can usually be derived from the characteristic length (L_s) and the characteristic velocity of the system under investigation. In this example, it is the free-stream velocity (U_b); hence $t_s = L_s/U_b$. A small Stokes number ($St \ll 1$) signifies that the particles are in near velocity equilibrium with the carrier fluid. For larger Stokes number ($St \gg 1$), particles are no longer in equilibrium with the surrounding fluid phase; they divert substantially from the fluid stream path.

Mean quantities are of utmost interest in engineering applications. For the gas and particle phases, the mean velocity, concentration, and fluctuation distributions along the bend are compared against measurements at the mid-plane along the spanwise direction of the duct geometry. All the values reported here (unless otherwise stated) are normalized using the inlet bulk velocity (U_b). Figure 7.9 shows the comparison of the numerical results against the experimental data for the mean streamwise gas velocity (the case for $St < 0.1$) along

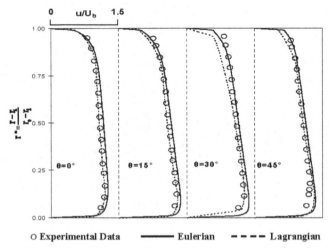

FIGURE 7.9   Mean streamwise particle velocities along the bend for Stokes number $St = 0.01$.

various sections of the bend. Both of the numerical models appear to be in qualitative agreement with the data reported by Kliafas and Holt (1987).

Enhanced understanding through research: As a research tool, CFD can be used to further enhance our understanding of particle behavior around the carrier gas phase by carrying out *numerical experiments* for the gas-particle flow. The following results demonstrate the use of this important tool.

To better understand the particle paths along the bend of different Stokes numbers, Figure 7.10 depicts the paths using Lagrangian tracking for each respective case. For a Stokes number of 0.01, it is clearly evident that the particles have a tendency to act as "gas tracers." In such a flow scenario, they are generally considered to be in equilibrium with the carrier phase. However, this phenomenon is less pronounced as the Stokes number is increased. A positive slip velocity exists between the particulate and gas phase at the outer walls, along with the gas velocities at the inner walls, due to the presence of a favorable pressure gradient. This "gas-tracing" property of the particles becomes less pronounced as they approach the bend exit, since the flow regains the energy it lost due to slip. It is observed that for flows with $St \geq 1$, the positive slip velocity between the particle and gas velocity decreases along with the bend radius and reverts to negative slip velocity at the bend exit, where the particles now lag the gas. They are unable to keep pace with the gas due to its own inertia as well as the energy lost through particle–wall collisions. As the Stokes number increases, the particles show a greater tendency to migrate toward the outer bend.

Conclusion: With increasing computing power and the availability of a viable discrete approach described by the Lagrangian model or a continuum approach described by the Eulerian framework, which entails the consideration of additional conservation equations, many complex multi-phase flow problems

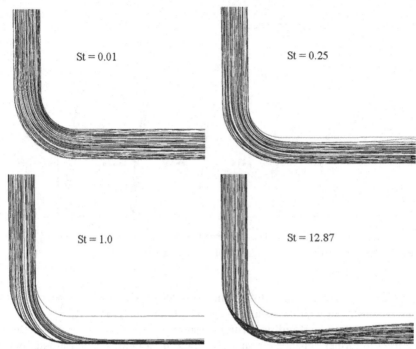

FIGURE 7.10   Lagrangian particle paths for varying Stokes numbers.

are being tackled using CFD. The feasibility of attaining qualitative and quantitative numerical results in this example represents not only the many rich physical insights that can be gained through the CFD methodology but also a testimonial to the rigorous advancements that have been made in state-of-the-art multi-phase flow research.

7.4   OTHER IMPORTANT APPLICATIONS

7.4.1   Heat Transfer Coupled with Fluid Flow

Another application of CFD is with heat transfer coupled to fluid-flow behavior. The proper handling of such flows is illustrated in the two examples presented below, which further elucidate pertinent features associated with modeling and procedures for obtaining meaningful practical solutions associated with this type of CFD problem.

7.4.1.1   Heat Exchanger

Heat exchangers are employed in numerous industries. Steam generation in a boiler, air cooling within the coil of an air conditioner, and automotive radiators are just some of the conventional applications of this mechanical system.

Of particular importance in the design of heat exchangers is the pivotal under-standing of heat transfer in flow across a bank of tubes. Tube banks, as used in many heat exchangers, can be arranged in in-line or staggered systems, as described in Figure 7.11. Here, the flow of cooler fluid flowing over the tubes containing the warmer fluid which flows perpendicular to the schematic draw-ing is considered. For design purposes, tube banks are usually characterized by a number of important dimensionless parameters, such as the transverse, longitu-dinal, and diagonal pitches. These parameters allow engineers to assess various heat-transfer augmentation methods in order to design special types of heat exchangers, such as compact heat exchangers. According to Žukauskas and Ulinskas (1988), these types of heat exchangers (the main focus in this CFD example) are categorically considered by the dimensionless transverse and

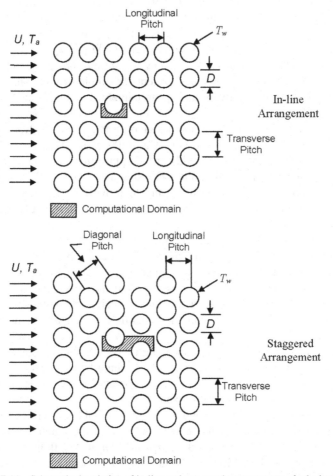

FIGURE 7.11   Schematic description of in-line and staggered arrangements of tube banks.

longitudinal pitches being less than 1.25 and have shown to provide higher heat transfer coefficients in *laminar flow* than in highly *turbulent flow*. Typically, the heat-transfer surface areas are substantially increased, while augmenting the flow, creating various secondary flows, and simultaneously destroying the hydrodynamic and thermal boundary layers created by the repetitive nature of the flow.

Clearly, long-standing practice in design and manufacture of heat exchangers has been dominated by experimental methods and semi-empirical integral approaches based on simple criteria relations for the heat-transfer coefficient. As early as the 1930s, Colburn (1933) proposed a simple correlation for the heat transfer in flow across banks of staggered tubes, while later others, such as Aiba et al. (1982a,b) and Žukauskas and Ulinskas (1988), reported extensive experimental heat transfer and fluid friction during viscous flow across in-line and staggered banks of tubes covering both isothermal and isoflux boundary conditions. Lately, Khan et al. (2006) provided analytical correlations that could be used with either in-line or staggered arrangements for a wider range of parameters. With the advent of commercial CFD packages, engineers are warming up to the possibilities for gaining insight into processes that are not easily amenable to analysis by measurements or simple overall computations, for assessing both qualitatively and quantitatively the performance, and for optimizing design and operation parameters through CFD. There are many advantages in the use of CFD for the design of heat exchangers. Recent publications by Witry et al. (2005), Hájek et al. (2005), and Wang et al. (2006) exemplify some of the current trends in the application of CFD to a variety of heat exchanger systems. With the increasing use of this methodology, it is therefore appropriate to demonstrate, and to provide some useful practical guidelines for, how CFD can be applied in solving the flow and heat-transfer characteristics of compact heat exchangers.

Problem considered: For the in-line and staggered arrangements of tube banks (which will be illustrated in Figure 7.17), fluid at a prescribed mass-flow rate or velocity U and an inlet ambient temperature T_a much lower than the wall temperature T_w enters the tube banks from the left and exits at the right. By taking advantage of special geometrical features, such as the inherent repetitive nature of the flow behavior, the computational fluid domain allows the possible exploitation of *symmetric* and *cyclic* (*periodic*) boundary conditions (as described in Chapters 2 and 6) in speeding up the computations and, in turn, enhancing the computational accuracy of the simplified geometries. The use of these boundary conditions, which are important features in practical CFD applications, is demonstrated in this example.

Cyclic (periodic) boundary conditions can be suitably prescribed to ensure that the characteristics of the fluid leaving and entering the domain are identical. They are usually imposed on walls *perpendicular* to the flow direction. In contrast, the rationale behind *symmetric* boundary conditions is to replicate the fluid flow within the solution region to adjacent regions containing the same flow structures. They are commonly applied on walls *along* the flow direction. The solution regions of the two boundary conditions are depicted by the shaded areas in Figure 7.11. Specification of the relevant boundary conditions and the computational meshes that have been generated are described in Figure 7.12.

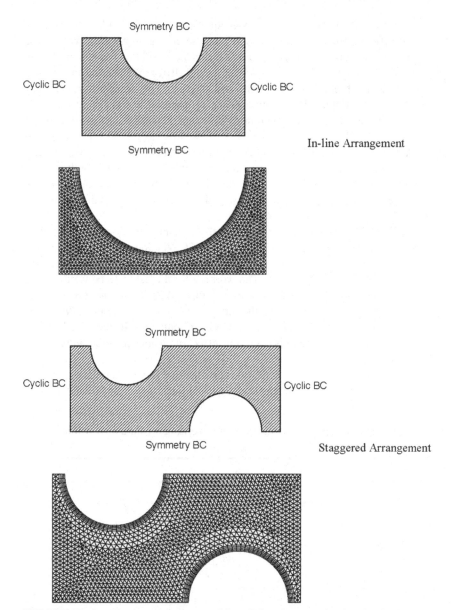

FIGURE 7.12   Specification of boundary conditions (BC) and computational meshes for the solution regions associated with the in-line and staggered arrangements.

CFD simulation: The governing equation for the two-dimensional laminar flow within the simplified computational domains can be written as

$$\frac{\partial(\rho u \phi)}{\partial x} + \frac{\partial(\rho v \phi)}{\partial y} = \frac{\partial}{\partial x}\left[\Gamma_\phi \frac{\partial \phi}{\partial x}\right] + \frac{\partial}{\partial y}\left[\Gamma_\phi \frac{\partial \phi}{\partial y}\right] + S_\phi \qquad (7.6)$$

By setting the transport property ϕ equal to 1, u, v, H $(= C_p T$ where C_p is the specific heat of constant pressure) and selecting appropriate values for the diffusion coefficient Γ_ϕ and source terms S_ϕ, we obtain the special forms presented in Table 7.2 for each of the partial differential equations for the conservation of mass, momentum, and energy. The solution is marched toward steady state, where convergence is deemed to have been reached when the iteration residual of enthalpy is reduced by six orders of magnitude (e.g., 1×10^{-6}).

A hybrid mesh (structured and unstructured elements) is generated for each of the computational domains represented later in Figure 7.18. For both cases, a boundary-layer stretched mesh of rectangular elements (structured) is concentrated near the tube walls while the rest of the internal flow domain is filled by triangular elements (unstructured).

CFD results: This example is calculated using the commercial finite-volume CFD computer code ANSYS-FLUENT, Version 6.1.22. The relevant data employed for the CFD calculations of the in-line and staggered arrangements are provided in Table 7.3. The working fluid is taken to be water. Fluid properties of water at the reference temperature of 25°C are also given in the table. Based on these properties, the dimensionless Prandtl number Pr can be evaluated; a value of 7.0 is obtained. For the computed results presented henceforth, it is assumed that there is negligible effect of change in temperature on the fluid properties.

As was shown in Chapter 2, the non-dimensional Reynolds number is a useful parameter in characterizing fluid flow. Another important dimensionless

TABLE 7.2 Conservation equations governing the fluid flow within the in-line and staggered arrangements

ϕ	Γ_ϕ	S_ϕ
ρ	0	0
u	μ	$-\dfrac{\partial p}{\partial x} + \dfrac{\partial}{\partial x}\left[\mu \dfrac{\partial u}{\partial x}\right] + \dfrac{\partial}{\partial y}\left[\mu \dfrac{\partial v}{\partial x}\right]$
v	μ	$-\dfrac{\partial p}{\partial y} + \dfrac{\partial}{\partial x}\left[\mu \dfrac{\partial u}{\partial y}\right] + \dfrac{\partial}{\partial y}\left[\mu \dfrac{\partial v}{\partial y}\right]$
H	$\dfrac{\mu}{Pr}$	0

TABLE 7.3 Data used for the in-line and staggered tube bank arrangements

Quantity	Dimension
Tube diameter D, mm	10.0
Longitudinal pitch, mm	12.5
Transverse pitch, mm	12.5
Tube wall temperature T_w, °C	70
Mass-flow rate, kg/s	0.01 – 0.4
Water properties:	
Thermal conductivity k, W/m · K	0.6
Density ρ, kg/m³	998.2
Specific heat C_p, J/kg · K	4182
Kinematic viscosity v, m²/s	1×10^{-6}
Prandtl number Pr	7.0

parameter that is commonly associated with heat transfer is the Nusselt number (Nu) (named after Wilhelm Nusselt), which may be obtained by multiplying the convective heat-transfer coefficient h with the ratio L/k, where L is the characteristic length scale and k is the thermal conductivity; that is,

$$Nu = \frac{hL}{k} \qquad (7.7)$$

A large Nu essentially signifies that the *convection heat transfer* supersedes the *conduction heat transfer* in the bulk fluid flow. For the design of heat exchangers, the diameter of the tube is usually taken as the characteristic length in the definition of the Nusselt number as well as the Reynolds number. The numbers can be defined as $Nu_D = D\ h/k$ and $Re_D = D\ U_{max}/v$. Instead of the usual inlet velocity being used as the reference velocity, the definition of the Reynolds number here differs by employing U_{max}, that is, the maximum velocity, in the minimum flow area to be determined directly from the CFD solution. It is noted that the dimensions given for the longitudinal and transverse pitches in Table 7.3 correspond to the configuration of compact heat exchangers, where the dimensionless longitudinal and transverse pitches are 1.25 (12.5 mm/10 mm). Figure 7.13 illustrates the streamlines at different Reynolds numbers for the in-line arrangement. There appear to be two distinct flow regimes within the spaces separating the tubes along the longitudinal and transverse distances at LRN flows. In between the spacing of the top and bottom

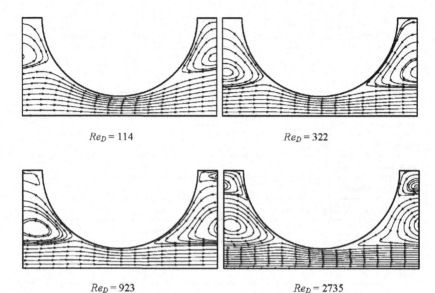

Re_D = 114   Re_D = 322

Re_D = 923   Re_D = 2735

FIGURE 7.13   Streamlines of in-line arrangement at different Reynolds numbers.

tubes lies the primary fluid flow of the system, which represents the multiple horizontal layers of fluid flowing along the longitudinal direction of the entire in-line tube bank. Sandwiched between these layers are two secondary recirculation vortices in between the spacing of the front and rear tubes. At sufficiently high-Reynolds-number flows, the breakdown of these larger secondary cells results in the formation of two smaller cells, further adding to the complexity of the fluid-flow and heat-transfer processes.

Figure 7.14 shows the streamlines predicted for the staggered arrangement, with boundary-layer separation prevailing downstream of the fluid flow (indicated by points "S" in the figure). Here, the flow patterns resemble a snake weaving around the tubes. Further increase of Reynolds numbers tends to elongate and intensify the fixed recirculation eddies, which clearly show the flow breaking away from the tube at high Reynolds numbers. Such observations are similar to the classical vortex shedding phenomenon, which represents an important aspect from the viewpoint of ensuring structural integrity. The shedding of these vortices gives rise to a lateral force acting on the tube. Since these forces are generally periodic, following the frequency of vortex shedding, the tubes may be subjected to forced vibration, sometimes called self-induced vibration. As also observed in Figure 7.14, the mainstream of the wavy flow layers in between the staggered spacing of the top and bottom tubes envelops the secondary elongated recirculation eddies residing in between the spacing of the front and rear tubes.

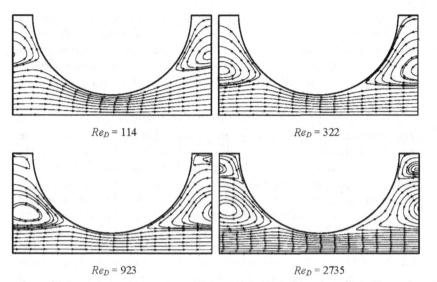

$Re_D = 114$ $\qquad\qquad\qquad\qquad$ $Re_D = 322$

$Re_D = 923$ $\qquad\qquad\qquad\qquad$ $Re_D = 2735$

FIGURE 7.14 Streamlines of staggered arrangement at different Reynolds numbers.

The higher heat-transfer rates experienced in a staggered arrangement than in an in-line arrangement of tube banks in compact heat exchangers is further confirmed by the plot of average Nusselt number against Reynolds number for the same dimensionless pitches in Figure 7.15. From thermal-hydraulic considerations, one plausible explanation is that, in a staggered bank, the path of the main flow is more tortuous and extends greater coverage over the surface areas

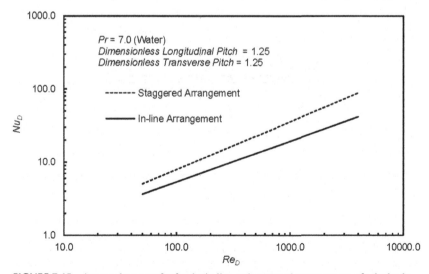

FIGURE 7.15 Average heat transfer for the in-line and staggered arrangements of tube banks.

of downstream tubes, thereby resulting in more efficient removal of the heat contained within the internal tubes of the compact heat exchanger. The in-line arrangement is the opposite: From the flow structures observed, the main flow path appears to glide over only the tips of the top and bottom tube surfaces, which may explain the reduced effectiveness of an in-line arrangement.

7.4.1.2 Conjugate and Radiation Heat Transfer[1]

Safety concerns and integrity issues are paramount in the removal of a single-plate molybdenum target (layered with a composite mixture of uranium and aluminum) from water into air after the target has been irradiated in the Open Pool Australian Light-water (OPAL) research reactor. During irradiation, owing to the bombardment of neutrons, heat is generated within the plate as a result of nuclear fission. After considerable time to allow the plate to cool sufficiently in the water, the plate experiences a decay power output due to residual fission heating. At low power levels, the plate is unloaded, placed in a holder, and then cooled in air by natural convection; process handling to extract molybdenum for nuclear medicine purposes proceeds thereafter. The schematic representation of the configuration of a typical plate layered with uranium and aluminum is shown in Figure 7.16. From thermal-hydraulic considerations, the key objectives are to ascertain at which power level the temperature inside the plate remains below the blistering temperature of 600°C (melting temperature of aluminum) in the air environment and to ensure that the structural integrity of the plate is not breached or compromised.

From a modeling viewpoint, this particular CFD example poses a number of challenges. First, heat being conducted out from the solid plate is radiated to the surroundings as well as being taken away by natural convective air currents.

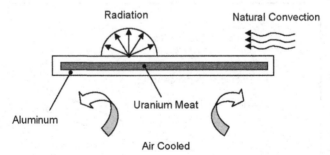

FIGURE 7.16 Schematic representation of a single-plate molybdenum target radiating heat and cooled by natural convective currents of air.

1. The materials in this section were provided by David Wassink and Mark Ho, working at the Australian Nuclear Science Technology Organization (ANSTO). The authors are indebted to them for kindly preparing the materials and for their useful input into the writing of this section in the spirit of the rest of the book.

This highlights the importance of considering the heat-transfer processes between solid and gas regions, which require *conservation of energy equations to be solved for each region*, in order to accurately predict the solid–gas interface and internal temperatures of the solid plate (*conjugate heat transfer*). Second, modeling of buoyant flow in air generally requires additional considerations particularly in the governing equations for the momentum and turbulence. Formulation of appropriate source terms is required to capture this essential flow characteristic. Third, this example demonstrates the requirement for using a suitable turbulence model as well as imposing appropriate boundary conditions in order to attain numerical results that closely mimic the actual flow behavior observed during experiments.

Problem considered: The experimental target plate designed at the ANSTO Water Tunnel Laboratory was constructed by carefully clamping two aluminum plates together to form a single plate with a thickness of 5 mm, as shown in Figure 7.17. Between the plates, matched grooves were machined into the matching faces of the plates to accommodate the heater wire and thermocouples. Electrical insulation of the heater wire to the target plate was achieved by anodizing the plates prior to assembly. The aluminum plates were held together with small screws, and a heat transfer compound was applied to the mating faces to ensure good thermal contact between the heater wire and thermocouples.

The case where the target plate is suspended horizontally in air is considered. Figure 7.18 depicts the experimental setup for the target plate being centrally supported in a box having open sides. During experiments, the outer boundaries of the box were closed to permit flow visualization for the convective air flow. Smoke seeding was used and the resultant flow streamlines were recorded by measuring the velocities at various points around the target plate using imaging and laser Doppler velocimetry (LDV) techniques. Internal temperatures in the solid plate were measured via thermocouples connected to the LabView Data Acquisition System. The plate power was manually controlled,

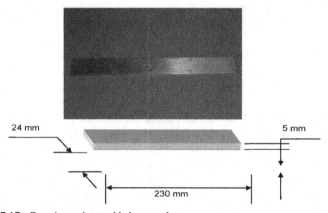

FIGURE 7.17   Experimental assembled target plate.

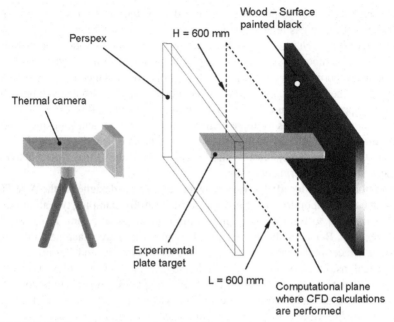

FIGURE 7.18   Experimental setup of an assembled target plate in an open box.

with the power level determined as a product of the *voltage* and *current* applied to the plate; losses in transmission were allowed to ensure accuracy of the specific plate power measured. In order to determine the effect of emissivity on the power radiated as heat from the plate—radiation heat transfer—data were obtained from the plate using different surface finishes, i.e, (i) black anodized; (ii) anodizing removed, surface burnished using 800-grit carborundum paper; and (iii) after development of an oxide layer by immersion in heated water. In order to enable the emissivity value to be estimated for CFD calculations, images of the target plate were obtained using a thermal camera.

CFD simulation: The two-dimensional "suspended" experimental target plate is modeled according to the computational domain shown in Figure 7.19 (see also Figure 7.18, where the computational plane—an area of 600 mm^2—is considered for CFD calculations) and solved using ANSYS-CFX, Version 10, commercial code. An orthogonal mesh totaling 280,000 elements covering the solid plate and surrounding air is generated for the computational domain. Fine-mesh resolution around and inside the solid plate, as shown by the exploded view in Figure 7.19, is aimed to resolve the formation of complex boundary layers due to the natural convective flow and internal heat conduction through the composite materials. To replicate the boundary conditions that enclose the physical experiment, all boundaries are simulated as wall boundaries at a uniform temperature of 25°C. The working fluid is taken to have the properties of air at a reference temperature of 25°C.

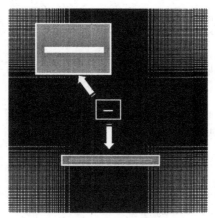

FIGURE 7.19   Computational model for the experimental assembled target plate in a closed environment.

As mentioned, buoyant air flow requires additional modeling effort. Here, the momentum equation should now include the body force resulting from buoyancy. For this particular example, gravity acts in the y direction; hence, the additional source term due to buoyancy needs to be incorporated in the y-momentum equation. The two-dimensional form of the y-momentum equation becomes

$$\frac{\partial(\rho v)}{\partial t} + \frac{\partial(\rho u v)}{\partial x} + \frac{\partial(\rho v v)}{\partial y} = \frac{\partial \tau_{xy}}{\partial x} + \frac{\partial \sigma_{yy}}{\partial y} - g(\rho - \rho_0) \qquad (7.8)$$

For completeness, the x-momentum and continuity equations can be expressed as

$$\frac{\partial(\rho u)}{\partial t} + \frac{\partial(\rho u u)}{\partial x} + \frac{\partial(\rho v u)}{\partial y} = \frac{\partial \sigma_{xx}}{\partial x} + \frac{\partial \tau_{yx}}{\partial y} \qquad (7.9)$$

$$\frac{\partial \rho}{\partial t} + \frac{\partial(\rho u)}{\partial x} + \frac{\partial(\rho v)}{\partial y} = 0 \qquad (7.10)$$

where the normal stresses σ_{xx} and σ_{yy} due to the combination of pressure p, and normal viscous stress components τ_{xx} and τ_{yy} acting perpendicular to the control volume, as well as the tangential viscous stress component $\tau_{xy} = \tau_{yx}$, formulated considering a Newtonian fluid, can be found in Appendix A. The additional source term $-g(\rho - \rho_0)$ in Eq. (7.8) represents the buoyancy term, with the reference density denoted by the variable ρ_0. In addition, besides the buoyancy term incorporated within the y-momentum equation, appropriate modifications to the two-equation k-ε turbulence model are also required. Like the production (P) and destruction (D) of turbulence in the two equations (as described in Chapter 3), the inclusion of buoyancy in the model follows similar considerations. For example, the k-equation now takes the form

$$\frac{\partial(\rho k)}{\partial t} + \frac{\partial(\rho u k)}{\partial x} + \frac{\partial(\rho v k)}{\partial y} = \frac{\partial}{\partial x}\left[\frac{\mu_T}{\sigma_k}\frac{\partial \phi}{\partial x}\right] + \frac{\partial}{\partial y}\left[\frac{\mu_T}{\sigma_k}\frac{\partial \phi}{\partial y}\right] \boxed{D}$$

$$\boxed{P} + \left\{2\mu_T\left[\left(\frac{\partial u}{\partial x}\right)^2 + \left(\frac{\partial v}{\partial y}\right)^2\right] + \mu_T\left(\frac{\partial u}{\partial x} + \frac{\partial v}{\partial y}\right)^2\right\} - \rho\varepsilon + G_{buoy} \tag{7.11}$$

where the generation term G_{buoy} relating to the buoyancy along the y direction can be formulated as

$$G_{buoy} = g\frac{\mu_T}{\sigma_T}\frac{1}{\rho}\frac{\partial \rho}{\partial y} \tag{7.12}$$

For the dissipation of turbulence kinetic energy ε, the modeled transport equation accounting buoyancy is given by

$$\frac{\partial(\rho\varepsilon)}{\partial t} + \frac{\partial(\rho u\varepsilon)}{\partial x} + \frac{\partial(\rho v\varepsilon)}{\partial y} = \frac{\partial}{\partial x}\left[\frac{\mu_T}{\sigma_k}\frac{\partial\varepsilon}{\partial x}\right] + \frac{\partial}{\partial y}\left[\frac{\mu_T}{\sigma_k}\frac{\partial\varepsilon}{\partial y}\right]$$

$$+ \frac{\varepsilon}{k}(C_{\varepsilon 1}P - C_{\varepsilon 2}D) + \frac{\varepsilon}{k}C_{\varepsilon 1}C_{\varepsilon 3}\max(G_{buoy}, 0) \tag{7.13}$$

In Eqs. (7.11) and (7.12), the adjustable constants σ_T and $C_{\varepsilon 3}$ are usually specified as having values of 0.9 and 1.0, respectively.

To account for the heat transfer in the solid plate and air, the appropriate conservation equations for the solid and gas phases are solved. For the solid plate, the heat-conduction equation in two dimensions can be expressed as

$$\rho_s C_{ps}\frac{\partial T_s}{\partial t} = \frac{\partial}{\partial x}\left[k_s\frac{\partial T_s}{\partial x}\right] + \frac{\partial}{\partial y}\left[k_s\frac{\partial T_s}{\partial y}\right] + \dot{S} \tag{7.14}$$

where T_s is the solid-phase temperature and $\dot{S}$ is the internal volumetric heat source. The variables ρ_s, C_{ps}, and k_s are the thermophysical properties of the solid plate. For the gas phase, it is generally preferable to solve the heat transfer through the thermal energy H. The two-dimensional governing equation for enthalpy is given by

$$\frac{\partial(\rho H)}{\partial t} + \frac{\partial(\rho u H)}{\partial x} + \frac{\partial(\rho v H)}{\partial y} = \frac{\partial}{\partial x}\left[\left(\frac{\mu}{Pr} + \frac{\mu_T}{Pr_T}\right)\frac{\partial H}{\partial x}\right]$$

$$+ \frac{\partial}{\partial y}\left[\left(\frac{\mu}{Pr} + \frac{\mu_T}{Pr_T}\right)\frac{\partial H}{\partial y}\right] \tag{7.15}$$

Here, μ and Pr are the laminar viscosity and Prandtl number, while μ_T and Pr_T are the turbulent counterparts, with values given as 0.7 and 0.9, respectively. The gas-phase temperature can be determined as a result of the solution obtained from the enthalpy equation, Eq. (7.15). At the solid plate's surface interfacing with the air, energy balance is invoked to obtain the surface or wall temperature, which entails

$$-k_s\frac{\partial T}{\partial n}\bigg|_s = -k_g\frac{\partial T}{\partial n}\bigg|_g + e\sigma\left(T_g^4 - T_w^4\right) \tag{7.16}$$

$$\underbrace{\qquad\qquad}_{solid\text{-}phase\ heat\ flux}\quad\underbrace{\qquad\qquad}_{gas\text{-}phase\ heat\ flux}\quad\underbrace{\qquad\qquad}_{radiation\ heat\ flux}$$

where σ is the Stefan-Boltzman constant. Appropriate values for the emissivity e are required to evaluate Eq. (7.16); they are determined through experimental recordings by the thermal camera.

Because of the fine resolution imposed around the solid plate, the shear stress transport (SST) turbulence model is utilized before other available turbulence models to resolve the low turbulent flow near the solid surface. The SST model actually represents a better alternative in bridging the wall with the bulk air flow than the default use of the logarithmic wall function coupled with the standard k-ε model. Another important consideration for this flow problem is the need to incorporate the *radiation heat transfer* between the different viewing orientations of the solid boundaries and outer walls of the computational domain. Here, the Monte-Carlo radiation model is applied to account for the radiation rays transmitted, primarily from wall to wall, within the domain.

CFD results: One important input parameter required for the present CFD calculations is a surface characteristic of the solid plate, the surface emissivity. During the experiments, a plate model at a low emissivity of 0.1 was exposed to water at a power of 25 W over 3 days. The growth in the aluminum oxide layer resulted in an increased emissivity to a range of 0.3 $\sim$ 0.5 as measured by the pyrometer. As a conservative approximation, an emissivity of 0.3 is adopted in the CFD simulations. Based on the power of 25 W, the internal volumetric heat source $\dot{S}$ appearing in Eq. (7.15), considering the region encompassing the heating elements inside the solid plate that comprises a rectangular prism with a volume of 2.76×10^{-5} m^3 (0.005 m [thickness] $\times$ 0.024 m [width] $\times$ 0.23 m [length]), can be evaluated to yield a power density of 9.058×10^{-6} W/m^3 (25 W/2.76×10^{-5} m^3). A transient simulation is performed and convergence is deemed to have been reached within each time step when the normalized residual falls below the criterion of 1×10^{-5}.

The streamline of smoke traces observed during experiments is shown by the photograph in Figure 7.20. In determining the velocities, the smoke seeding used to permit visualization of the convection flow allowed velocities to be measured at the specific locations around the plate using imaging and LDV techniques (as will be shown in Figure 7.28). Table 7.4 shows the comparison of the measured velocities and the predictions made by the computational model. The close agreement between the measured and predicted velocities validates and verifies the numerical models for resolving this flow problem. The adopted models also provide confidence in the usage of CFD methodology in tackling the complex processes associated with conjugate and radiation heat transfer.

Figure 7.21 illustrates the temperature contours predicted inside the solid plate and the surrounding air enveloping the solid plate accompanied by the

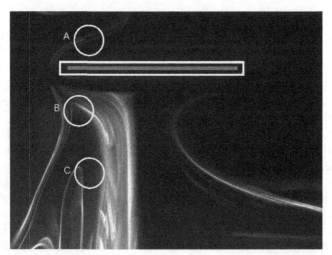

FIGURE 7.20   Experimental streamline of smoke traces.

TABLE 7.4 Comparison between predicted and measured velocities at the specific locations highlighted in Figure 7.20

Location	Measurement	Prediction
A	100 mm/s	108.4 mm/s
B	35 mm/s	32.4 mm/s
C	15 mm/s	12.2 mm/s

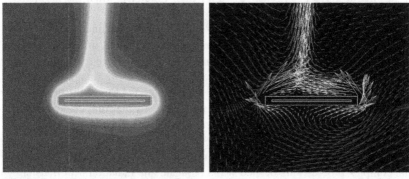

FIGURE 7.21   Temperature contours showing the surrounding and internal temperatures and velocity vectors around the solid plate.

velocity vectors around the solid plate. For the latter, the predicted flow pattern shows remarkable resemblance to the observed streamline in Figure 7.20, where the flow appears to be biased toward the left of the solid plate. Cool air is seen being entrained from some distance away onto the solid plate and is heated while flowing around the solid plate. A strong upwardly buoyant thermal plume is generated, as confirmed by the temperature contours and velocity vectors. Since aluminum is a very good conductor, there is essentially no appreciable difference of temperatures that can be found within the solid plate and on the solid walls. For the current CFD simulation, a uniform temperature of 164.5°C is predicted for the entire solid region. The surrounding temperature is depicted at 25°C.

Calculations without the consideration of radiation heat transfer were also performed. The simulation revealed a solid temperature of around 300°C, which is considerably higher than the case with radiation heat transfer. *Which of these two temperatures is the correctly predicted value and, most importantly, which of them actually reflects reality?* Let us ponder the implication of the two computational results. During experiments, rigorous measurements have repetitively been found to err toward a lower temperature, approximately 159°C. This makes it clear that there is a need to incorporate additional sophistication into the CFD modeling by accounting for radiation heat transfer in order to yield CFD results corresponding to the actual physical flow processes. Also, radiation heat loss measured during experiments accounts for almost 40% of the total heat emission from the solid plate, which further emphasizes the need to aptly incorporate the effect of radiation heat transfer into the CFD modeling.

Additional CFD calculations on different power settings, ranging as high as 100 W, with the same emissivity value of 0.3 are illustrated in Figure 7.22. The close agreement between the predicted temperatures and the experimentally measured data, as well as the analytical total heat loss expression based on Incropera and DeWitt (1985), further confirms the application of models to adequately resolve the flow and heat-transfer characteristics of this particular flow problem.

7.4.2 A Buoyant Free-Standing Fire

As evidenced by experiments performed by McCaffrey (1979) and Cox and Chitty (1980), buoyant fires have three distinct regions: *a persistent flame, an intermittent flame*, and *a buoyant plume*. Figure 7.23 exemplifies a typical three-zone flame structure for a buoyant fire (Cheung et al., 2007).

Buoyant fires are generally characterized by a very low initial momentum and they are strongly governed by buoyancy effects. It is also recognized that such fires exhibit an oscillatory behavior. The occurrence of this oscillation, generally known as the "puffing" effect, stems from the presence of coherent

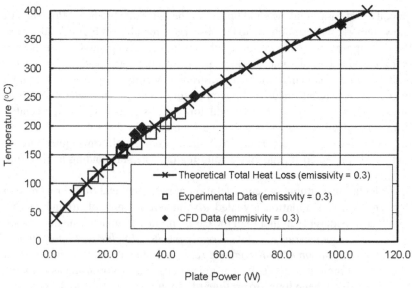

FIGURE 7.22   Distribution of temperature inside and around the solid plate.

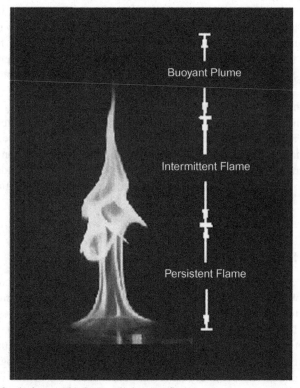

FIGURE 7.23   A photographic image of a buoyant fire.

structures above a fire plume. These structures are a consequence of the developing buoyancy-driven instabilities, which subsequently lead to vortex shedding, especially through the formation of large flaming vortices that rise up until they burn out at the top of the flame. The pulsating characteristics of such fires are strongly governed by the rate of air entrainment into the flame, flame height, combustion efficiency, and radiation heat output of the flames. Simulating such fires can be challenging.

According to McCaffrey (1983), the "puffing" mechanism can be described as follows. The initial momentum and buoyancy of combustion gases in stagnant air set a large toroidal vortical structure. The entrainment of air is assisted into the reaction zone, which causes the combustion of gases to accelerate, creating the characteristic "necking." As the vortex propagates upward, it leaves an area of low pressure that is immediately filled by the combustion gases, creating a "bulge." This bulge rises, resulting in the formation of another toroidal vortex at the base of the fuel source. The vortex below the bulge shifts the plume surface outward in the radial direction while the bulge above drags the plume surface inward. Hence, the bulge structure is maintained. The rotational motion in the upper vortex causes the plume to be stretched in the axial direction. This periodic oscillation is usually referred to in the literature as the pulsation frequency of the flame-flickering phenomenon. Pulsation frequencies for buoyant diffusion flames have been reported by many researchers (Portscht, 1975; Zukoski et al., 1984), with Malalasekera et al. (1996) providing an excellent review.

CFD simulation: The turbulent buoyant fire is solved using an in-house large eddy simulation (LES) computer code to examine the coupled turbulence, combustion, soot chemistry, and radiation effects. The three-dimensional, Favre-filtered, compressible mass, momentum, energy, and mixture fraction and its scalar variant conservation equations are closed using the LES Smagorinsky sub-grid-scale (SGS) turbulence model. The numerical method is based on a two-stage predictor-corrector approach for low Mach number compressible flows to account for the strong coupling between the density and fluid-flow equations. The infinitely fast chemistry approach is adopted as the combustion model. A combination of a presumed beta-filtered density function and a conservation equation for the scalar variance is used to account for the SGS mixture fraction and scalar dissipation fluctuations on the filtered composition and local heat-release rate. A soot model incorporating nucleation and surface-growth agglomeration is considered. The radiation heat transfer has been accommodated through the discrete ordinates model. More details regarding this model can be found in Cheung et al. (2007).

A three-dimensional computation is performed on a designated 3 m × 3 m × 3 m domain. The gaseous fuel is injected through a floor-level porous square burner with dimensions of 0.3 m × 0.3 m. A structured mesh totaling 96 × 96 × 96 cells is generated for the computational domain. Transient analysis is preformed with about 27,500 time steps. To ensure that the analysis achieves stability and a pseudo-steady-state status, a simulation of 35 seconds in real time is used. By employing a Pentium IV PC with speed of 3.0 GHz and 2.0 Gbytes RAM, the CPU time for a real 35-second simulation is on the order of 400 hours.

CFD results: Figure 7.24 illustrates a series of frames capturing a flickering period. Macroscopic predictions using the LES approach are compared against the photographic images obtained from McCaffrey (1979) experiments. The naturally flickering behavior demonstrates the flow undergoing a phase shift along the axial direction. Numerically predicted images in Figure 7.25 correspond to three isometric surfaces of temperatures at 800 K (visible flame), 450 K, and 310 K (cold smoke). Two distinct regions of the flame, the upper flame separated from the lower persistent flame depicted in frames 1 and 8 by the numerical model, show remarkable resemblance to the associated frames of the photographic images. The upwardly surging flame structures in frames 6 and 7 before flame separation in frame 8 are also adequately captured by the model when compared to the corresponding images of the developing fire observed during experiments.

The radiative cooling due to soot is demonstrated through the instantaneous soot distribution accompanied by the temperature contours and velocity vectors shown in Figure 7.25. Buoyant diffusion flame behaves differently from jet diffusion flame. At the beginning of the puffing period, soot is formed at the outer fringes of the fire bed, where the meeting of the fuel and air is ideal. Owing to the self-excited toroidal vortex motion above these fringes, the soot being formed is entrained in the continuous flame region, which is then carried upward during the middle and end of the puffing periods. The peak soot content

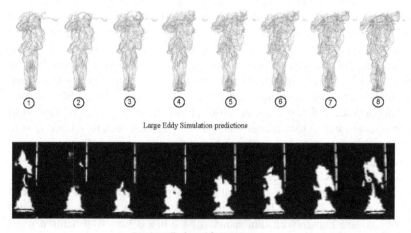

Large Eddy Simulation predictions

Experimentally observed flame structures

FIGURE 7.24 Demonstration of the puffing effect during one flickering cycle.

Beginning of Puffing Period

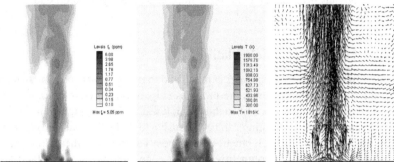

Middle of Puffing Period

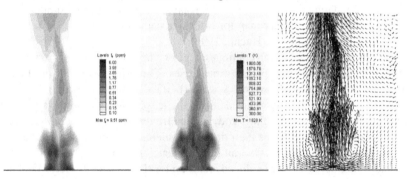

End of Puffing Period

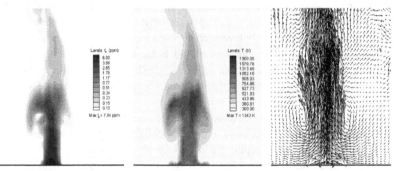

FIGURE 7.25 Instantaneous soot distribution accompanied by the temperature contours and velocity vectors during one puffing period.

is located predominantly in the middle portion of the continuous flame region for the buoyant fire; the temperature is thus subsequently lowered due to soot radiation.

7.4.3   Flow over Vehicle Platoon

In order to meet surging consumer demand and to reduce costs and time to market, automobile manufacturers are progressively aiming to develop more economical, safer, and more comfortable vehicles at an increasingly rapid pace. Traditional wind-tunnel testing can be expensive to build and road-testing techniques may incur long development-cycle times. To overcome these difficulties as well as to maintain a competitive market edge, automotive manufacturers need to focus on computational techniques, such as the emergence of CFD. Nowadays, CFD and scale-model tests are often used in car development, with full-size wind tunnels used for validation and refinement in global simulation of the entire flow field, rather than for extensive parametrical studies as found in current industrial practice (Hucho, 1996).

As the human population continues to grow, the need for more practical and efficient methods of transportation increases. The primary aim of the Future Generation Intelligent Transport System (FGITS) is to establish a platoon of vehicles with significantly reduced inter-vehicle spacing on the highway. In this system, it is expected that the aerodynamic efficiencies of the individual vehicles will be improved, since their drag coefficients are expected to be reduced with close vehicle spacing. Overall fuel efficiencies of the vehicles in a platoon can also be improved, with reduced emission levels and added safety. Vehicles would be equipped with Intelligent Transport Systems (ITS), such as distance sensors, adaptive braking systems (ABS), global positioning systems (GPS), and other ITS systems that would enable the vehicles to travel close together *safely* (e.g., less than one car length apart), so that they closely resemble a convoy of train cars. However, before such systems can be implemented, a detailed understanding of how the vehicles would behave aerodynamically when in close proximity to one another is required. CFD can play an important role in conducting significant aerodynamic research on various forms of vehicle-to-vehicle interference.

Problem considered: The Ahmed model geometry, as shown in Figure 7.26, is a generic automotive bluff body with the backlight (slant) region configurable through various angles. Despite the relatively simple geometry, the model is capable of replicating dominant flow structures pertinent to those generated about the C-pillar of a practical road car.

The experiment representation of two vehicles in tandem considered in this example is illustrated in Figure 7.27. The vehicles are placed inside a closed-jet, fixed-ground wind-tunnel facility, having a test section 2 m high, 3 m wide, and 9 m long. Two Ahmed models, a lead model (75% scale model) and a trailing model (100% scale model) with 30° rear slant angles, are positioned at different

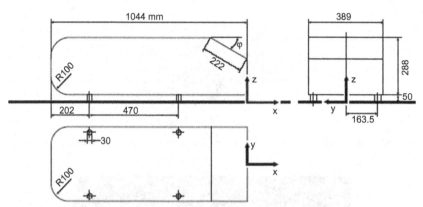

FIGURE 7.26   The Ahmed car geometry.

FIGURE 7.27   Two Ahmed vehicles in tandem inside a wind-tunnel facility.

inter-vehicle spacing, which is varied according to the length of the 100% scale model ($L = 1.044$ m). The inter-vehicle spacing is varied between $0.25\,L$ and $2\,L$. The free-stream velocity is set at 30 m/s, corresponding to a Reynolds number of 1.55×10^6 based on 75% of the scale Ahmed model length. Free-stream turbulence intensity is maintained at 1.8%, and the blockage ratio is less than 2%, which is not corrected for in the study.

To obtain the correlation between the flow structure and forces exerted on the trailing model, time-averaged lift and drag measurements are taken at each model position. This is achieved through a JR3 six-component force-balance sensor that has been mounted inside the lead (75% scale) test model. The arrangement is shown in Figure 7.28. More details on the experimental setup as well as the CFD model discussed below can be found in Rajamani (2006).

CFD simulation: The computational domain that comprises a section of the wind-tunnel configuration is shown in Figure 7.29. Within this figure, the

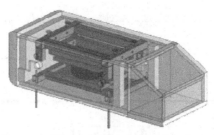

FIGURE 7.28   Internally mounted force-balance arrangement for the lead model.

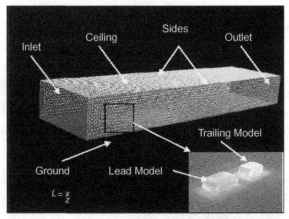

FIGURE 7.29   Computational domain comprising the leading and trailing Ahmed vehicle models inside a section of the wind tunnel.

enlarged view illustrates the lead and trailing Ahmed model vehicles. The total length of the test domain is 13 L. The inlet is positioned 4 L from the front of the lead model. The outlet was positioned 8 L from the rear of the trailing model. The sides were 2 L on either side from the full-scale model while the ceiling was 2.5 L from the top of the full-scale model. The lead model is located 0.0375 m from the ground and the trailing model was positioned 0.05 m from the ground in both the experimental and computational investigations.

An unstructured mesh consisting of tetrahedral elements totaling over 1.8 million cells is generated to accommodate all the necessary geometrically intricate details. The simulation of the turbulent flow is handled through the commercial CFD code ANSYS-FLUENT, Version 6.1.22. The realizable k-ε model is chosen as the preferred turbulence model for this particular flow problem. To satisfy mass conservation, the SIMPLE algorithm is adopted to establish the linkage between the velocity and pressure, while a multi-grid solver is employed to accelerate the numerical calculations toward the steady-state solution. Convergence is deemed to have been reached when the residual mass falls below a value of 1.0×10^{-5}.

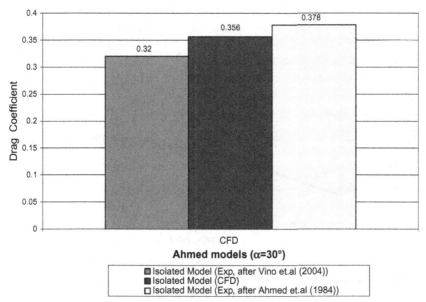

FIGURE 7.30   Comparison of drag coefficients for an isolated Ahmed model.

CFD results: The case for the critical rear slant angle of 30°, after Ahmed et al. (1984), is used to validate the CFD results for the case of an isolated model. For the two models in tandem, however, this case has been compared against the experimental data of Vino et al. (2004). The predicted drag coefficient for the isolated Ahmed model shown in Figure 7.30 is seen to agree rather well with experimental measurements by Ahmed et al. (1984) as well as Vino et al. (2004). Despite some discrepancies between the predicted and measured drafting effects on the coefficient of drag, especially at low inter-vehicle spacing for the case of two vehicles in tandem, as illustrated in Figure 7.31, the predicted values are nonetheless still in close agreement with experiments.

The 30° rear slant angle of the Ahmed model is often considered the most critical angle (Ahmed et al., 1984), since any slight decrease or increase in the angle would considerably reduce the drag coefficients. In this particular configuration, the separation bubble, which is caused by flow separation at the slant side, reaches its maximum size at this critical angle—one of the main reasons for its high drag coefficient. This high drag regime is mainly characterized by the presence of strong counter-rotating C-pillar vortices that are rich in kinetic energy, as shown in Figure 7.32, where the color spectrum (shown in shades of gray) indicates the highest value and the lowest value.

At very close inter-vehicle spacing, as demonstrated by the case of $x/L = 0.25$ in Figure 7.33, the drag coefficient of the lead model is considerably lower than that of the trailing model. This effect is caused by the apparent presence of another model in its wake. The mere presence of the trailing model increases the base

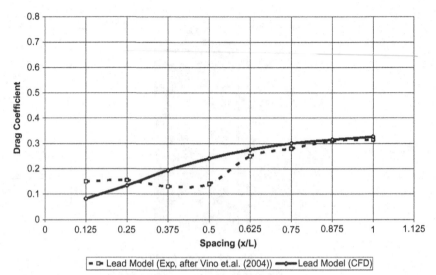

FIGURE 7.31   Effects of inter-vehicle spacing on the lead model.

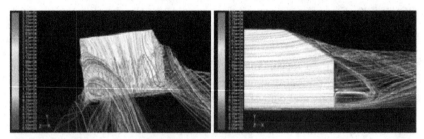

FIGURE 7.32   High drag configuration: critical angle of 30°.

pressure of the lead model, which subsequently reduces the drag of the lead model. The flow that separates from the leading edge of the rear slant of the lead model also creates a flow impingement on the front portion of the trailing model. This effect increases the drag coefficient of the rear model, which can be observed from the pathlines of pressure in Figure 7.33. At an inter-vehicle spacing of $x/L = 1$ (see Figure 7.33), the trailing model's drag coefficient reaches its peak value; the increase in drag is caused by the very high flow impingement.

CFD is also well suited for analysis of a wide range of shape options. These simulations are particularly useful in predicting trends in considering different design configurations that can affect the flow-field features of vehicles. A CFD simulation also permits investigations that in some circumstances cannot be realistically duplicated in a wind tunnel. Let us consider the 25% scale model of a Ford Falcon EXT 2003, with the model dimensions given in Figure 7.34. The model is considered to be a bluff model that produces the basic flow structures of a pickup truck. Information on the side rearview mirrors, bull bars,

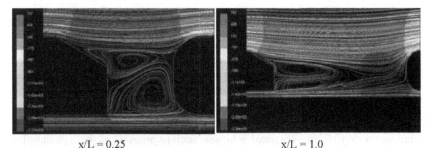

x/L = 0.25   x/L = 1.0

FIGURE 7.33   Pressure pathlines within inter-vehicle spacing.

Model Information

Length = 1.2 m
Height = 0.12 m
Width = 0.432 m
Ground Clearance = 0.06 m

FIGURE 7.34   Pickup truck model information.

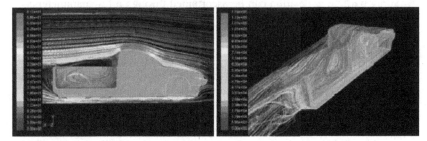

FIGURE 7.35   Recirculation region of open-tub configuration of pickup truck.

engine components, and under-body is not included. The test velocity is 40 m/s with a model-length-based Reynolds number of 3.3×10^6.

The effects on the drag coefficient are analyzed. The rear tub is left open for this case. A drag coefficient of 0.44 is predicted, which shows an immediate increase from a value of 0.38 if the rear tub is closed. With the rear bed exposed, the formation of a large recirculation flow region behind the cab is evident, as illustrated in Figure 7.35. The flow pathlines are colored (shown in shades of gray) according to the static pressure. The highest and lowest pressure are depicted. Along the symmetry plane represented in the same figure, the pressure drop is comparatively high in the vicinity of the cab-bed region.

7.4.4   Air/Particle Flow in the Human Nasal Cavity

An alternative to oral and injection routes for delivery of systemic drugs is nasal drug delivery. The many advantages associated with nasal drug delivery are well documented, and one of them is treatment of respiratory ailments, such as congestion and allergies. One remaining concern about the prescription of such treatment is the availability of useful particle-deposition information for the human nasal cavity, which could significantly improve the effectiveness of a nasal sprayer device in delivering the drug to specifically targeted sites within the nasal cavity.

Current *in vivo* and nasal cavity replica methods are usually limited in scope because of their intrusiveness, as well as because of their implementation's being very time consuming and expensive. Nevertheless, the availability of computers to perform numerical analyses, such as analysis of the repeatability and accuracy of a nasal spray injection released from the same location, with quick turnaround times, and possibly extending to a wider range of investigations, has certainly revolutionized how medical research can be carried out nowadays. Advanced application of CFD is gaining enormous interest, as evidenced in Keyhani et al. (1995), Yu et al. (1998), and Hörschler et al. (2003). With the aid of graphical representations of the local particle deposition sites, particle- and air-flow paths, velocity contours, and vectors at any location, detailed critical assessments can be obtained on the effectiveness of delivery of a particular nasal sprayer device being tested.

Spray particle deposition can be ascertained through a few parameters:

1. Gas-phase flow field, such as velocity and turbulence effects
2. Deposition mechanisms involving the interaction between particles and their continuum
3. Material properties of the particle and initial spray conditions, such as particle density and size and spray-cone angle.

Knowledge of the fluid-flow field allows prediction of particle dispersion and deposition. The construction of the complex nasal cavity geometry needs to be appropriately managed for CFD simulations before calculations can proceed. This has been addressed in detail by Inthavong et al. (2006), and some pertinent practical aspects of the geometry construction are discussed below.

Problem considered: A schematic drawing of the human nasal cavity is shown in Figure 7.36. Constructing the complex geometrical model of the human nasal cavity is a very challenging task. Slices of three-dimensional CT images are taken at various positions from the entrance of the nasal cavity to the anterior of the larynx. The image analysis data obtained consist of *xyz* coordinates of the airway perimeters for cross-sections spaced 1 to 5 mm apart, depending on the topography of the anatomy.

Through the solid modeling tool GAMBIT—the pre-processor facility of the CFD commercial code ANSYS-FLUENT, Version 6.1.22—a computational mesh is generated enveloping the entire anatomical geometry, as illustrated in Figure 7.37. The reader should note that the mesh-generation step is the most labor-intensive component of the entire simulation process because of the requirement to appropriately mesh the intricate surface topology of the airway geometry. A denser mesh distribution is concentrated near the surface boundaries to better resolve the large velocity gradients of the air flow.

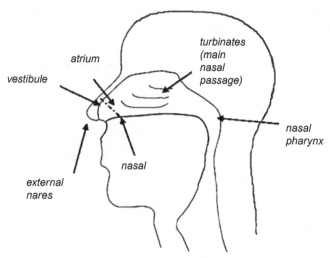

FIGURE 7.36   Schematic drawing of the human nasal cavity.

FIGURE 7.37   Construction of surface topology of the human nasal cavity through CT scan.

A preliminary model is created with a total mesh cell count of 82,000. This coarse mesh provides a basis whereby enhancements are further performed to address grid quality criteria, such as cell skewness, resolution in the vicinity of the cavity wall, and cell-to-cell volume change. Three further meshes are generated—286,000 cells, 586,000 cells, and 822,000 cells —to check for grid

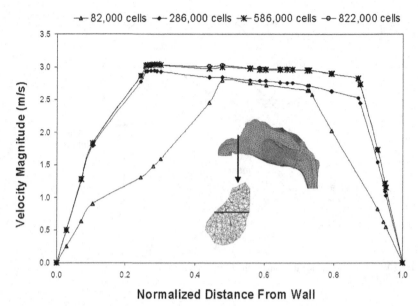

FIGURE 7.38   Velocity profiles of a coronal section near the nasal valve region for the four different nasal cavity models.

independence. The reader should be aware that accuracy of computed results increases with finer mesh but at higher computational costs. Carrying out a grid-independence test allows an optimal mesh grid size to be ascertained. The Navier–Stokes equations for the gas phase are solved at a flow rate of 20 l/min. Figure 7.38 shows the velocity profiles for different mesh densities. It is observed that a mesh of 822,000 cells exhibits no further improvement from a much less dense mesh of 586,000 cells. The nasal cavity model with 586,000 cells is thus accepted to be sufficient for current simulation purposes and is used for generating the results below.

   CFD simulation: The solution of the flow field representing the continuum gas phase under steady-state conditions is achieved by solving the full Navier–Stokes equations. These equations are discretized using the finite-volume approach. The *QUICK* scheme is used to approximate the momentum equation, while the pressure–velocity coupling is realized through the SIMPLE method. According to Tu et al. (2004a), the flow velocity inside the human nasal cavity can reach as high as 22 m/s. This suggests that the flow is turbulent, which is also confirmed by Keyhani et al. (1995) and Yu et al. (1998), where the k-ε turbulence model has been used for flows around 200–300 ml/s. This two-equation approach is the most commonly used turbulence model, based on the assumption of isotropic turbulence and a single eddy viscosity for all three components of the velocity vector. This can nonetheless cause inaccuracies for flows with high swirling action where the turbulence is apparently anisotropic. For flows in the nasal cavity, which exhibit low levels of swirl, the k-ε approximation should be sufficient, given its low computational costs compared with more sophisticated approaches, such as large eddy simulations and the Reynolds stress model. In this example, the realizable k-ε model proposed by Shih et al. (1995) is adopted. For the particle dispersion, a Lagrangian particle-tracking method is used to track the migration of particles within the nasal cavity.

The human nasal cavity is subjected to constant air-flow rates of 20–40 l/min. The internal walls are modeled using an enhanced wall-treatment function to consider the no-slip condition on the air flow. The particles are assumed to be spherical water droplets, since most drug formulations can be considered to be diluted with water. Initial particle conditions are taken from analytical methods due to lack of experimental data. The conditions for the release of particles into the constant flow rate differ by the parameter under investigation and are elucidated later. The internal walls of the nasal cavity are set for a "trap" condition, meaning that particles that touch a wall deposit at that location.

Appropriate boundary conditions for the particles are specifically chosen to simulate a number of realistic situations. Physically speaking, the initial particle velocity can be controlled in many ways, such as changing the nozzle diameter and the actuation mechanism. The full spray cone angle, β, characterizes the internal atomizer type of the spray device. A small β generally refers to a narrow spray, which is typical of plain orifice atomizers. However, a larger β depicts a wider spray, which belongs to the family of pressure swirl atomizers. This angle is analyzed to simulate the dispersion of particles exiting from the nozzle tip of a nasal spray device to observe the physical differences with changing β. Monodispersed particles are released and the average deposition within the left and right nasal cavity is recorded. Particles are released at 10 m/s from an internally fixed location with a diameter of 0.8 mm and a range of β between $20°$ and $80°$.

CFD results: The air-flow characteristic within the human nasal cavity with a total air-flow rate of 20 l/min through both nostrils is illustrated in Figure 7.39. The air flow increases at the nasal valve, where the cross-sectional area is known to be smallest, and reaches a maximum area-averaged velocity of 2.28 m/s from an initial uniform inlet velocity $u_{in} = 2.1$ m/s. The velocity subsequently decreases as the nasal cavity expands. It can be observed that the bulk of the air flow remains within the middle and lower regions of the nasal cavity, close to the septal walls, rather than diverging outward. Recirculation appears in the expanding region near the top of the olfactory region. At the nasal pharynx, the velocity increases due to the smaller flow area. Such complex flow patterns for the single-phase air flow have been confirmed through a number of prominent numerical studies, as evidenced in Zwartz and Guilmette (2001) and Hörschler et al. (2003). The reader may wish to refer to these important works as supplements.

For study of the different full spray-cone angle β affecting particle distribution, the smaller range of particles, which follow the gas-phase velocity more

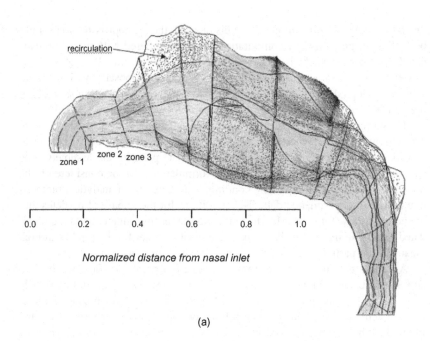

(a)

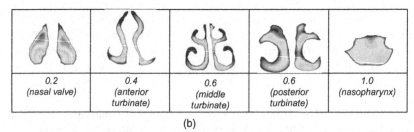

(b)

FIGURE 7.39 Velocity distribution within the human nasal cavity: (a) vector field in the horizontal plane and (b) velocity contours at different cross-sections.

readily, are optimized when released with a narrow β, rather than a wider β, because the latter gives rise to a larger range of dispersion of particles due to the nature of a 360° spray cone. The larger dispersion creates a low ratio of favorably dispersed particles (those pointing with the flow) to those being dispersed away from the curvature (i.e., in the opposite direction) of the gas flow. The effect becomes more apparent as the particle size increases. Figure 7.40 shows the flow for 15 μm being centralized when $\beta = 20°$ and an increase in deviation from the center when $\beta = 80°$. Deposition for $\beta = 20°$ remains along the roof of the nasal cavity and near the septal walls, with 28% deposit in the first two zones. At $\beta = 80°$, deposition in the two frontal zones has increased to 47%. Note that the color spectrum (shown in shades of gray) in Figure 7.40 and Figure 7.41 of the particle flow trajectories indicates the velocity magnitude.

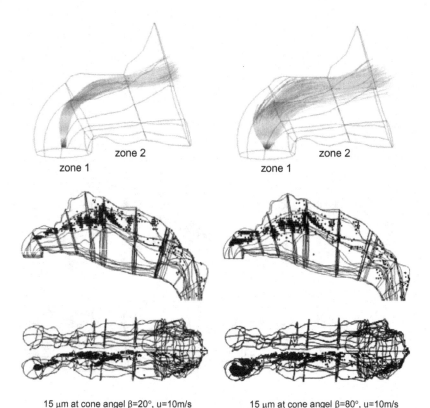

15 µm at cone angel β=20°, u=10m/s   15 µm at cone angel β=80°, u=10m/s

FIGURE 7.40 Deposition patterns for 15-µm particles released at 10 m/s from a small internal diameter at 20° and 80° spray-cone angles.

Figure 7.41 compares the two deposition patterns for particles of 20 µm at β = 20° and β = 80°. A number of parametric investigations have revealed that 20-µm particles result in almost 100% deposition in the two frontal zones. By increasing β, more particles are dispersed favorably by projecting the particles into the already curved streamlines, thereby allowing them to travel further downstream along the nasal cavity. At β = 20°, impaction occurs at the roof of the vestibule directly above the injection release point in a concentrated area. When β = 80°, a wider area of deposition is observed in the frontal zones where the particles projected favorably toward the nasal valve are able to travel beyond the 90° bend. However, deposition eventually occurs within the middle sections of the nasal cavity due to the higher particle inertia.

7.4.5 High-Speed Flows

All the examples considered thus far have involved the application of CFD only to *subsonic* flows—flows below the velocity of sound. The velocity of sound, or sonic velocity, is an important consideration in fluid mechanics.

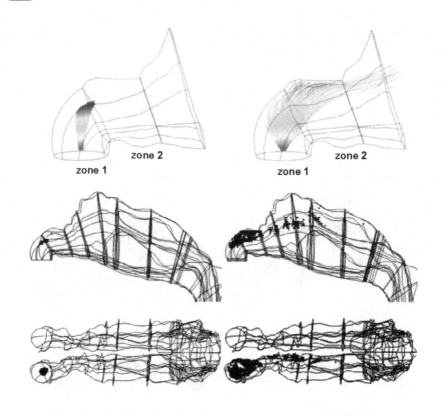

20 μm at cone angel β = 20°, u = 10 m/s   20 μm at cone angel β = 80°, u = 10 m/s

FIGURE 7.41   Deposition patterns for 20-μm particles released at 10 m/s from a small internal diameter at 20° and 80° spray-cone angles.

At *subsonic* velocities, small pressure waves can be propagated from both upstream and downstream. Nevertheless, when the velocity of fluid exceeds the sonic velocity, the fluid flow becomes *supersonic* and small pressure waves cannot be propagated upstream. Many numerical investigations in the aerodynamic and aerospace fields, the genesis of most CFD methods and techniques, involve *supersonic* fluid flows. The ratio of the fluid velocity to the sonic velocity is generally known as the dimensionless Mach number (*Ma*). If the Mach number is > 1, flow is *supersonic*; if the Mach number is < 1, flow is *subsonic*.

With the aim of providing guidelines similar to those provided for the examples of subsonic flow above, two examples of supersonic flows (flow over a simple flat plate geometry and flow over a complex NACA0012 wing configuration) are given in this section. Salient aspects, particularly in obtaining practical solutions for high-speed flows, are described.

7.4.5.1   Supersonic Flow over a Flat Plate

Supersonic flow over a flat plate is a classical *boundary-layer* fluid dynamics problem. Despite its simplicity, an exact analytical solution for the fluid flow over this simple geometry remains *absent*. Consider the supersonic flow over a thin, sharp, flat plate at zero incidence and length L, as sketched in Figure 7.42. As the free-stream fluid approaches the flat plate, a boundary layer develops at the leading edge of the flat plate. It is noted that the boundary layer is taken as the region of fluid close to the surface immersed in the flowing fluid. Away from the leading edge, the free-stream fluid no longer "views" the sharp flat plate; instead, a fictitious curvature is formed due to the presence of the viscous boundary layer. If the Reynolds numbers are based on the distance from the leading edge of the plate, it can be appreciated that, initially, the value is low, so the fluid flow close to the surface may be categorized as *laminar*. However, as the distance from the leading edge increases, so does the Reynolds number, until a point must be reached where the flow regime becomes *turbulent*. In practice, the transition does not occur at one well-defined point, but a transition zone is established between the laminar and turbulent flow regimes, as was shown in Figure 7.42. As also illustrated in the same figure, a curve-induced shock wave is generated at the leading edge and the *shock layer* encapsulates the inviscid and viscous flows. In addition, *viscous dissipation* (dissipation of kinetic energy within the viscous flow) can cause high flow-field temperatures; high heat-transfer rates are thus attained within the boundary layer.

Problem considered: In this particular example, the reader will gain an understanding of the flow physics through solving the complete Navier–Stokes equations. The flow over a flat plate is arguably the easiest application and it is packed with interesting fluid phenomena. This example consists of two parts.

First, we begin by simply solving a laminar flow in two dimensions. Of course, this means that the length of the plate needs to be extremely small

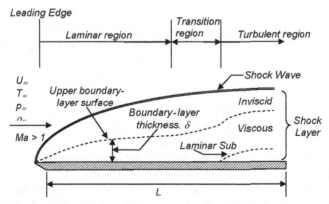

FIGURE 7.42   Illustration of supersonic flow over a sharp leading edge flat plate at zero incidence and development of boundary layer along the plate.

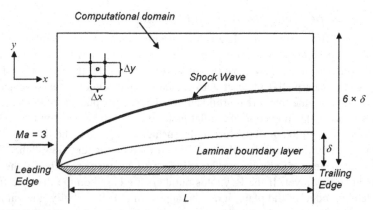

FIGURE 7.43   Illustration of two-dimensional computational domain.

for the Reynolds number to be low. More important, the main aim of this problem is to demonstrate the practical use of a commercial computer code to sufficiently capture the desired flow physics. Figure 7.43 depicts the size of the computational domain for the numerical calculations of $Ma = 3$ over the flat plate. The length of the plate has been chosen to be $L = 2.85 \times 10^{-5}$ m, which yields a Reynolds number of around 2000 based on the gas properties at reference sea-level values. A uniform-structure mesh of rectangular elements totaling 80×80 cells overlies the computational domain. The step size in the x direction (Δx) is 3.5625×10^{-7} m (0.000285/80). In the y direction, it is imperative that the height of the domain encapsulate the shock wave in order to obtain an accurate description of the fluid flow. As predicted by a Balsius calculation at the trailing edge, it is reasonable to assume that the computational domain is at least six times the height of the boundary layer δ, where δ is given by

$$\delta = \frac{5L}{\sqrt{Re_L}} = 3.186 \times 10^{-6} \text{ m} \tag{7.17}$$

The step size in the y direction (Δy) is thus 2.3898×10^{-7} m ($6 \times 3.186 \times 10^{-6}/80$). As a consequence of the stronger gradients in the direction normal to the plate, it makes perfect sense that the step size in the y direction is smaller than that in the x direction in order to better capture the flow field, especially near the flat plate surface.

Second, with the current availability of improved computational speeds and resources, such as the advent of computer clusters, employing direct numerical simulation techniques (see Chapter 6) is within reach for predicting the onset of the transition to turbulence in the fluid flow over this simple geometry. Here, the supersonic flow at $Ma = 4.5$ is solved. The three-dimensional complete Navier–Stokes equations are solved to fully describe the laminar–turbulent transition of the supersonic boundary layer, as shown by the schematic drawing in Figure 7.44. Based on the local boundary-layer length scale $l_s = 3.0486 \times 10^{-4}$

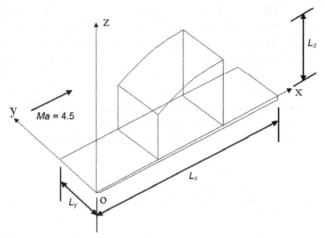

FIGURE 7.44   Illustration of three-dimensional computational domain.

TABLE 7.5 Parameters employed for each computational box

Box no.	Re	$N_x \times N_y \times N_z$	Points/Wavelength
1	400~1800	864 × 16 × 132	60
2	1600~2146	864 × 32 × 132	90
	1400~2099	1024 × 32 × 160	96
3 (Transition)	1968~2540	1080 × 48 × 160	96
	1968~2649	1590 × 102 × 160	112
	1968~2649	1590 × 256 × 160	112

Note: The range of Reynolds numbers evaluated above is based on the local boundary-layer length scale l_s. N_x, N_y, and N_z represent the total number of grid nodal points along, the x, y, and z directions.

m, the non-dimensional dimensions of the computational domain are, respectively, $L_x/l_s = 3259$, $L_y/l_s = 65.6$, and $L_z/l_s = 438 \sim 587$. Owing to the increasing local resolution requirements for the streamwise evolution of the non-linear instability wave, the direct numerical simulation is performed in multiple stages involving overlapping sub-domains or boxes (see details in Table 7.5) to better manage the highly demanding computational resources. (Note the requirement for the increasing number of grid nodal points to resolve the wave-propagation characteristics in the last column of the table at high Reynolds numbers). The last box contains a total of approximately 65 million grid nodal points, a resolution that is deemed to be sufficient in capturing the structure of a fully developed turbulent flow. More details on the simulations carried out for the flow on this particular geometry can be found in Jiang et al. (2006).

CFD simulation: The advantage of employing the complete Navier–Stokes equations extends to investigations that can be carried out on a wide range of flight conditions and geometries, and, in the process, the location of the shock wave as well as the physical characteristics of the shock layer can be precisely determined. We begin by describing the three-dimensional forms of the Navier–Stokes equations below. (Note that the two-dimensional forms are just simplifications achieved by the omission of the component variables in one of the coordinate directions.) Neglecting the presence of body forces and volumetric heating, the three-dimensional Navier-Stokes equations are derived as

Continuity:

$$\frac{\partial \rho}{\partial t} + \frac{\partial (\rho u)}{\partial x} + \frac{\partial (\rho v)}{\partial y} + \frac{\partial (\rho w)}{\partial z} = 0 \tag{7.18}$$

x-momentum:

$$\frac{\partial (\rho u)}{\partial t} + \frac{\partial (\rho u u)}{\partial x} + \frac{\partial (\rho v u)}{\partial y} + \frac{\partial (\rho w u)}{\partial z} = \frac{\partial \sigma_{xx}}{\partial x} + \frac{\partial \tau_{yx}}{\partial y} + \frac{\partial \tau_{zx}}{\partial z} \tag{7.19}$$

y-momentum:

$$\frac{\partial (\rho v)}{\partial t} + \frac{\partial (\rho u v)}{\partial x} + \frac{\partial (\rho v v)}{\partial y} + \frac{\partial (\rho w v)}{\partial z} = \frac{\partial \tau_{xy}}{\partial x} + \frac{\partial \sigma_{yy}}{\partial y} + \frac{\partial \tau_{zy}}{\partial z} \tag{7.20}$$

z-momentum:

$$\frac{\partial (\rho w)}{\partial t} + \frac{\partial (\rho u w)}{\partial x} + \frac{\partial (\rho v w)}{\partial y} + \frac{\partial (\rho w w)}{\partial z} = \frac{\partial \tau_{xz}}{\partial x} + \frac{\partial \tau_{yz}}{\partial y} + \frac{\partial \sigma_{zz}}{\partial z} \tag{7.21}$$

Energy:

$$\frac{\partial (\rho E)}{\partial t} + \frac{\partial (\rho u E)}{\partial x} + \frac{\partial (\rho v E)}{\partial y} + \frac{\partial (\rho w E)}{\partial z} = \frac{\partial \left(u \sigma_{xx} + v \tau_{xy} + w \tau_{xz} \right)}{\partial x}$$

$$+ \frac{\partial \left(u \tau_{yx} + v \sigma_{yy} + w \tau_{yz} \right)}{\partial y} + \frac{\partial \left(u \tau_{zx} + v \tau_{zy} + w \sigma_{zz} \right)}{\partial z} \tag{7.22}$$

$$+ \frac{\partial}{\partial x} \left(k \frac{\partial T}{\partial x} \right) + \frac{\partial}{\partial y} \left(k \frac{\partial T}{\partial y} \right) + \frac{\partial}{\partial z} \left(k \frac{\partial T}{\partial z} \right)$$

Assuming a Newtonian fluid, the normal stresses σ_{xx}, σ_{yy}, and σ_{zz} can be taken as combinations of the pressure p and the normal viscous stress components τ_{xx}, τ_{yy}, and τ_{zz} while the remaining components are the tangential viscous stress components whereby $\tau_{xy} = \tau_{yx}$, $\tau_{xz} = \tau_{zx}$, and $\tau_{yz} = \tau_{zy}$. Expressions for the normal and shear stresses can be found in Appendix A. For the energy conservation for *supersonic* flows, the specific energy E is solved, instead of the usual thermal energy H applied in *subsonic* flow problems. In three dimensions, the specific energy E (see Eq. A.26 in Appendix A) is repeated below for convenience:

$$E = \underbrace{e}_{\text{internal energy}} + \underbrace{\frac{1}{2}\left(u^2 + v^2 + w^2\right)}_{\text{kinetic energy}} \qquad (7.23)$$

It is evident from above that the kinetic energy term contributes greatly to the conservation of energy because of the high velocities that can be attained for flows where $Ma > 1$. Equations (7.18)–(7.23) represent the form of governing equations that are adopted for *compressible flows*.

The solution to the governing equations nonetheless requires additional equations to close the system. First, the equation of state for the assumption of a perfect gas is employed, i.e.,

$$p = \rho RT$$

where R is the gas constant. Second, assuming that the air is calorically perfect, the following relation holds for the internal energy:

$$e = C_v T$$

where C_v is the specific heat of constant volume. Third, if the Prandtl number is assumed to be constant, and which is approximately 0.71 for calorically perfect air, the thermal conductivity can be evaluated by the following:

$$k = \frac{\mu C_p}{Pr}$$

Sutherland's law is typically used to evaluate the viscosity μ, which is provided by

$$\mu = \mu_0 \left(\frac{T}{T_0}\right)^{1.5} \frac{T_0 + 120}{T + 120} \qquad (7.24)$$

where μ_0 and T_0 are reference values at standard sea-level conditions.

Figure 7.45 illustrates the boundary conditions for the high-speed laminar fluid flow in two dimensions. At the left-hand side and upper boundaries of the computational domain, Dirichlet boundary conditions on the velocity, pressure, and temperature are imposed at their respective free-stream values. At the right-hand side, it is imperative to note that the outflow condition in supersonic flows is not influenced by downstream conditions and thus differs from the usual Neumann boundary condition in subsonic flows whereby all properties are now calculated based on an extrapolation from upstream in contrast to imposing the *zero* normal gradient constraint. Finally, the no-slip condition is prescribed for all the velocity components ($u = v = 0$). One of the most significant advantages of CFD is the ability to conduct numerical experiments in order to gain an understanding of the implications of changing a flow parameter. The interaction of a constant wall temperature and an adiabatic wall will be assessed in the numerical calculations. For the constant-temperature

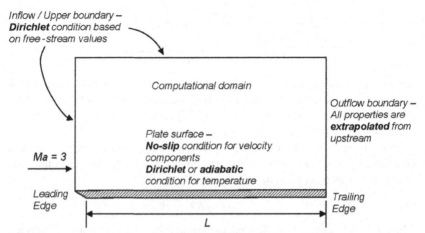

FIGURE 7.45   Application of boundary condition for the two-dimensional laminar supersonic flow.

wall boundary condition, it is assumed that the temperature is prescribed at the free-stream value.

One of the many challenging aspects of direct numerical simulation is the requirement to impose appropriate flow conditions at the inflow boundary. To represent the actual characteristics of flow through this boundary, different single pairs of oblique first-mode disturbances have been introduced to parametrically study the sensitivity of the chaotic inflow conditions entering the computational domain influencing the internal flow. For the outflow boundary, all properties are extrapolated as for the two-dimensional case. For the side boundaries encasing the flow along the spanwise *y* direction (see Figure 7.44), *periodic* boundary conditions, such as those exemplified for the compact heat exchanger problem above, are employed. At the upper boundary, the *free-slip* condition is imposed, while the usual no-slip condition is prescribed with an adiabatic condition for the temperature at the surface of the plate.

CFD results: The relevant data, such as the free-stream conditions and several thermodynamic constants for the laminar boundary-layer simulation, were detailed earlier in Table 7.4. For the particular case considered, the commercial code ANSYS-CFX, Version 10, is used to obtain all the numerical results of the laminar flow condition.

Figure 7.46 shows the steady-state temperature contours and velocity vectors for the supersonic flow over the flat plate. Note that the range of colors (shown in shades of gray) depicted by the velocity vectors has been chosen to correspond to the temperature distribution. For both the adiabatic and constant-wall-temperature cases, the distinctive regions occupied by the inviscid and viscous flows are clearly identified within the shock layer by the CFD methodology. The profile of the curve-induced shock wave separating the silent zone (upstream free stream) from the shock layer is also well established.

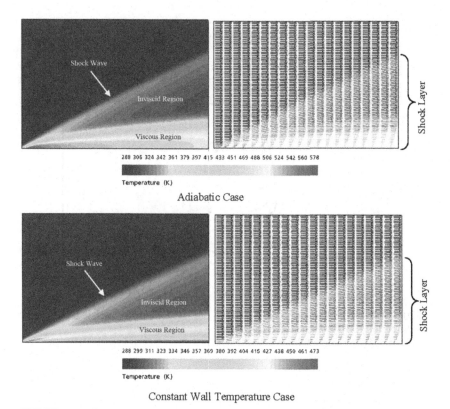

Adiabatic Case

Constant Wall Temperature Case

FIGURE 7.46   Temperature contours and velocity vectors describing the distinct regions of the inviscid and viscous flows within the shock layer for the adiabatic and constant-wall-temperature conditions for $Ma = 3$.

An adiabatic wall is seen to increase the boundary-layer temperature by a substantial margin above the constant wall. This is primarily due to the relatively lower density, which leads to a thicker boundary layer and consequently a broader shock layer, as illustrated in Figure 7.46.

The normalized temperature and u component velocity profiles at the trailing edge are plotted in Figures 7.47 and 7.48 to further confirm the aforementioned observation. In plotting the profiles, the normalized y distance is adopted as suggested by Van Driest (1952), i.e., $\bar{y} = y\sqrt{Re}/x$. The normalized temperature (T/T_∞) profiles adequately capture the leading edge shock wave as well as the classical boundary-layer behavior near the plate surface. Since the temperature gradient is zero at the wall for the adiabatic case, temperatures within the thermal layer are expected to be higher than those of the constant-wall-temperature case. Essentially, a colder wall temperature suppresses the boundary-layer flow. Similarly, a closer examination of the normalized velocity (U/U_∞) profiles near the surface, as exemplified in Figure 7.48, also shows a thicker boundary layer for the adiabatic case. The local Mach number at the

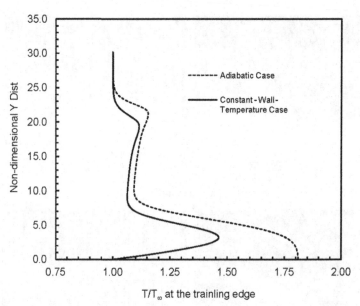

FIGURE 7.47 Normalized temperature profiles for the adiabatic and constant-wall-temperature conditions for $Ma = 3$ at the trailing edge.

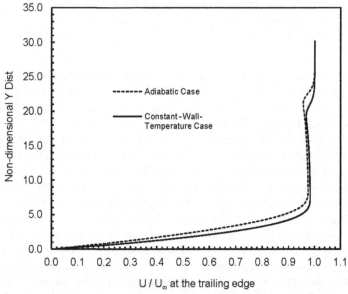

FIGURE 7.48 Normalized u component velocity profiles for the adiabatic and constant-wall-temperature conditions for $Ma = 3$ at the trailing edge.

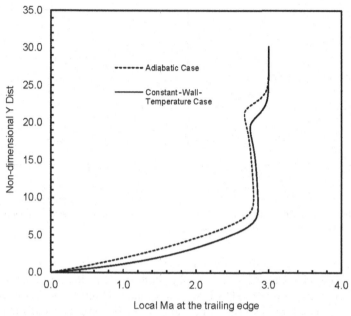

FIGURE 7.49   Local Mach number profiles for the adiabatic and constant-wall-temperature conditions for $Ma = 3$ at the trailing edge.

trailing edge is graphically presented in Figure 7.49, in which the relative strength of the two leading edge shock waves clearly illustrates a stronger shock wave migrating across the flat plate when the adiabatic condition is imposed at the surface of the plate.

Nowadays, increasing computational capabilities offer immense possibilities for actually revisiting a number of classical CFD problems, such as the boundary-layer problem typified in this example, through the use of DNS techniques. Particularly, the laminar flow investigation described above can be further extended to study the onset of flow instability and to better understand the transition from laminar flow to turbulent flow. For this flow problem, an in-house DNS computer code developed at the University of Texas at Arlington is employed to predict the non-linear evolution of instability waves and the onset of breakdown to turbulence in a Mach 4.5 flat-plate boundary-layer flow after the occurrence of shock. To verify the DNS predictions, they are also subjected to intense scrutiny against the Parabolic Stability Equation (PSE) calculations of NASA Langley, work based on the investigation performed by Jiang et al. (2006). For the remainder of this section, results extracted from work on the laminar-turbulent transition investigation are presented and discussed.

One of the many challenges in DNS is to prescribe suitable inflow boundary conditions. Since DNS is a fully deterministic approach where all spatial and temporal scales are required to be computed, the specification of appropriate

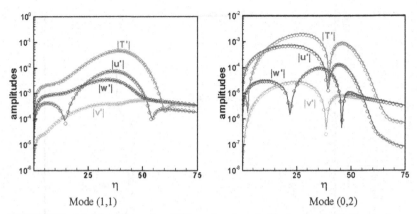

FIGURE 7.50 Modal profiles at $Re = 1500$ as a function of η (the local scaled wall-normal coordinate) subjected to two disturbance modes.

inflow conditions is of paramount importance. The comparison of wave-propagation characteristics prior to the onset of transition based on various disturbance modes predicted by DNS is illustrated in Figure 7.50. In this figure, the DNS modal profiles of the fluctuating velocity components and temperature subject to different disturbance modes as a function of the local scaled wall-normal coordinate are shown to be in excellent agreement with the results determined through the PSE calculations. For each of the disturbance modes affecting the numerical calculations, a successful cross-validation of the DNS solutions for the linear and early non-linear stage of disturbance growth is clearly demonstrated, which strongly suggests the independent influence of the initial boundary conditions on the onset of laminar–turbulent transition.

It is noted that the prediction of the onset of breakdown to turbulence after the non-linear interaction is beyond the capability of PSE. However, since DNS solves the full unsteady Navier–Stokes equations without any *ad hoc* assumptions, the present methodology is able to provide solutions across a broad band of fluid dynamics, ranging from the laminar state to the early and later stages of transition as well as even to capture the manifestation of chaotic structures in a fully developed turbulent flow. To gain an insight into the complex flow structures occurring along the flat plate, the evolution of the contours for the density fluctuation and wall-normal vorticity along an increasing distance along the flat plate is illustrated in Figures 7.51 and 7.52. As observed from these two figures, the flow remains laminar for $x < 4600$. Flow instability or transition nonetheless begins to creep in at about $x = 4800$, with the increasing undulation persisting over the flow patterns after this point. Shortly after some distance downstream at $x = 5400$, the onset of breakdown of the flow to turbulence prevails and the fluid flow remains turbulent thereafter by the establishment of substantial density fluctuations within the fluid and vigorous coherent structures of wall-normal vorticities at $x = 6800$.

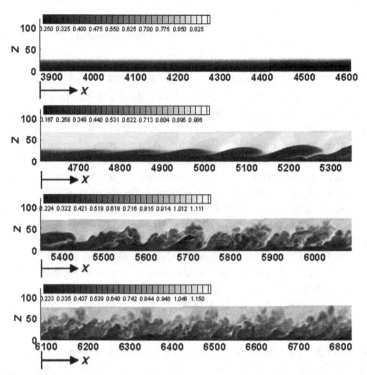

FIGURE 7.51 Contours of density fluctuation in the *x-z* plane midway along the spanwise direction.

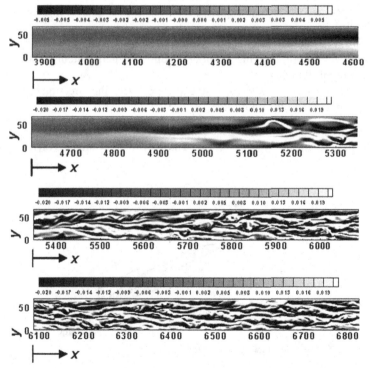

FIGURE 7.52 Contours of wall-normal vorticity in the *x-y* plane at a vertical non-dimensional distance of $z^+ = 12.21$.

7.4.5.2 Subsonic and Supersonic Flows over a Wing

An airfoil can be defined as a streamlined body designed specifically to produce lift. It will also experience the counteracting influence of drag while placed in a fluid stream. The primary purpose for the construction of a streamlined airfoil is to minimize the drag imposed on the body. As a measure of an airfoil's usefulness, for example, as a wing section of an aircraft, the ratio of lift to drag must be sufficiently large that it is capable of producing high lift at a small penalty of drag. For an aircraft to remain in the airspace, the creation of lift on the wing surface is of paramount importance. Nevertheless, what is necessary to move the craft forward is the drag that absorbs the engine power.

Problem considered: In order to illustrate CFD application to high-speed flows, the fluid flowing past an NACA0012 airfoil is considered herein. This particular geometry has been specifically chosen because of the numerous aerodynamic investigative studies that have been carried out in research and design practices. There are a number of accepted terms related to an airfoil; familiarization with them is necessary in order to understand the discussion of the flow past such geometry.

Figure 7.53 is a schematic drawing of airfoil geometry. Some terms relating to an airfoil cross-section that are used throughout the discussion are

- *Leading edge*—the front, or upstream, edge, facing the direction of flow
- *Trailing edge*—the rear, or downstream, edge
- *Chord line*—a straight line linking the centers of curvature of the leading and trailing edges
- *Chord, c*—the length of chord line between the leading and trailing edges
- *Span, b*—the length of the airfoil in the direction perpendicular to the cross-section of the wing
- *Angle of attack (incidence), α*—the angle between the direction of the relative motion and the chord line

The CFD example in this section considers the subsonic and supersonic flows past an infinitely long airfoil. The Reynolds-averaged Navier–Stokes

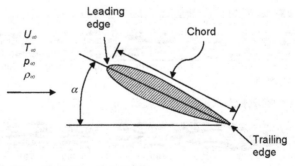

FIGURE 7.53 Schematic drawing of an airfoil.

(RANS) equations are solved along with the shear stress transport (SST) turbulence model for supersonic flow ($Ma = 2.5$) over the wing geometry. Because only time-averaged results are of primary interest, especially in RANS simulation, and since the length b is infinite, such conditions or assumptions mean that the flow is truly two-dimensional; there is negligible spanwise variation of flow patterns and forces for a constant-chord airfoil. Numerical calculations are thus carried out in a two-dimensional fluid domain, and the calculations also include parametric investigations of the influence of different angles of attack on the expanding shock layers as the fluid travels past the airfoil. At flows that are just below the speed of sound ($Ma = 0.2$ is considered for the present problem), currently available computational hardware permits the use of DNS techniques to study the onset of flow instability and subsequent transition to turbulence subject to different inlet conditions. Here, the complete three-dimensional Navier–Stokes equations are solved to accommodate the spectrum of varying length scales that exists within the complicated fluid phenomena along the streamwise (x), spanwise (y), and vertical (z) directions.

CFD simulation: One challenging aspect of the CFD example of fluid flowing over an airfoil is the generation of appropriate meshes surrounding the geometry. Two different grid topologies that could be specifically used to solve the problem are illustrated in Figures 7.54 and 7.55. One approach that can be suitably considered, especially for RANS simulations, is the block-structured or multi-block mesh (see Chapter 6) shown in Figure 7.54. The entire fluid region is subdivided into six contiguous blocks. This approach offers immense flexibility in generating different grids in each block, particularly the increased mesh

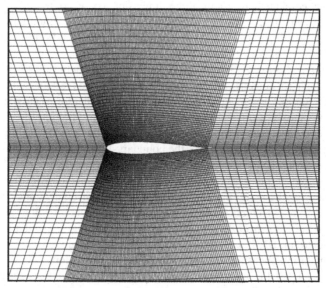

FIGURE 7.54   A multi-block approach for the mesh surrounding the airfoil for RANS calculations.

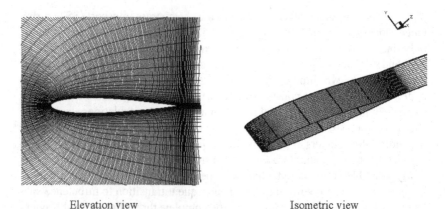

Elevation view   Isometric view

FIGURE 7.55   A C-type mesh scheme adopted for DNS.

densities in blocks 2 and 5 capturing the developing boundary layer along the chord line between the leading and trailing edges and the expanding shock layers along the airfoil. From the viewpoint of practicality and general application, the attractiveness of multi-block meshing is the ease of porting the mesh information into any commercial computer package. Alternatively, it is also possible to employ a single-block mesh to fit the entire fluid domain. The three-dimensional structured C-grid type encapsulating the airfoil seen in Figure 7.55 typifies such grid topology, and it has been purposefully employed for the DNS calculations. Numerical solutions are obtained in a mesh totaling 22,000 rectangular elements for the two-dimensional case and 1200 (streamwise) × 32 (spanwise) × 180 (vertical) hexahedral elements for the three-dimensional case. For the latter, the large overlay of elemental volumes ensures that the mesh distribution based on the non-dimensional spacing along the streamwise, spanwise, and vertical directions is below $\Delta x^+ < 13$, $\Delta y^+ < 15$ and $\Delta z^+ < 1$.

The *full compressible* Navier–Stokes equations, namely Eqs. (7.18)–(7.23), are applicable for consideration of both the subsonic and the supersonic flows in this CFD example. (Note again that the two-dimensional forms are obtained by simplifying the three-dimensional governing equations; the simplification is the omission of the component variables in one of the coordinate directions.) Boundary conditions like those for the boundary-layer flow over a sharp-edged flat plate in the previous example can also be employed in the present CFD example. In other words, they are the *no-slip* condition for the velocity components on the wing surface, the *free-slip* condition on the upper and lower boundaries, the *Dirichlet* condition based on the free-stream values at the inlet boundary, and an *extrapolation* of the upstream values at the outlet boundary. With regard to DNS, *periodic* boundary conditions are employed for the side boundaries encasing the three-dimensional flow along the spanwise y direction. For all the numerical results attained, *adiabaticity* is assumed for the wing surface.

TABLE 7.6 Data used for the laminar flow case

Quantity	Dimension
Plate length L, m	0.0000285
Free stream air properties at sea level	
Speed of sound U_∞, m/s	340.28
Pressure p_∞, N/m^2	101325.0
Temperature T_∞, K	288.16
Density ρ_∞, kg/m^3	1.225
Dynamic viscosity μ, kg/m · s	1.7894×10^{-5}
Specific gas constant R, J/kg · K	287.0
Specific heat C_p, J/kg · K	1005.4
Prandtl number	0.71
Ratio of specific heats γ (C_p/C_v)	1.4

CFD results: For the RANS simulations, the commercial CFD code ANSYS-CFX, Version 10, is utilized to obtain all the numerical results of the supersonic flow condition. The relevant input data, such as the free-stream conditions and several thermodynamic constants, are the same as those tabulated in Table 7.6.

Figure 7.56 illustrates the contours of the local Mach number and pressure for the fluid flow past the wing geometry based on a free-stream $Ma = 2$ subject to 0° and 5° angles of attack. From these results, the formation of shock waves above and below the wing surface is clearly identifiable as the fluid travels downstream from the leading edge to the trailing edge of the airfoil. Because of complete flow symmetry, it is not entirely surprising in the case of the 0° angle of attack that the growth of the boundary layer is identical at the top and bottom surfaces; one would not expect any pressure gradient to be developed and hence the flow vorticities should be of equal strength and opposite rotation. However, at the 5° angle of attack, it is seen that the fluid flowing over the bottom surface of the airfoil is slowed down, as evidenced by the local Mach number contours. The pressure is significantly increased from the case of the 0° angle of attack at the same flow region, which means that the pressure gradient is favorable, the boundary layer thickness is small, and hence the vorticity in it is also small. The opposite prevails over the top surface, and the boundary layer is thicker, the pressure gradient is adverse, and the vorticity in it is larger. This pressure difference gives rise to an upward resultant force, namely lift, which is of primary importance for an airfoil when it is placed in a fluid stream.

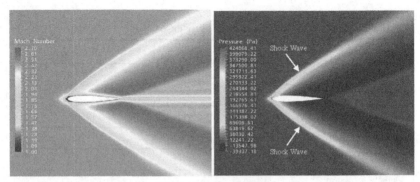

Angle of attack – 0°

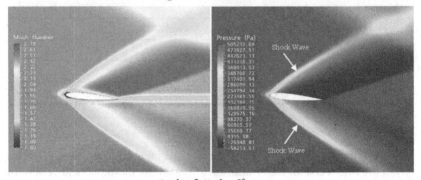

Angle of attack – 5°

FIGURE 7.56 Local Mach number and pressure contours for supersonic flow based on the free-stream $Ma = 2.5$ subject to 0° and 5° angles of attack past the wing geometry.

The foregoing discussion indicates a strong dependence of lift upon the incidence angle. Well-documented observations have shown that there is also a compelling linkage between the creation of lift and the flow characteristic surrounding the airfoil. Figure 7.57 shows the velocity vectors for the two cases of the 0° and 5° angles of attack with the spectrum of colors (shown in shades of gray) associated with the pressure distribution. For these *small* angles of attack, there appears to be no separation of the boundary layer near the trailing edges. Nevertheless, a threshold limit exists whereby any further increase of incidence no longer produces an increase of lift. At this instance, significant flow separation occurs at the top surface and subsequently widens the wake behind the trailing edge of the airfoil. This is known as the *stall* position and it constitutes a critical angle of attack for aerodynamic design since the lift drops rapidly thereafter.

Upstream, the flow away from the leading edge or stagnation point as indicted by "P" on the temperature contours in Figure 7.58 is usually designated as the *silent zone*. Because the flow is supersonic, the disturbance generated at "P" is not communicated to any part of the zone. Further downstream, the zone depicting the significant changes of the fluid within the shock layer is

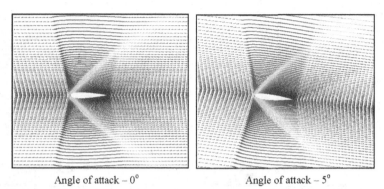

Angle of attack – 0°   Angle of attack – 5°

FIGURE 7.57 Velocity vectors of supersonic flow based on the free-stream $Ma = 2.5$ subject to 0° and 5° angles of attack past the wing geometry.

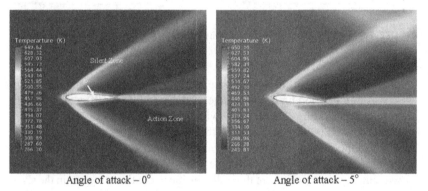

Angle of attack – 0°   Angle of attack – 5°

FIGURE 7.58 Temperature contours of supersonic flow based on the free-stream $Ma = 2.5$ subject to 0° and 5° angles of attack past the wing geometry.

sometimes known as the *action zone*. Here again, the upper and lower temperature distributions are identical at the top and bottom of the wing surface for the case of 0° angle of attack because of complete symmetry. For the case where the angle of attack increases to 5°, the symmetry is broken, registering a temperature gradient across the action zone. Temperatures are significantly higher over the bottom wing surface, while lower than expected temperatures are recorded over the top wing surface when compared to the case of 0° angle of attack.

The remainder of this section presents the non-linear evolution of instability waves and onset of breakdown to turbulence for the subsonic flow over the airfoil based on investigations performed by Shan et al. (2005) and Deng et al. (2007) using DNS techniques. The boundary conditions for the inflow and outflow are non-reflecting conditions based on the analysis of one-dimensional characteristic equations on the time-dependent Euler equations.

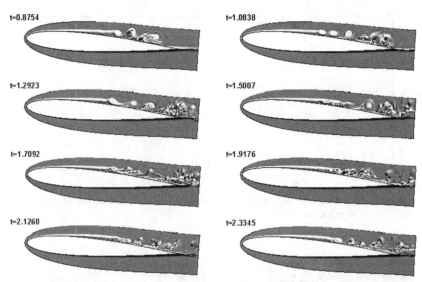

FIGURE 7.59 Instantaneous spanwise vorticity at different times (t is non-dimensionalized by c/U_∞).

Figure 7.59 shows a snapshot of segments representing instantaneous spanwise vorticity developing in the middle of the x-z plane at different times for fluid flow over the airfoil at $Ma = 0.2$ with an angle of attack of $4°$ investigated by Shan et al. (2005). As shown by the instantaneous spanwise vorticity at $t = 0.8794$, the presence of instability waves propagating downstream above the wing surface clearly establishes the onset of flow transition. As time progresses, these waves, which are three-dimensional, continue to grow in the shear layer and they subsequently cause the vortex to break down near the trailing edge, as indicated by the chaotic three-dimensional fluid flow at the later stages of the DNS calculations. In its fully turbulent state, the presence of vortex shedding behind the trailing edge is associated with the rapid growth of the strong non-linear interactions of velocity fluctuations within the wake region.

Isosurfaces of instantaneous vorticity along the x, y, and z directions are further illustrated in Figure 7.60 to gain a better understanding of the flow transition. The appearance of undulating vorticity generated along the spanwise direction suggests that the fluid flow is three-dimensional and also depicts the propagation of instability waves along the top surface of the airfoil. In general, the results clearly show the simultaneous breakdown of the rolled-up shear layer along the chord line and that the vortices are shed from the separated shear layer and become distorted while traveling downstream. More specifically, the shed vortex, which is quickly deformed and stretched, actually conforms to a negative vortex being induced by the prime vortex that is situated alongside the negative vortex, and eventually the prime vortex breaks down into smaller fragments corresponding to the flow transitioning to

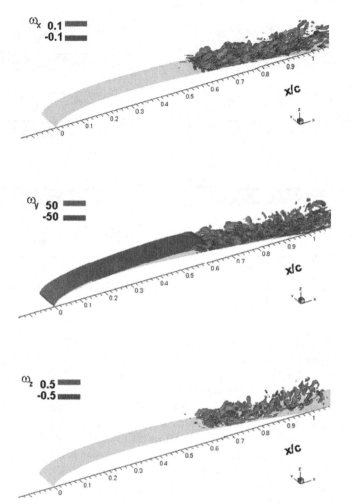

FIGURE 7.60 Isosurfaces of instantaneous vorticity components (from top to bottom: streamwise, spanwise, and vertical components).

turbulence. The interaction between the streamwise and spanwise vortices leads to the development of a λ-shaped vortex, which then rolls up and breaks down. The boundary layer becomes fully turbulent after re-attachment of the flow behind the trailing edge.

Recent results from Deng et al. (2007) in the DNS performed on flow separation control subject to pulsed blowing jets over the airfoil at the same angle of attack are presented below as a case for comparison to the one above. Here again, a snapshot of the instantaneous spanwise vorticity being developed in the middle of the x-z plane at different times is illustrated in Figure 7.61. The unsteady blowing enforced before the separation point with

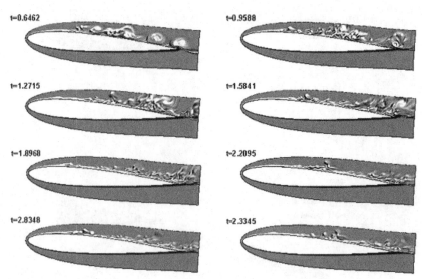

FIGURE 7.61   Instantaneous spanwise vorticity at different times with pulsed blowing jets (t is non-dimensionalized by c/U_∞).

the creation of a large vortex shedding prevails close to the leading edge of the airfoil triggers the early transition of the boundary layer. Iso-surfaces of vorticity along the respective Cartesian directions in Figure 7.62 further exemplify the breakdown of the separated shear layer and the development of the vortex structure. The fluid phenomenon reveals a number of interesting features. From the results obtained, it is clear that the pulsed blowing jets imposed at the inlet significantly affect the re-attachment of the boundary layer shortly after separation. A shorter separation zone is developed, resulting in a much-reduced separation bubble, as evidenced by the vortical structures plotted in Figure 7.62.

7.5   SUMMARY

The examples in this chapter were purposefully selected to demonstrate the feasible application of CFD methodology across a wide range of engineering disciplines and to provide some useful guidelines for handling some of these challenging flow problems in practice.

In the first example, detailed investigation of the appropriate turbulence models for ventilation system design exemplifies how these models can be put to good use as a *design tool* in such areas as architecture and civil or construction engineering.

The ongoing relevance of CFD as a *research tool* is considered in the second example. For gas-particle flows, particles can be treated either as another continuum medium through additional conservation equations by the Eulerian

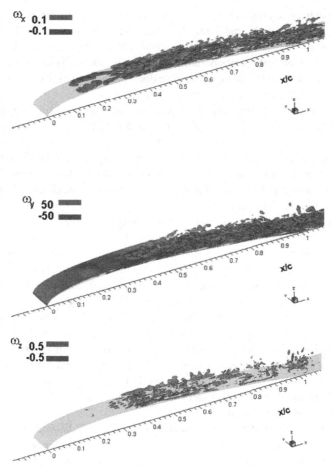

FIGURE 7.62 Isosurfaces of instantaneous vorticity components with pulsed blowing jets (from top to bottom: streamwise, spanwise, and vertical components).

approach or as an ensemble of discrete solid particulates tracked by the Lagrangian approach within the parent-phase fluid conservation equations. These approaches represent the currently adopted state-of-the-art methods in handling such flows. The reader should take note that the so-called multi-phase models, with appropriate constitutive closure relationships, are not restricted to gas-particle or gas–solid flows. They can also be readily applied to resolve the transport of gas–liquid, liquid–solid, or even gas–liquid-solid mixtures.

By the consideration of an additional equation for the conservation of energy coupled with appropriate source terms in the momentum and turbulence equations, fluid flows coupled with *heat transfer* are illustrated through two practical examples. The wealth of information obtained from CFD in analysis of the fluid flow within the in-line or staggered arrangement of a compact heat exchanger allows mechanical engineers to determine the improvements needed to augment the performance of the mechanical system. More important, CFD plays an important role in helping nuclear engineers to address important radiation safety and structural integrity issues pertaining to the exposure of a molybdenum plate with residual nuclear fission heating in an open-air environment (*conjugate heat transfer as well as radiation heat transfer*).

The fifth example presents the practicality of adopting LES instead of the usual RANS in resolving the combustion and radiation processes associated with a free-standing fire. Capturing the fluctuating characteristics of a buoyant fire through LES signifies the advances in knowledge benefiting fire engineers as well as researchers from a combustion and science background, particularly knowledge about the environmental consequences that could result from accidental large-scale fires in oil fields.

Flow over a vehicle platoon, the sixth example, further establishes CFD as an effective design tool in improving the aerodynamic efficiencies of automobiles. The potential of this methodology also presents automotive designers and engineers with improvement strategies to use in working toward the construction of more environmentally friendly vehicles that will consequently have very low fuel consumption and significantly reduced emissions.

Mature development of CFD meshing and numerical models, as demonstrated in the seventh example, represent the state of the art for probing the many complex flows inside a human nasal cavity. CFD is increasingly being considered a viable tool in the field of biomedical engineering, and the application of the methodology is steadily growing, especially with the enormous interest in studying the deformation of vessel walls that affects arterial blood flow inside the human vascular system.

Despite the long-standing application of traditional theoretical boundary-layer-solution techniques in aircraft design in the aerodynamic industries, *supersonic* flows over varying aircraft geometries and flight conditions can now be handled through efficient CFD methods and models. For the two fluid-dynamics examples shown for this engineering discipline, complete two- and three-dimensional solutions using the full Navier-Stokes equations under supersonic conditions can be obtained nowadays with commercial CFD computer packages. In a research-oriented investigation, application of DNS to predict the onset of flow instability and transition to turbulence indicates a future direction for CFD in enhancing understanding of basic fluid dynamics and in all classical physics, especially for the prediction of turbulence.

The windows of opportunity for CFD are indeed plentiful. CFD is aptly being considered as the critical technology for fluid-flow investigation in the 21st century. While relishing the many successful applications of CFD achieved thus far, the reader should also be well informed about some emerging innovative techniques that will continue to revolutionize CFD methodology. These are discussed in the next chapter.

REVIEW QUESTIONS

7.1 Why do engineers prefer to use the standard k-ε turbulence model as a starting point in solving their design problems, instead of using more complex models, such as the LES approach? When would an LES approach be used?

7.2 For the internal pipe flow geometry shown below, formulate appropriate answers to the following questions:

(a) Define and state all the boundary conditions for this problem.

(b) What would be an approximately suitable length for L_1 and L_2 to achieve a fully developed flow at the outlets ($d=0.2$ m) for this case?

(c) Show where a fine mesh would be needed.

(d) From your understanding of fluid mechanics, sketch roughly what the flow streamlines will look like for this flow problem.

(e) Discuss what design ideas may be used to reduce the head loss in this pipe flow and how this can be implemented using CFD.

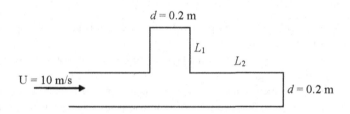

7.3 Explain the *Lagrangian* description of a fluid motion.

7.4 What is the *Eulerian* description of a fluid motion? How does it differ from the *Lagrangian* description?

7.5 A streak of dye is released into an internal flow and its motion is tracked and recorded. Is this a *Lagrangian* or a *Eulerian* measurement?

7.6 A stationary pitot tube is placed in a fluid flow to measure pressure. Is this a *Lagrangian* or a *Eulerian* measurement?

7.7 The flow in a 90° bend is shown below. Formulate appropriate answers to the following questions:

(a) Would a structured or unstructured mesh be better for this geometry? Describe how you would create the mesh.

(b) From your knowledge of the conservation equations of motion, sketch the expected flow streamlines for a flow that has a Reynolds number of 100 and a flow that has a Reynolds number of 10,000.

(c) If a particle with a Stokes number of 0.1 is released in a passive manner from the inlet under laminar conditions, sketch the expected particle trajectory throughout the pipe. How is the trajectory different if the flow is turbulent?

(d) What would the expected particle trajectory be if the particle Stokes number was instead equivalent to 20 for a laminar flow and a turbulent flow?

(e) If you solved this problem in the *Eulerian* reference, what kind of measurements could be made?

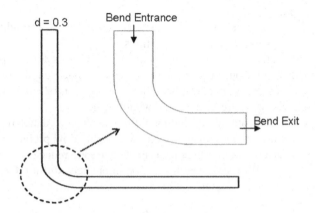

7.8 For heat transfer coupled with fluid flow, formulate appropriate answers to the following questions:

(a) Why is a compact heat exchanger different from a conventional heat exchanger?

(b) Symmetric and cyclic boundary conditions can usually be employed to simplify the flow geometry of a compact heat exchanger. Indicate the boundary conditions for the in-line and staggered arrangements shown below.

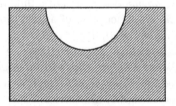

In-line arrangement

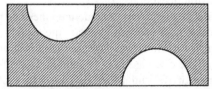

Staggered arrangement

(c) The Nusselt number is a ratio of two fluid properties. What are they?

(d) Explain why higher heat-transfer rates are experienced in a staggered arrangement than in an in-line arrangement.

(e) What is *conjugate heat transfer*?

(f) Why is it important to incorporate the buoyancy effect in natural convection? What additional modeling effort is required for the transport equations?

(g) What is *radiation heat transfer*? How does it affect the surface solid temperature (see section 7.4.1.2)?

7.9 The Ahmed model is often used to model bluff car bodies. The schematic for one model is shown here, along with two cars traveling in line, together with the mesh geometry.

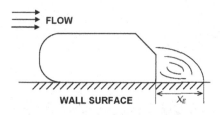

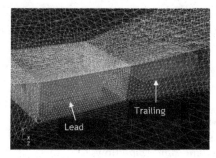

Analyse this problem by answering the following questions:

(a) Explain why there is a fine mesh surrounding the near regions of each car body.

(b) Indicate the regions of high and low pressure encountered around the car bodies.

(c) What is the general effect that the lead car model has on the drag coefficient of the trailing model?

(d) What happens to the drag and lift coefficients if the two models are in close proximity (i.e., $x/L \approx 0.25$, where x is the distance and L is the length of the models)?

(e) When using two different k-ε turbulent models, one model over-predicts while the other under-predicts the turbulent kinetic energy k. Explain the difference of the flow behaviour against the experimental value X_E shown in the figure.

7.10 Two particle-trajectory profiles are shown below for mono-sized particles released from the bottom inlet. The trajectories are colored (shown in shades of gray) by the particle-velocity magnitude. The initial particle velocity is a factor of ten of the air-flow velocity, i.e., $U^* = 10 = U_{particle}/U_{air\ flow}$.

(a) What would be the approximate particle Stokes number for the flow path in a and for the flow path in b?

(b) What are the velocities of the particles in flow path a when they impact the upper walls? Compare them with the velocities of particles in flow path b at the same location.

(c) Explain the difference between particle trajectories obtained for flow path a (straight) and for flow path b (curve). (Hint: compare the velocity profiles.)

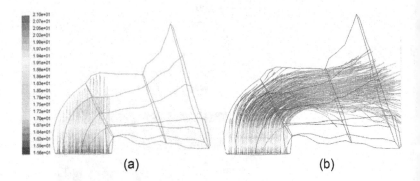

(a)   (b)

7.11 Formulate appropriate answers to the following questions about high-speed flows:

(a) Why is the direct numerical solutions (DNS) technique very useful for studying *supersonic* flows?

(b) How does the outflow condition in *supersonic* flows differ from the usual *Neumann* boundary condition in *subsonic* flows?

(c) For the following supersonic flow over a flat plate, discuss what types of boundary conditions are suitable to capture the flow features.

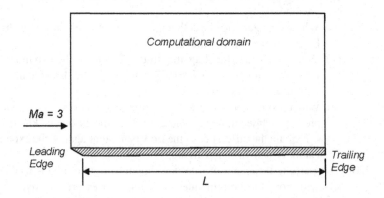

(d) Discuss the effects of using a constant wall temperature instead of an adiabatic wall in terms of the boundary-layer development and its effect on the temperature profile in *supersonic* flows.

(e) Discuss the advantages and disadvantages of using DNS methods over theoretical approaches, such as the Parabolic Stability Equation (PSE) calculations.

(f) For the following flow over an airfoil, discuss why the following grid is necessary.

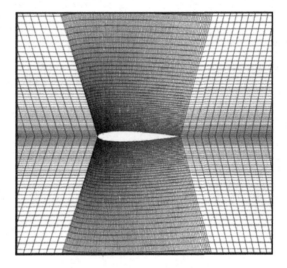

(g) When the angle of attack for the NACA0012 airfoil is at 5°, what happens to the pressure over the bottom surface and how does this affect the temperature? (See section 7.4.5.2.)

Chapter 8

Some Advanced Topics in CFD

8.1 INTRODUCTION

Over the last three decades, considerable progress has been made in the development of CFD. In many areas of applications, the field of CFD is attaining a stage of maturity where most of the basic methodologies are well established and many have been implemented in a number of commercially available computer packages. Despite many significant achievements, there is still a need to develop and to advance CFD to meet the demands emanating from various engineering industries for resolving complex flow problems, as well as from research communities, especially in the area of biomedical research.

The materials presented in this book thus far have served as an introduction to the uses and applications of CFD. For students who are keen to pursue a research career or for those who are currently involved in research and development activities, this chapter is written specifically to provide the latest developments in CFD and to address some of the important issues and challenges that are currently faced by many CFD researchers.

8.2 ADVANCES IN NUMERICAL METHODS AND TECHNIQUES

8.2.1 Incompressible Flows

Incompressible flow, by definition, is an *approximation* of flow where the flow speed is insignificant compared to the speed of sound. In accordance with this definition, the majority of fluid and associated flow we encounter in our daily lives belongs to the incompressible category. Mathematically, the incompressible flow formulation poses unique and challenging issues not present in compressible flow equations because of the incompressibility requirement. Physically, incompressible flow is characterized by an elliptic behavior of the pressure waves, whereby the speed in a truly incompressible flow is *infinite*, which imposes stringent requirements on computational algorithms for satisfying incompressibility. Inherently, the major difference between an incompressible and a compressible Navier–Stokes formulation is in the continuity equation. The incompressible formulation can usually be viewed as a singular limit of the compressible one where the pressure field is just considered part of the solution. The primary issue in solving the set of governing equations is to appropriately satisfy the mass conservation equation.

Many engineering applications dealing with the flow of air and water and the flow of bio-fluids are generally in the incompressible flow domain. As flow devices are required to be designed with more sophistication, CFD techniques are being adopted at a staggering frequency to tackle such types of flows. Because of the escalating use, this book has been specifically written from a practical viewpoint to present the fundamental development of the conservation equations accompanied by suitable numerical methods. Focusing further on the advances achieved to date and a possible research perspective, the authors aim to examine the challenges and level of complexities being confronted in fluid engineering for incompressible and low-speed flow, as well as the current and future status of development, which are discussed herein.

Based on the approach employing primitive variables, the two most commonly adopted methodologies are (1) methods based on pressure iteration and (2) the artificial compressibility method.

For the various methodologies based on pressure iteration, the marker-and-cell (MAC) method developed by Harlow and Welch (1965) represents the first primitive variable method employing a derived Poisson equation for pressure. In this method, the pressure is used as a mapping parameter to satisfy the continuity equation. Reviewing the Navier–Stokes mass and momentum equations with constant density (see Eqs. (3.13), (3.24), and (3.25)), they can be expressed in the general form as

$$\frac{\partial u_i}{\partial x_i} = 0 \tag{8.1}$$

$$\frac{\partial u_i}{\partial t} + u_i \frac{\partial u_i}{\partial x_i} = -\frac{1}{\rho} \frac{\partial p}{\partial x_i} + v \frac{\partial^2 u_i}{\partial x_i^2} \tag{8.2}$$

Taking the divergence of the momentum equation, the Poisson equation for the pressure can be obtained according to

$$\frac{1}{\rho} \frac{\partial^2 p}{\partial x_i^2} = \frac{\partial h}{\partial x_i} - \frac{\partial}{\partial t} \frac{\partial u_i}{\partial x_i} \tag{8.3}$$

where

$$h_i = v \frac{\partial^2 u_i}{\partial x_i^2} - u_i \frac{\partial u_i}{\partial x_i} \tag{8.4}$$

Ever since its introduction, numerous variations of the MAC method have surfaced and successful computations have been made.

Another pressure-based method that is also commonly used is the fractional-step procedure (Chorin, 1968; Yanenko, 1971). It is usually performed in two steps: (1) the auxiliary velocity field is obtained by solving the momentum equation, in which the pressure-gradient term can be entirely excluded or computed from the pressure in the previous time step; and (2) the pressure is computed through a Poisson equation similar to the MAC method that maps the auxiliary velocity onto a divergence-free velocity field. The method with the exclusion of the pressure-gradient term is as follows:

$$\frac{u_i' - u_i^n}{\Delta t} = -u_i^n \frac{\partial u_i^n}{\partial x_i} + v \frac{\partial^2 u_i^n}{\partial x_i^2} \tag{8.5}$$

$$\frac{u_i^{n+1} - u_i'}{\Delta t} = -\frac{1}{\rho} \frac{\partial p^{n+1}}{\partial x_i} \tag{8.6}$$

where u_i^n is the velocity at the previous time level n, u_i' is the intermediate velocity, and u_i^{n+1} is the velocity at the current time level $n + 1$. By taking the divergence of Eq. (8.6), the Poisson form of the pressure can be expressed as

$$\frac{1}{\rho} \frac{\partial^2 p^{n+1}}{\partial x_i^2} = \frac{1}{\Delta t} \frac{\partial u_i'}{\partial t} \tag{8.7}$$

One important aspect of the fractional-step method the reader should heed is that care needs to be exercised in evaluating the boundary conditions for the intermediate variables (Orszag et al., 1986). A logical way to overcome such difficulty is to employ the physical boundary conditions for the intermediate steps. This aspect has been thoroughly discussed by Rosenfeld et al. (1991).

The artificial compressibility approach involves modifying the continuity equation by adding a time-derivative of the pressure term with an *artificial compressibility parameter* β in the form of

$$\frac{1}{\beta} \frac{\partial p}{\partial t} + \frac{\partial u_i}{\partial x_i} = 0 \tag{8.8}$$

Together with the unsteady momentum equations, this forms a hyperbolic-parabolic type of time-dependent system of equations. A series of pre-conditioning methods (Choy et al., 1993; Turkel, 1999) and numerical schemes, based on the alternating direction implicit (ADI) procedure by Briley and McDonald (1977) and upwind differencing schemes implemented by Rogers et al. (1991), have been developed in conjunction with the compressible Navier–Stokes equations to solve the incompressible flow. Artificial compressibility relaxes the strict requirement to satisfy mass conservation in each step. Nonetheless, to utilize this convenient feature effectively, it is essential that the nature of artificial compressibility both physically and mathematically be properly understood. Useful guidelines in choosing the artificial compressibility parameter can be found in Chang and Kwak (1984).

In general, it is the authors' opinion that the above methods are well established and that they have been employed successfully to solve a wide range of fluid-flow problems. Most of these developed methods are commonplace in many daily fluid-engineering problems. Many review articles and books on

CFD have been written on state-of-the-art computational methods for viscous incompressible flow. Interested readers are strongly encouraged to refer to Gunzburger and Nicolades (1993), Hafez (2002), Loner et al. (2002), Gustafsson et al. (2002), and Kiris et al. (2002) for the latest trends and advances in these methods. Extensions of these basic formulations targeting a number of specific applications have also been reported. For example, a semi-implicit scheme has been formulated by Najm et al. (1998). A zero-Mach-number formulation of the compressible conservation equations coupled with detailed chemistry is constructed based on the predictor-corrector treatment with additive fractional-step procedures to stabilize computations of reacting flows with large density variations.

Despite the many significant advances achieved for incompressible CFD methodologies, one aspect the authors believe still deserves some concerted attention is *computational efficiency*, which is directed primarily toward the solution of the Poisson equation for pressure. The choice of pressure solver can be a crucial consideration, particularly in solving a large physical domain with high mesh density and/or in marching the solution through small time steps for a transient flow problem. In light of the computations in rectangular-type regions, one possible consideration is to employ a direct (non-iterative) fast Fourier transform (FFT)-based solver (Sweet, 1973) for the pressure equation, which is easily accessible through a computer package called CRAYFISH-PAK.[1] However, the application of this direct Poisson solver for general three-dimensional coordinates is not straightforward and normally is considered impractical; iterative solvers still remain the only feasible option to resolve such complex geometries. Traditional iterative solvers are well known to be inefficient due to their slow convergence, which therefore requires the development of more advanced solution methods. Currently, parallel computing presents one possible way of accelerating the computations. To maximize the computing speed, iterative matrix solvers, such as the parallelized version of the Generalized Residual Minimal Equation Solver (GRMES) by Saad and Schultz (1985), which essentially belongs to the family of Krylov methods, is a viable choice for solving the pressure equation. Another avenue that is also worth exploring is the application of the multi-grid acceleration technique, as suggested by Kwak et al. (2005).

8.2.2 Compressible Flows

Historically, numerical methods for solving equations for compressible flows were developed because of their importance and relevance in aerodynamics and aerospace. In air flow around an aircraft, enormously high speeds can be achieved as the aircraft travels through the atmosphere, thus resulting in a very

---

1. CRAYFISHPAK, a vectorized form of the elliptic equation solver FISHPAK, was originally developed at the National Center for Atmospheric Research (NCAR) in Boulder, Colorado.

high Reynolds number; the turbulence effect is mainly concentrated within the thin boundary layers. Ignoring the frictional drag, the flow may be attributed to the presence of the *wave* drag due to shocks and the *pressure* drag that is essentially inviscid in nature. The latter property has indeed necessitated much of the revolutionary and evolutionary effort focused on special methodologies designed specifically to obtain the solution for the inviscid Euler equations. As reviewed by Fujii (2005), CFD was first applied to simulation of transonic flows in the 1970s. Embedded shock waves were adequately captured and the design process for commercial aircraft has drastically changed ever since then. In the middle of 1980s, CFD was employed to simulate hypersonic flows associated with space flight, including re-entry vehicles. Most of these can be readily solved through the inviscid Euler equations. Practical flow simulations in aerospace employing the fully compressible Navier–Stokes equations first appeared in Fuiji and Obayashi (1987a,b). For high-speed flows, the use of these equations to resolve the thin, viscous, boundary layers around airfoils still presents enormous challenges. The fine-mesh requirement is unavoidable in adequately predicting the near-wall turbulence for flow analysis and eventual control of flows. Finely distributed mesh resolution near walls greatly limits the time-step size for computation; this in turn dramatically increases the computational burden needed to achieve a solution.

With regard to some specific methods for compressible flows, the earliest scheme was the method of MacCormack (1969), which is based on an *explicit* method and central differencing. It is still being used extensively even today. To avoid the problem of oscillations, especially due to the discontinuity at the shock front, the concept of artificial dissipation was introduced into the equations. A fourth-order dissipative term is the most common addition, but higher-order terms have also been successfully applied. Another effective numerical method for solving the equations for compressible flow is the *implicit* method developed by Beam and Warming (1978), which is based on the approximate factorization of the Crank-Nicolson method. Like the MacCormack method, the addition of the explicit fourth-order dissipative term to the equations is imperative due to the central-differencing consideration.

Roe (2005), in his review paper "CFD—Retrospective and Prospective," identified some issues key to the seminal developments in CFD algorithms. One of them is to overcome the deficiency in shock capturing when the shocks are very strong. Enormous interest has been placed on developing a class of upwind schemes of greater sophistication to produce a well-defined discontinuity without introducing an undue error into the smooth part of the solution elsewhere. In the next section, the authors provide a survey of the latest developments in high-resolution schemes. The survey doesn't dwell on the fine details of construction of these schemes, so interested readers are strongly encouraged to refer to the literature for more in-depth understanding. Another approach of significant interest in capturing the unsteady moving shock front is the use of adaptive meshing, which is also discussed below.

8.2.2.1   High-Resolution Schemes

Consider the simple illustration of the inviscid Euler equation based on the one-dimensional wave equation in the non-conservative form:

$$\frac{\partial u}{\partial t} + c\frac{\partial u}{\partial x} = 0 \qquad (8.9)$$

where c is the advective velocity describing the propagation of a wave in the direction of the x axis. There is a discontinuity in the velocity u across the wave, as described in Figure 8.1. Assuming that c is positive, properties at grid point i should depend on the upstream flow-field properties at grid point $i-1$. Grid point $i+1$, on the other hand, should not physically influence the point at i; hence the choice of numerical scheme must reflect the flow physics. If the gradient $\partial u/\partial x$ is approximated using central differencing, such as the traditional approach described above for MacCormack (1969), the velocity u profile results in an oscillatory behavior near the discontinuous wave front. In some circumstances, the numerical procedure can lead to an unstable and chaotic solution. The common remedy, as mentioned, is to introduce an artificial dissipation term. Despite the numerical result exhibiting a *monotone* variation (no oscillations), the diffusive property remains an undesirable element, as illustrated for the numerical representation in Figure 8.1.

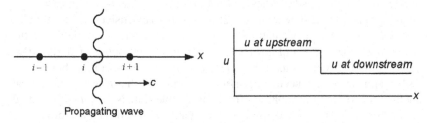

Physical Representation

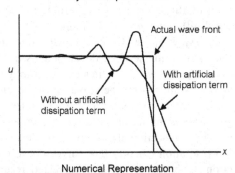

Numerical Representation

FIGURE 8.1   Schematic physical representation of a propagating wave in the positive x direction accompanied by the flow field at a given instant of time and numerical representation with and without an artificial dissipation term in the vicinity of the discontinuous wave interface.

To eliminate this undesirable property, some mathematically elegant algorithms have been developed during the past decade or two. These modern algorithms have included flux limiters by Sweby (1984) and Anderson et al. (1986) in the MINMOD and SUPERBEE schemes and slope limiters due to the Monotone Upwind Scheme for Conservation Laws (MUSCL) family of methods that can be found in Van Leer (1974, 1977a,b, 1979), Godunov (1959), and approximate Riemann solvers (Toro, 1997). Among the many shock-capturing techniques in the literature, the total variable diminishing (TVD) algorithm is considered to be well suited for capturing shock waves. Essentially, the integrated quantity (total variation) of the gradient $\partial u/\partial x$ in Eq. (8.9) is now discretized according to

$$\mathrm{TV}(u) = \sum_i |u_{i+1} - u_i| \tag{8.10}$$

To achieve monotonic variation, the condition $\mathrm{TV}(u^{n+1}) \leq \mathrm{TV}(u^n)$ is imposed. In general, the design of a numerical scheme that can both represent small-scale structures with minimum numerical dissipation and capture discontinuities without spurious oscillations is a formidable task. The first-order upwind scheme (Appendix B), which does not result in oscillations in the vicinity of discontinuities, can be readily shown to obey the TVD condition. Nevertheless, it still suffers the introduction of the artificial dissipation term as being very diffusive in the vicinity of discontinuities.

The greatest challenge in ongoing research in this area is therefore to develop schemes that are *highly accurate in smooth regions of the flow* and have *sharp non-oscillatory transitions at discontinuities*. Colella and Woodward (1984) first attempted to overcome particularly the latter aspect by introducing a piecewise parabolic method (PPM), which is a four-point centered stencil to define the interface value; the formulation of this value is then limited to control the oscillations. Leonard (1991) later combined this limiting approach with a higher-order (up to ninth-order) interface value. It is noted that the PPM can be considered an extension of Van Leer's MUSCL scheme, and MUSCL in turn is an extension of the Godunov approach. These limiting procedures nonetheless still cause the solution accuracy to degenerate because of the first-order approximation near extrema.

Along a different line of thought, ever since the development of the third-order essentially non-oscillatory (ENO) schemes by Shu and Osher (1988, 1989) and Harten et al. (1989), improved schemes, such as the fifth-order weighted-ENO (WENO) by Liu et al. (1994) and Jiang and Shu (1996), have surfaced to better define the interface value as a weighted average of the interface values from all stencils. The weights have been designed so that very high accuracy is achieved in smooth regions. The WENO schemes, however, still show diffusive behavior in the vicinity of discontinuities; they smear the interface front as much as the ENO schemes. The approach by Suresh and Huynh (1997) has been developed to enlarge the TVD constraint for a better representation near extrema. The reconstructed value at the interface is limited in order to preserve both monotonicity and high-order accuracy by using local geometrical considerations to relax the monotonicity constraints near extrema. This scheme, referred to as MP, where MP stands for "monotonicity preserving," is fifth order and was developed to address the chronic problem of upwind methods, namely the narrow stencils that cannot distinguish between shocks and extrema.

As a demonstration of the above schemes, some results extracted for the one-dimensional advection equation, Eq. (8.9), which appears in the article by Suresh and Huynh (1997), are shown in Figure 8.2. Not surprisingly, the

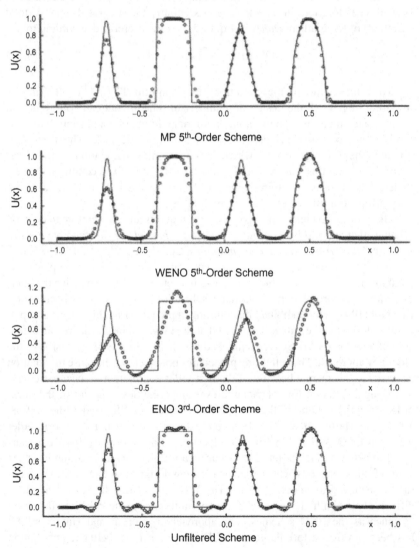

FIGURE 8.2 Results extracted from Suresh and Huynh (1997) for the unsteady one-dimensional advection equation.

unfiltered approach generally exhibits unwanted oscillations—a violation of the TVD condition—near the discontinuity locations. The ENO scheme, although preserving the monotone variation of the velocity u, fares no better than the unfiltered scheme in exceeding the limiting bounds—the realizability condition. Higher-order schemes, such as the fifth-order methods of the WENO and MP schemes, clearly satisfy not only the monotonic variation of velocity u but also the realizability condition near the extrema. The superiority of MP over WENO is succinctly demonstrated, especially near the discontinuities of the velocity u profile, where the latter scheme still shows some diffusion of the velocity u at these locations.

Recently, Daru and Tenaud (2004) extended the MP schemes of Suresh and Huynh (1997) to seventh order for the solution of Euler and Navier–Stokes equations using local linearization and dimensional splitting in the multi-dimensional case. They commented that the schemes, despite their lack of preservation of the formal high accuracy, still gave very accurate results that compared well with the high-order WENO schemes, at a lower cost. They also pointed out that the classic drawback associated with dimensional splitting, similar to that of the fractional method for incompressible flows, is the treatment of boundary conditions for the intermediate step boundary in bounded viscous flow calculations. Interested readers may wish to further investigate this pertinent aspect to improve these MP schemes.

From this brief survey, it is apparent that the development of high-resolution TVD schemes for the numerical simulation of unsteady compressible flows remains an area of intense research. The trend in research at the time of this writing appears to be the development of higher-order schemes to better capture the strong shock profiles. In the next section, the authors illustrate another cutting edge of research, where the moving shock front is progressively tracked throughout the computational domain, in addition to the usual approach of resolving the shock front in a fixed-grid environment, such as has been described for the methodology discussed herein.

8.2.2.2 Adaptive Meshing

For a strong migrating shock front, dynamic grid adaptation may be required to resolve the sharp discontinuities that exist within the flow domain as the solution changes continuously through time. A number of adaptation categories have recently been discussed, especially in Thompson et al. (1999) and Kallinderis (2000). Interested readers are encouraged explore the literature for a better understanding of the various state-of the-art techniques that are being applied.

One method that has shown to perform well is the *r-refinement* technique. The basic idea is to devise an algorithm to physically relocate the original non-adapted mesh to the time-varying changes of the solution in the domain, based on some criteria for adaptation while maintaining the identity and data structure of the domain being solved. Many successful means are available for relocating nodes or re-meshing in order to perform adaptation. However, the success of any procedure depends on the criteria used to properly guide the node location. Benson and McRae (1991) employed a gradient-based algorithm to obtain a weight function. One such measure is the local gradient (curvature) of the dependent variable for resolution evaluation. Note that there are other approaches that could also be adopted for resolution evaluation. Nevertheless, let us consider the two-dimensional multi-block dynamic adaptive algorithm developed by Ingram et al. (1993) for the purpose of illustrating shock transition. By tracking the evolution of the outer shock from right to left and observing the grid orientation, topology, cell volume, etc., the mesh clearly reflects the continuous changes of the solution process, as depicted in Figure 8.3. The sharp discontinuities are precisely captured through the concentrated mesh surrounding the shock front.

Another refinement method that is also prevalent in the literature but not often applied to unsteady problems is the enrichment or *h-refinement* technique. The development of an adaptive re-meshing scheme for application to two-dimensional triangular and three-dimensional tetrahedral meshes has been reported in Hassan and Probert (1999). Here, the cell volume and shape are changed by the *insertion* or *deletion* of mesh nodes from the data structure, resulting in an overall *increase* or *decrease* in the number of cells. It is certainly a more natural technique, since the algorithms used for reconnecting the nodes are the same as those used during grid generation to improve distribution or to achieve some predetermined conditions for the grid. Generally speaking, h-refinement is not strictly confined to only one type of mesh structure. This flexible technique permits a combination of both structured and unstructured meshes. Dynamic adaptation can also take place through a series of mesh layers fitted to the body.

Adaptive meshing is still in its infancy in CFD. Many research issues remain unresolved (McRae, 2000). First, there is a further requirement to investigate the relationship between cell shape/size/orientation, etc., in particular for the r-refinement on a structured mesh concerning the grid skewness and aspect ratio in the vicinity of the shock front, as illustrated by the problem in Figure 8.3. Highly skewed cells can pose enormous problems in numerical methods; the sharp response of the flow behavior may lead to divergence of the solution procedure. Second, it must be recognized that even the best adaptive techniques with the best criteria (weight function) may not provide the same degree of resolution for all portions of the solution. A recent article by Soni et al. (2001) stated about the adaptive weight function that "determination of this function is one of the challenging areas of adaptive grid generation." Third, the defining characteristic of the unstructured triangular/tetrahedral meshes for h-refinement is their ability to resolve solutions based on the orientation of the cells' surfaces to the solution, which can be quite random locally. This does not pose any problems in smooth regions but can create serious resolution difficulties in regions where the solution varies rapidly.

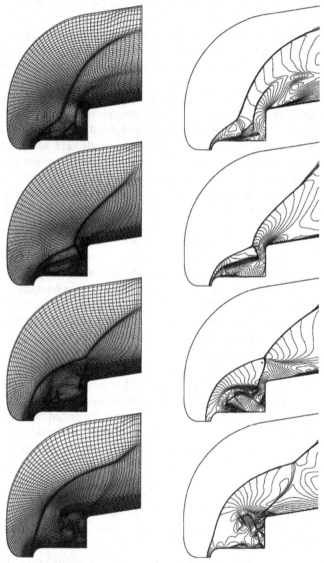

FIGURE 8.3   Results extracted from Ingram et al. (1993) for the grid and density contours produced by *r-refinement* dynamic adaptation.

8.2.3   Moving Grids

Moving grids exist in many engineering applications. Take for example the simulation of the flow in a diesel internal combustion engine in Chapter 1, where deforming grids have provided the means of simulating the piston and valve motion to gain better insight into features that are present within in-cylinder flows. Another important example in Chapter 1 is the simulation of the rotating impeller that is common in gas-sparged stirred tanks. To resolve such a problem, one part of the grid can be taken to be attached to the impeller and can be allowed to move freely in time, while the other encompasses the impeller and remains stationary. It is worth noting that such an approach can also be applied to the rotor-stator interaction in turbomachinery. As suggested by Lilek et al. (1997) and Demirdzic et al. (1997), the moving grid is allowed to *slide* along the interface without deformation between the part of the grid that is attached to the static stator and the other part that is attached to the rotating rotor. The grids do not have to match at the interface; this allows flexibility in employing different kinds of meshes and/or achieving the desired fineness in the respective domains. Except for difficulties in ensuring exact conservation, there are essentially no limitations on the applicability of this approach. Recently, Farhat (2005) reviewed the application of CFD to the prediction of flows past flexible and/or moving deforming bodies in the aeronautical area. The driving application has centered on the non-linear computational aeroelasticity to address problems associated with local transonic effects, limit-cycle oscillation, high-angle-of-attack flight conditions, and buffeting. Despite the achievements in the formulation and discretization of turbulence models and wall laws on dynamic meshes (Koobus et al., 2000), as well as successful simulations of vortex-dominated flows around maneuvering wings over a wide range of Mach numbers (Kandil and Chung, 1988), most successes of moving-grid applications according to Farhat (2005) have concentrated so far on either *complex geometries with inviscid flows* or *viscous flows with simple geometries*. Hence, to fill this gap, it is the authors' opinion, along with Farhat's, that further research into CFD on moving grids needs to be undertaken to perfect the robustness of mesh-motion algorithms and to combine them with the possibility of automatic partial regridding.

In all the problems discussed above, the movement of the domain is predetermined by external effects. For another class of problems, such as free surface flows, the movement must be calculated as part of the solution process where the grid has to move with the boundary. In the majority of these cases, the free surface interface can be a boundary between liquid and gas, such as an ocean separated from the surrounding air, or liquid and water, such as the freezing of water into ice. For such problems, the critical issues for this approach are the efficiency and stability of the numerical algorithm for the movement of the interface. As each grid point on the interface is moved through time, the surface grid may be rather irregular due to an uneven distribution and unconstrained movement. Some grid-generation algorithms that retain the curvature information of the interface are generally required to redistribute the grid points to ensure numerical integrity and stability of the solution. For illustration, some results in a numerical study of three-dimensional natural convection and freezing of water in a cubical cavity by Yeoh et al. (1990) are discussed below. The distorted ice and water domains at a particular instant of time are shown in Figure 8.4. The ice–water interface dramatically deforms under the influence

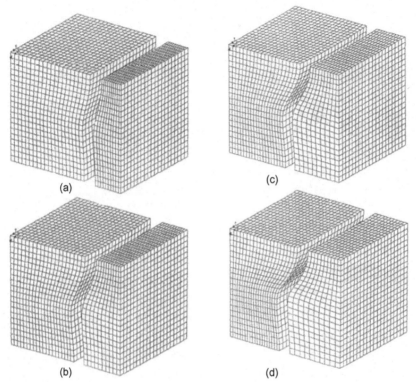

(a)

(c)

(b)

(d)

FIGURE 8.4 Results extracted from Yeoh et al. (1990) for the distorted water (*left*) and ice (*right*) domains at (a) 12.7 min, (b) 25.5 min, (c) 38.2 min, and (d) 54.1 min.

of natural convection effects that are present in the water. The appearance of two counterclockwise flow regions shown in Figure 8.5 is distinct for freezing of water because of its density extremum at 4 °C. As expected, the flow inversions around this temperature significantly influence the development of the ice–water interface. Simulating such flow problems usually requires very small time steps to advance the interface as well as generation of a new mesh using body-fitted grids in the respective domains at each time step. Both of these steps negatively affect the simplicity, robustness, and computational cost of the solution procedure. Tezduyar (2001) indicates that the computational efficiency degenerates dramatically for cases involving large motion. Development of a more robust method is thus still required to fulfill the need for better treatment of the class of flow problems with moving interfaces or boundaries.

8.2.4 Multi-Grid Methods

Chapter 4 discusses the use of the multi-grid method to accelerate the convergence of the iterative process for large systems of non-linear algebraic equations, while Chapter 5 presents the simplest strategy for achieving computational

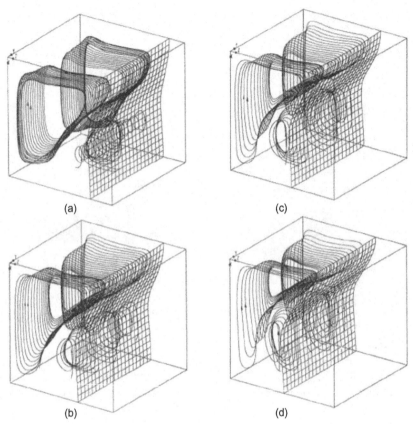

(a)   (c)

(b)   (d)

FIGURE 8.5   Results extracted from Yeoh et al. (1990) for the isometric views of interface locations and particle tracks at (a) 12.7 min, (b) 25.5 min, (c) 38.2 min, and (d) 54.1 min.

efficiency in solving such a system of equations through the V-cycle with five different grid levels. These are by no means a comprehensive description; the authors have aimed mainly to provide the reader with a bird's eye view of such an approach. However, because of its surging importance in areas of modern CFD, the basic philosophy behind the multi-grid method is further explored within this section.

As with the V-cycle, calculations are initially carried out on a fine grid and these results are progressively transferred downward to a series of coarser grids; the results on the coarsest grid are obtained and transferred upward back to the fine-grid level. The process is repeated until satisfactory convergence is achieved. On a mathematical basis, the advantage of the multi-grid method is the consideration of the enhanced damping of numerical errors through the flow field. A whole spectrum of errors can propagate throughout the numerical solution of such a flow field. For a discrete grid increment Δ_m, the *high-frequency errors* can be represented by the smallest value $\lambda_{min} = 2\Delta_m$.

The majority of iterative procedures, such as the Jacobi and Gauss-Siedel (Chapter 4), are efficient in removing high-frequency errors in a few iterations. It is the removal of *low-frequency errors* that causes the slow convergence of iterative methods on a fixed grid. Nonetheless, let us imagine that, after carrying out a few iterations on the fine grid, the immediate results are transferred to a coarser grid. The high-frequency errors are now essentially lost or hidden in the coarse grid and the solution procedure begins to damp at a more rapid rate than would have taken place in the fine grid, because of the larger Δ_m. Hence, by progressively moving the intermediate results to coarser grids, the low-frequency errors are essentially damped; when these results are transferred back to the fine grid, the low-frequency errors are indeed much smaller than they would have been for an equal number of sweeps performed on the fine grid itself.

It is clear from above that the multi-grid method is more of a strategy than an application of a particular solution method to solve the algebraic equations for a given flow field. Research is still ongoing to optimize the solution process, particularly to improve the numerical methods in solving the algebraic equations at each grid level. Many challenges remain in designing the "best" multi-grid strategy, whether it is ascertaining a suitable combination of simple or advanced iterative methods on fine grids accompanied by direct methods to obtain solutions on the coarse grids or employing different cycling strategies using either a series combination of the V-cycle with the W-cycle or the implementation of a *full-multi-grid* F-cycle. Extensive efforts are also being made to extend the multi-grid method to parallel computing to achieve quicker computational speed than is possible with traditional single- or dual-processor computations. Interested readers can refer to Mavriplis (1988), Wesseling (1995), Timmermann (2000), and Thomas (2003) for trends and developments in this subject area.

8.2.5   Parallel Computing

Central processing units (CPUs) in most single- or dual-processor computers (PCs and workstations) are getting faster and more compact. Currently, most computer chips are attaining amazing speeds. To gain any significant increase in speed, future high-performance CFD computing ultimately will involve parallel computing.

The basic idea of parallel computing stems from the simple desire to perform simultaneous operations in multiple computational tasks on a computer system. Commercially available parallel computers generally comprise thousands of processors, gigabytes of memory, and computing power measured in megaflops or even gigaflops. The advantage of parallel computers over vector supercomputers is *scalability*. These systems usually allow the accommodation of standard computer chips and therefore are cheaper to produce. Parallel computing, particularly in CFD, is a broad field of research and development.

The details of the taxonomy of parallel computing architectures and programming paradigms are beyond the scope of this introductory book. Interested readers can refer to numerous articles in journals, such as the *International Journal of High Speed Computing*, *The Journal of Supercomputing*, or the *Journal of Parallel Programming*, and the book by Simon (1992) for more in-depth study, particularly on the aforementioned aspects and the principal issues behind the utilization of parallel computing in CFD.

Here we will highlight outstanding issues that still need to be adequately addressed. For the effective utilization of parallel computing, we can first concentrate on the issues of *domain decomposition* and *load balancing*. The idea of partitioning data and computational tasks among multiple processors is commonly denoted as *domain decomposition*. The principal objective of domain decomposition is to maintain uniform computational activities on all processors and is known as *load balancing*. Load balancing may seem straightforward, but there are a number of factors that can complicate it. For example, combustion flows that involve chemical reaction rate source terms are computed only where the static temperature exceeds some threshold value. Another example is particle tracking, where particles can accumulate in a particular sub-region. These pose serious challenges in parallel computing that are being actively researched.

Another key concern is the impact of scaling on a parallel computer architecture with increasingly larger numbers of processors. An important question arises: How does the *efficiency* of the computation depend on the number of processors? Generally speaking, the performance of parallel computing may be influenced by the cost of scheduling processors, communication between processors, and synchronization of time (i.e., the time required to allow processors to reach a common point following execution of the parallel section of the code). Lastly, regarding the issue of *portability*, significant effort has been devoted to the development of standardized environments for the development of parallel computer codes. In recent years, rigorous development of the standard computer language FORTRAN into High Performance FORTRAN (HPF) (Forum, 1993; Koelbel et al., 1994), of standard heterogeneous, network-based parallel computing environments such as Parallel Virtual Machine (PVM) (Sunderam et al., 1990; Mattson, 1995), and of standard message-passing interface (MPI) (Forum, 1994) has been undertaken to expand the appeal and broaden the flexibility of parallel computing.

At the present time, many CFD researchers may still regard parallel computing with some reservations, judging it as

1. Lacking a decisive performance advantage over conventional serial (and vector) computers in many instances
2. Difficult to program efficiently
3. Grossly lacking in portability

It appears that all these factors are likely to diminish in the not too distant future. It is also highly probable that modern parallel computers can achieve speed that equals or exceeds, for example, the performance of even the largest multi-processor Cray supercomputers. Currently, the aerospace industry has taken the lead in applying parallel computing to practical analysis and design. Other engineering areas will eventually follow as advancements in parallel computers make them more attractive for practical use and for more realistically resolving complex fluid-flow problems.

8.2.6   Immersed Boundary Methods

Immersed boundary methods were first introduced by Peskin (1972) to simulate cardiac mechanics and associated blood flow. One distinguishing feature of this approach is the ability to perform the entire simulation on a fixed Cartesian grid. A novel procedure was developed by directly imposing the effect of the immersed boundary on the flow; the requirement for the grids to conform to the complex geometrical structure of the heart was thus avoided. Since its inception, numerous modifications and refinements have been proposed and a number of variants of this approach now currently exist.

Consider the schematic drawing of the simulation of flow past a solid body in Figure 8.6(a). The conventional approach would be to employ structured or unstructured grids conforming to the particular body shape. This is achieved by first defining the surface grid covering the boundary Γ_b. The internal grids encompassing the regions occupied by the fluid Ω_f and the solid Ω_s are then generated afterward. For the finite-difference method, the partial differential governing equations are transformed into a curvilinear coordinate system that aligns with the gridlines. They are then discretized and solved in the

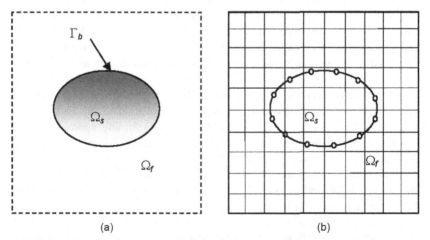

(a)   (b)

FIGURE 8.6   (a) Schematic drawing showing flow past a generic body. The solid body occupies the volume Ω_s with boundary Γ_b. The volume of the fluid is denoted Ω_f. (b) Schematic of a body immersed in a Cartesian grid on which the governing equations are discretized.

computational domain with relative ease. For the finite-volume method employing a structured grid, the integral form of the governing equations is discretized and the geometrical information regarding the grid is incorporated directly into the discretization. If an unstructured grid is used instead, either the finite-volume or finite-element methodology can be adopted by incorporating all the relevant local cell geometry into the discretization. Neither of these approaches has to resort to grid transformations.

On the other hand, let us consider the non-conformal body on a Cartesian grid in Figure 8.4b. Here the immersed boundary would still be represented through some means of the surface grid covering the boundary Γ_b, but the Cartesian grid would be generated with no regard to this surface grid. Therefore, the solid body cuts through the Cartesian volume grid. Because the grid does not conform to the solid boundary, the immediate task is to incorporate appropriate boundary conditions by modifying the equations in the vicinity of the boundaries. The governing equations can therefore be discretized using any technique, whether it is finite difference, finite volume, or finite element, without the need to resort to coordinate transformation or complex discretization operators. This is essentially the essence of the immersed boundary method.

Clearly, the imposition of suitable boundary conditions is not straightforward in immersed boundary methods. As indicated by Mittal and Iaccarino (2005), there are two approaches that are amenable to precisely accommodating the effects of the boundary condition on the immersed boundary. The modification can take place in the form of source terms (or forcing functions) in the governing equations in order to reproduce the effect of a boundary. In the first approach, which we can call *the continuous forcing approach*, the forcing function is incorporated into the continuous equations applied to the entire domain before discretization, whereas in the second approach, which can be called *the discrete forcing approach*, the "forcing" is introduced after the equations are discretized. The continuous forcing approach is very attractive for flows with immersed elastic boundaries. For such flows, the method has a sound physical basis and is very simple to implement. Many successful applications in biology, such as for sperm motility studies (Fauci and McDonald, 1994) and for multiphase flows (Unverdi and Tryggvason, 1992), testify to the viability of such an approach where elastic boundaries abound. On the other hand, the discrete forcing approach is not as practical as the continuous forcing approach, but it enables a sharp representation of the immersed boundary and this is especially desirable for high-Reynolds-number flows. It also allows direct control over the numerical accuracy, stability, and discrete conservations of the solver. This method has been used successfully for simulating compressible flow past a circular cylinder and an airfoil (Ghias et al., 2004), flow through a rib-roughened serpentine passage (Iaccarino et al., 2003), flapping foils (Mittal et al., 2002), and objects in free fall through a fluid (Mittal et al., 2004). It is apparent that the decision to employ either the continuous or the discrete forcing approach is very dependent on the particular fluid-flow process. More details on the

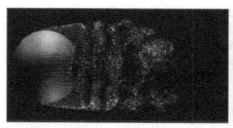

FIGURE 8.7 Results extracted from Yun et al. (2006) for the flow structures in the near wake behind a sphere: (a) Reynolds number $Re = 10^4$ (immersed boundary simulation) and (b) Reynolds number $Re = 1.5 \times 10^4$ (experiment).

mathematical basis and numerical considerations in both of these approaches can be found in the article by Mittal and Iaccarino (2005).

As a demonstration of the application of the immersed boundary method, Figure 8.7 illustrates the flow structures in the near wake behind a sphere. Numerical simulations performed by Yun et al. (2006) are compared with the experimental flow visualization. The large eddy simulation (LES) results are shown to be in good agreement with the experiment, thereby validating the fidelity of the immersed boundary method.

The popularity of immersed boundary methods is increasing at a tremendous rate. The few applications highlighted above do not even begin to scratch the surface of the many simulations using these methods. In due time, concerted ongoing improvements in accuracy and efficiency and the development of innovative and creative approaches, such as adaptive grid meshing with the immersed boundary method (Roma et al., 1999), will yield increased applications in new areas of complex turbulent flows, fluid–structure interaction, and multi-material and multi-physics simulations, as well as in well-established areas, such as biological and multi-phase flow applications.

8.3 ADVANCES IN COMPUTATIONAL MODELS

8.3.1 Direct Numerical Simulation

Turbulence in fluids is a non-linear phenomenon and it comprises a wide spectrum of spatial and temporal scales. Turbulent flow poses significant challenges to modelers, as already evidenced by the range of turbulence models that have been developed and applied. Decades of active research have certainly transformed some of these models into commonplace usage in many practical and design analyses. Lately, one significant area of activity in modern simulation strategies for turbulent flow is the return to direct simulation of turbulence to embrace increasing realism (complexity and Reynolds numbers) in the flow as computers increase in performance.

As described in Chapter 6, direct numerical simulation (DNS) refers to computations where all relevant spatial and temporal scales are adequately resolved *for the given application.* A valid simulation must accommodate the entire range of length scales, including the smallest scales in which the viscosity is active. This means that it is important to capture all of the kinetic energy dissipation within the turbulent flow. Estimates for the smallest scales are available from the so-called Kolmogorov micro-scales. From dimensional analysis, assuming dependence only upon viscosity v and dissipation rate of kinetic energy ε, the length microscale η can be expressed as

$$\eta = \left(\frac{v^3}{\varepsilon}\right)^{1/4} \qquad (8.11)$$

If the integral length and velocity macro-scales of the problem are respectively L and u and we further assume the dissipation scales in the same way as production, i.e., u^3/L, it can be shown according to Tennekes and Lumley (1976) that

$$\frac{L}{\eta} = Re_L^{3/4} \qquad (8.12)$$

Here $Re_L = uL/v$ is a Reynolds number based on the magnitude of the velocity fluctuations and the integral length scale. The difference between the largest and smallest length scales in turbulence thus increases as the Reynolds number increases; the number of grid points required to resolve turbulence increases as $Re_L^{9/4}$. If the time-step restriction is factored in, the computational cost scales as the cube of the Reynolds number. Clearly, flows with very high Reynolds numbers are impossible to simulate. Nevertheless, some phenomena in turbulence appear to be asymptotic in the range of high Reynolds numbers asymptote, such as free shear layer rates, near-wall behavior, etc., and so numerical simulations that can be feasibly performed at "high enough" Reynolds numbers are still required to capture these phenomena in order to enhance understanding and contribute to model development.

Some numerical issues pertaining to DNS simulations are discussed here. The key issue that typically dictates a DNS simulation is the mesh resolution within the flow domain. Some applications, such as those requiring statistics involving higher derivatives, will definitely need greater resolution than others. Detailed grid-refinement studies have shown that there are some "rules of thumb" that can be adopted in DNS calculations. One rule of thumb that the user may adopt is that grid spacing away from the wall should be on the order of 5η, which has been found to be sufficient for most purposes in the prediction of the mean flow, second moments of turbulence, and all the terms in the kinetic energy transport equation. Another rule of thumb, for example in calculations being performed for free shear layer flows, is that the ability to preserve six decades of roll-off in the energy spectrum could allow good second-moment turbulence statistics to be attained during the DNS calculations.

Another consideration for DNS is the need to implement suitable numerical methods to obtain accurate realization of a flow that contains a broad spectrum of time and length scales. DNS generally requires an accurate time history. Since small time steps are naturally adopted, *explicit* methods, which are usually stable because of the strict CFL requirements, have been the preferred time-advancement methods in most simulations, instead of *implicit* methods. There are, of course, notable exceptions, especially fine meshes being concentrated near solid walls to capture the essential small-scale flow structures that may cause instability, due to the explicit determination of the viscous terms in these regions. This can be easily overcome by treating these terms in an implicit manner to promote stability during transient calculations. The most commonly used time-advance methods have been the second-order Adams-Bashforth and the third- or fourth-order Runge-Kutta methods. The reader may wish to refer to Rai and Moin (1991) for a greater understanding of the use of these methods in the direct simulation of turbulent flows in a channel. In practice, Runge Kutta methods allow a larger time step for the same accuracy and thus compensate for the increased amount of computation. To treat the viscous terms, the Crank-Nicolson method (see Appendix C) is often applied in conjunction with these time-advance explicit methods.

DNS also requires appropriate *discretization* methods. Unquestionably, the prevalent method for DNS is the spectral method. However, this method places restrictions on the type of geometry and grids that can be efficiently handled. In consideration of the commonly adopted finite-difference or finite-volume approaches, it is imperative to reiterate the importance of employing an energy-conservative spatial differencing scheme. In many Reynolds-averaged Navier–Stokes (RANS) calculations, upwind differencing schemes are commonly advocated, since the dissipation that is introduced tends to stabilize the numerical methods and bound the scalar solutions. However, when these schemes are used in DNS, the dissipation produced is often much greater due to the physical viscosity, resulting in a solution that has little connection to the flow physics of the problem being solved. To overcome the issue of false diffusion, "diffusion-free" central-differencing schemes are used instead in many DNS simulations. Accuracy ranging from second to fourth order has been applied. Nonetheless, Rai and Moin (1991) have found that the fifth-order upwind-biased scheme has proven to yield good first- and second-order statistics that agree well with experimental data. It appears that the high-order accurate upwind-biased scheme is a good candidate for direct simulations of turbulent flows with complex geometries, provided that the mesh is sufficiently fine. However, the problem of false diffusion may still resurface due to inadequate resolution of the flow domain.

Generating initial and boundary conditions for the flow domain in DNS can also be challenging. In comparison to RANS computations, where only mean conditions are required, DNS must contain all the details of the three-dimensional velocity field, including the complete velocity field on a plane (or surface) for the inflow conditions of a turbulent flow at each time step. Since the memory of the initial and boundary conditions may well be required through some considerable time, they can have a significant effect on the results at the later stages of the flow simulations. Certain boundary conditions that are applicable to RANS cannot be readily applied in DNS, such as symmetry boundary conditions. For outflow boundaries, it is important to impose boundary conditions that do not allow the pressure waves to be reflected off these boundaries and back into the interior of the domain. Despite all attempts to prescribe initial and boundary conditions that are as realistic as possible, a DNS simulation must take its course for some lengths of time so that the flow develops with the correct characteristics of the physical flow. The best way to ascertain the flow development is to actually monitor some flow quantity; the choice depends on the flow that is being simulated. Initially, the quantity may reveal some systematic decreasing or increasing trends, but when the flow is fully developed, the value will show statistical fluctuations with time. At this stage, statistical averaging over time can thus be performed on the results (for example, the mean velocity and fluctuations).

Lately, more sophisticated DNS applications, such as bypass transition, in which passing wakes trigger turbulence spots in a flat-plate boundary layer, or vortex structure identification, have been performed (Guerts, 2001). The reader should be aware that such simulations usually consume hundreds of computational hours to arrive at a solution. It is not entirely surprising that the number of grid nodal points that can be used is limited by computer processing speed and memory. Nonetheless, as computers become quicker and larger memories are made available, solutions to more complex and higher-Reynolds-number flows are becoming attainable. DNS allows the provision of acutely detailed information about the flow. This wealth of information covering a wide range of length and time scales can be used to attain a better understanding of the flow physics, such as the mechanisms of turbulence production, energy transfer, and dissipation in turbulent flows or the effects of compressibility on turbulence. Through DNS, significant insights into the physics of flow can be realized, many of which may not have been possible through experiments.

8.3.2   Large Eddy Simulation (LES)

Large eddy simulation (LES), also briefly described in Chapter 6, is essentially a simulation that directly solves the large-scale motion but approximates the small-scale motion. The establishment of the LES method has its roots in the prediction of atmospheric flows since the 1960s, and, like DNS, the method has grown in importance as computers have increased in size and performance.

As mentioned, LES requires a flow field where only large-scale components are present. This can be achieved through a *filtering* process. After the small-scale eddies are eliminated by localized filter functions like either a Gaussian or top-hat filter, only the large or resolved scale field remains to be solved. The instantaneous flow variables are then decomposed into filtered or resolvable components (essentially a local average of the complete field) and a *sub-grid-scale* component that accounts for the scales not resolved by the filter width. In the finite-volume method, it is sensible and natural to define the filter width as an average of the grid volume. In a rough sense, the flow eddies larger than the filter width are *large eddies,* while eddies smaller than the filter width are *small eddies* that require modeling. When filtering is performed on the incompressible Navier–Stokes equations, a set of equations very similar to the RANS equations of momentum and energy discussed in Chapter 3 are obtained. Similar to RANS, there are additional terms where a modeling approximation must be introduced. In the context of LES, these terms are the sub-grid-scale turbulent stresses and heat fluxes, which require sub-grid-scale (*SGS*) models to close the set of equations.

The most widely used sub-grid-scale model is the one proposed by Smagorinsky (1963). Since it is prescribed through the eddy-viscosity assumption, it shares many similarities with the formulation of the Reynolds stresses obtained through the RANS approach. The unresolved sub-grid-scale turbulent stresses are modeled accordingly as

$$\tau_{ij} - \frac{\delta_{ij}}{3}\tau_{kk} = -2\nu_T^{SGS}\overline{S}_{ij}, \quad \overline{S}_{ij} = \frac{\partial \overline{u}_i}{\partial x_j} + \frac{\partial \overline{u}_j}{\partial x_i} \tag{8.13}$$

where ν_T^{SGS} is the sub-grid-scale kinematic viscosity and $\overline{S}_{ij}$ is the strain rate of the large-scale or resolved field. The form of the sub-grid-scale eddy viscosity μ_T^{SGS}, noting that $\nu_T^{SGS} = \mu_T^{SGS}/\rho$, can be derived by dimensional arguments and is given by

$$\mu_T^{SGS} = C_s^2 \rho \Delta^2 |\overline{S}_{ij}|, \quad |\overline{S}_{ij}| = \sqrt{2\overline{S}_{ij}\overline{S}_{ij}} \tag{8.14}$$

where Δ is denoted by the grid filter width and the model constant C_s varies between 0.065 and 0.3 depending on the particular fluid-flow problem. There is a difference in the way the turbulent viscosity is evaluated between the LES and RANS approaches. From Eq. (8.14), LES determines the turbulent viscosity directly from the filtered-velocity field. However, referring to Eq. (3.42) in Chapter 3, RANS requires the turbulent viscosity to be evaluated through the flow field containing two additionally derived variables, the turbulent kinetic energy k and its rate of dissipation ε values. Over the past decades, other basic sub-grid–viscosity models, such as the Structure Function model by Métais and Lesieur (1992) and the Mixed Scale model by Sagaut (2004), which exhibits a triple dependency on the vorticity of the resolved scales, have also been proposed in addition to the basic Smagorinsky model. The reader should realize that all of these models have been designed assuming that the simulated flow is turbulent, fully developed, and isotropic, and therefore they do not incorporate any information related to an eventual departure of the simulated flow from these assumptions. To obtain an automatic adaptation of the models for inhomogeneous flows, simulations of engineering flows are more likely to be based on the dynamic formulations of the basic versions of these models, which are briefly discussed below.

One possibility for designing a self-adaptive SGS model is the dynamic procedure originally developed by Germano et al. (1991). All the basic developments rest upon the use of the Germano et al. (1991) identity[2]:

$$\widetilde{L_{ij}} = \overline{\widetilde{u}_i \overline{u}_j} - \widetilde{\overline{u}}_i \widetilde{\overline{u}}_j = \tau_{ij}^T - \widetilde{\tau}_{ij} \tag{8.15}$$

where $\widetilde{\tau}$ represents a second filtering operation, called a "test" filter, with a larger filter width than τ, and τ^T, which indicates a quantity computed using the test-filtered velocity. The dynamic procedure can in principle be applied to any sub-grid model. By equating L_{ij} to the term $C_s^2 M_{ij}$, where M_{ij} is given by

$$M_{ij} = -2\left(\widetilde{\Delta}^2 |\,\widetilde{\overline{S}}_{ij}\,|\, \widetilde{\overline{S}}_{ij} - \widetilde{\Delta^2 |\overline{S}ij|\overline{S}_{ij}} \right) \tag{8.16}$$

with $\widetilde{\Delta}$ as the test filter width, and combining with the procedure of Lilly (1992), the model constant C_s can be obtained by minimizing the square of the error term $L_{ij} - C_s^2 M_{ij}$ to yield

$$C_s^2 = \frac{\langle L_{ij} M_{ij} \rangle}{\langle M_{ij} M_{ij} \rangle} \tag{8.17}$$

The procedure described above is the basic formulation of the dynamic Smagorinsky model. Following similar arguments, other dynamic models, such as the mixed Smagorinsky scale-similarity models developed by Zang et al. (1993) and Shah and Ferziger (1997) and the Lagrangian dynamic model of Meneveau et al. (1996), have been used with considerable success and have certainly broadened the accessible applications of LES.

The numerical models and boundary conditions required for LES are essentially the same as those used in DNS. In regions close to the solid surfaces, the dynamic Smagorinsky model is well suited for such flows since it automatically decreases the model constant in the correct manner near the solid wall. Nonetheless, it is also still possible to employ wall functions of the kinds used in RANS modeling while adopting the original Smagorinsky model or other basic SGS models in some turbulent flow problems. The van Driest (1956) wall-damping function that reduces the near-wall eddy viscosity according to the normal distance from the wall has proven to be a successful recipe. A variation of the van Driest wall-damping function, such as that formulated by Piomelli et al. (1987), can also be used to suppress the near-wall eddy viscosity. Both of these models are available in the commercial code ANSYS-CFX. Another approach that is based on the renormalization group (RNG) theory by Yakhot and Orszag (1986), where the eddy viscosity reduces according to the sub-grid-scale Reynolds number, has also been proposed. This model is adopted in the commercial code ANSYS-FLUENT. The use of wall functions in LES has been shown to work well for attached flow problems (Piomelli et al., 1989).

---

2. This identity is also known as the Leibniz identity in classical mechanics.

Like DNS, most LES schemes that model real flows correctly are still very computationally expensive. Owing to computational limitations, a possibly damaging aspect prevalent in most LES schemes is that the simulations are still not being performed on a fine enough grid and consequently do not capture some important dynamics at high wave numbers, which have not been filtered or modeled. Numerical truncation errors begin to overcome the SGS model and the original physics may be lost. Lately, a way to address some of the drawbacks has been to utilize the numerical truncation errors directly to act as a SGS model implicitly instead of the explicit models proposed above. This model is known as the monotone integrated large eddy simulation, better recognized by the acronym MILES. The key to success of this approach is constructing a numerical method that will depict the proper SGS model, and recent investigations have shown it is feasibile to employ high-resolution shock-capturing methods (see Section 8.2.2.1). At this particular juncture, it is not the authors' intention to present the formal procedure behind the MILES algorithm, which is beyond the scope of an introductory level. Interested readers are strongly encouraged to refer to Garnier et al. (1999), Sagaut (2004), and Hahn and Drikakis (2005) for a greater understanding of the theory behind MILES and its respective applications in turbulent flow simulations.

8.3.3 RANS-LES Coupling for Turbulent Flows

The idea behind this methodology is to address the need to solve high-Reynolds-number flows, especially for wall-bounded flows where the requirement for a very fine mesh near walls still precludes a full LES simulation. It is important that the near-wall turbulent structure in the viscous and buffer sub-layers consisting of high-speed in-rushes and low-speed ejections (often called the streak process) be properly resolved within the near-wall region. At low to medium Reynolds numbers, this streak process generates the major part of the turbulence production; these structures must be fully accommodated in LES to obtain an accurate representation of the phenomena. For wall-bounded flows at high Reynolds numbers of engineering interest, such as bluff bodies, the computational resource requirement of accurate LES can be prohibitively large. Because of the enormous amount of work that has been devoted to the development of RANS turbulence models for near-wall predictions, it is therefore possible to take advantage of these models, which are less expensive, to capture the near-wall turbulent structures instead of full resolution through LES. Nevertheless, it is well known that the RANS approach is unable to account for the spectrum of unsteadiness that is associated with the turbulent fluctuations away from the wall. LES, because of its explicit cut-off frequency, can choose the degree of accuracy for the description of the turbulent fluctuations; accurate results can be obtained at an affordable cost without imposing a very-fine-mesh resolution. Hence, the idea of combining both methods, where the inner near-wall (the unsteady RANS) region is handled by the RANS approach while the outer region of the bulk flow is solved through

LES, makes sense. In the LES region, the mesh resolution is dictated mainly by the requirement to resolve the largest turbulent scales rather than the near-wall turbulent processes.

The RANS-LES coupling for turbulent flows can be handled through a number of different approaches. For attached flow problems using LES as described in the previous section, wall law functions employed to reduce the computational cost are in some way representative of the zonal use of RANS and LES. Nonetheless, the application of these functions to simulate separated flows presents enormous problems and they also tend to be lacking in universality when complex geometries are handled. The use of a RANS model in the near-wall zone by adopting a reduced RANS model (e.g., of mixing-length type), by solving the turbulent boundary-layer equations in an embedded near-wall mesh (Balaras et al., 1996; Wang and Moin, 2002), or even by applying a full RANS model designated a hybrid RANS-LES, are the possible alternatives to circumvent the problems associated with wall functions. Recently, the non-linear disturbance equations (NLDE) approach, initially proposed by Morris and his co-workers (1997, 1998) and recently reviewed by Labournasse and Sagaut (2004), represents another RANS-LES technique that can be applied to handle a range of complex flows. Like the hybrid RANS-LES, it solves the near-wall flow field through the RANS model and the energy and dynamics of the main unsteadiness flow in the far-wall flow field through LES, but with an additional sub-grid model to handle the unresolved fluctuations (in the zones of interest). The remainder of this section focuses on the basic concepts and advancements made to the hybrid RANS-LES models.

Hybrid RANS-LES models can be divided into two categories. The first category is the zonal decomposition into the physical space, where the computational domain is divided into sub-domains. Some of these sub-domains are computed using a RANS approach; the others, using LES. In the unsteady RANS region, two-equation models, whether they are the *standard k-ε model*, the *k-ω model*, or other developed turbulent models, can be employed. Davidson and Peng (2003) applied the k-ω model in the near-wall flow field. The transition between the domains is prescribed through a predefined switching plane, which is explicitly provided. The location of this plane or interface can be determined in many ways. It can be chosen using some blending function (sharp or smooth) (Davidson and Peng, 2003; Temmerman et al., 2005), by comparing the unsteady RANS and LES turbulent scales, by computing from turbulence/physics requirements (Tucker and Davidson, 2004), or by solving different partial differential equations to automatically locate the interface (Tucker, 2003). Although encouraging results have been obtained for selective comparative cases with the application of this zonal hybrid RANS-LES, proper treatment of the interface bridging the RANS and LES regions is crucial to the success of this model and remains an area of active research. For the second category, hybrid computations are performed without predefining a switching plane. This approach appears to be preferable to the first category of models. However, the possibility that a sub-grid model with a RANS is required limits this method. Similar to the NLDE, where the same numerical scheme is used throughout the whole domain, instability modes such as wiggles can appear in the RANS zone, which is a main disadvantage of this approach; thus, no clear advantage can be found in adopting this second category over the first category of models. Examples of these models are Speziale's very large eddy simulation (VLES) (1998), detached eddy simulation (DES) (Spalart et al., 1997), and the renomalization group approach (de Langhe et al., 2005a,b).

The hybrid RANS-LES coupling is an excellent candidate for massively separated flow problems. Typical successful applications that have been found in Strelets (2001) and Squires and Constantinescu (2003) are flow around a sphere or cylinder, NACA0012 airfoil, backward-facing step, and landing gear truck. For flow over periodic hills, overall energy and mean velocity profiles are well predicted with 20 times fewer grid points than resolved using LES. Generally speaking, the mesh density in hybrid RANS-LES is comparable to those used in RANS and it provides better results than LES on the same coarse grids. It is also superior to a full RANS simulation in predicting the energy of the main unsteadiness in the flow. The current research focus for this methodology is turbulent synthetization. Interested readers can refer to Battan et al. (2004) for a better understanding of the required turbulent closures of this issue.

8.3.4 Multi-Phase Flows

In Chapter 7, we observed how CFD can enhance the understanding of physical processes associated with gas-particle flows in simple or complex geometries, such as in a $90°$ duct bend or the human airway system. There are many other engineering applications that often involve multi-phase flows; examples are gas bubbles in liquid or liquid droplets in gas, fluidized bed combustion, liquid fuel injected as a spray in combustion machines, etc.

Multi-phase flow investigations are certainly at the crossroads of intense research in the climate of significant advancements being achieved in computing power and performance. Two commonly adopted approaches for computing two-phase flows are the Eulerian–Eulerian and Eulerian–Lagrangian methods, briefly described in Chapter 7. The Eulerian approach is usually adopted for the carrying- or continuous-phase fluid. The distinction results in the dispersed phase, which may be handled either through the discrete approach in the Lagrangian framework or the continuous Eulerian approach similar to the carrying-phase fluid. When the mass loading is small, this phase can be represented by a finite number of particles whose motion can be computed in a Lagrangian manner. These particles are injected into the flow domain and their trajectories are determined from the pre-computed velocity field of the background fluid. This is the *one-way* coupling approach. When the mass loading is substantial, the influence of particles on the fluid motion needs to be properly accounted for. A *two-way* coupling requires the computation of particle trajectories and fluid flow to be carried out simultaneously; each particle contributes to the momentum, mass, and energy of the parent fluid in each control volume cell it passes through. For the range of Eulerian–Eulerian and Eulerian–Lagrangian investigations of gas-particle flow applications, the reader may wish to refer to the current works by Tian et al. (2005, 2006).

Interaction between particles, such as collision or agglomeration, and between particles and walls can both occur, and appropriate models to account for all these effects are areas of ongoing research. These matters are certainly complicated and the reader can refer to the book by Crowe et al. (1998) for a broader understanding. Recently, attempts have been made to address particle impaction particularly on curved wall boundaries, and some encouraging results are found in articles by Morsi et al. (2004) and Tu et al. (2004b). Ongoing research continues to address many of the complex issues associated with particle–wall behavior. Robust models for practical applications remain elusive at the present time.

For flows with large mass loadings when phase change occurs, the two-fluid model (Eulerian–Eulerian approach) is applied to both phases. Here, both phases are treated as continua with separate velocity and temperature fields. To account for inter-facial effects between phases, appropriate constitutive relationships for the inter-phase exchange terms are required to close the system. The principles of two-fluid models are described in detail in texts by Kolev (2005) and Ishii and Hibiki (2006) on application to gas–liquid flows. In the context of a phase-change multi-phase flow, sub-cooled boiling flow presents enormous challenges in modeling and simulation. The two-fluid model that is applied to handle both continua phases occupied by the gas bubbles and liquid requires additional wall models to predict the nucleation of bubbles generated from heated walls. Based on a number of experimental observations, bubbles departing from the heated walls have been found to form bigger or smaller bubbles by merging, shearing off, or disappearing due to condensation in the bulk sub-cooled liquid flow. The bubble mechanistic behaviors of coalescence and breakage handled through the *population balance approach* form an integral part, in addition to the two-fluid model, of determining the bubble size in the dispersed phase. The accurate determination of bubble diameter is required because it strongly governs the drag as well as non-drag characteristics of the fluid flow. Much research has been performed on this particular type of multi-phase flow and the reader can refer to the development of models in Yeoh and Tu (2005, 2006a,b). The methods required to compute these flows are similar to those described in this book except for the addition of models like the wall and bubble mechanistic models and boundary conditions.

Last, it is also worth mentioning another class of multi-phase flows, flows with free surfaces, like channel flows, flows around ships, mold filling including solidification or melting, etc. The free surface is generally treated in addition to the *interface-tracking* methods (moving grids) by another category of methods known as *interface-capturing*. These methods have some similarities, like the immersed boundary methods where the computation is also performed on a fixed grid; the shape of the free surface is determined by solving an additional transport equation for the void fraction of the liquid phase, such as the volume-of-fluid (VOF), which was first introduced by Hirt and Nichols (1981) to indicate whether the control volume cell is filled by liquid and/or gas. The critical issue in this type of method is the discretization procedure for the advective term in the equation. Low-order schemes have a tendency to smear the interface and introduce artificial mixing of the two fluids. Like shocks in compressible flows, the smearing effect degrades the profile, and high-resolution schemes are generally preferred. For a sharper interface, the interface-capturing methods based on the *level-set formulation* are another alternative. Here, a stepwise variation of the fluid properties is enforced. Nevertheless, this causes problems when computing viscous flows and arbitrarily introducing some finite thickness across the interface region is necessary to promote smooth but rapid change of the properties. More details on the level-set methods can be found in Sethan (1996).

8.3.5   Combustion

The interaction between turbulence and combustion is an area of great research interest, and its complexity defies an analytic solution or even a basic understanding in most cases. This is especially true in the case of turbulent non-premixed or diffusion flames in which turbulence and chemical reactions are strongly coupled within the thin reaction zone of the flame.

There are many difficulties associated with handling turbulent combustion flows. First, there are various degrees of mixing and consequently combustion, since turbulence has a wide range of time and length scales. Second, the chemical reactions of combusting species are highly non-linear; they strongly affect the flow variables, such as temperature and density, which can change rapidly by a factor of eight through the flame region. Third, a combustion process usually comprises a large number of species partaking in many reaction steps, which may result in very different, complicated transport properties. Fourth, the combustion time and length scales are typically smaller than the smallest turbulent scales, making them difficult to resolve by numerical or experimental means.

Nevertheless, all types of combustion are governed by the same set of conservation laws of mass, momentum, energy, and chemical species at the fundamental level. If these governing equations can be solved numerically without any approximations, then a full set of information about the flow and combustion is known. This is the basic idea behind the direct simulation of combusting flows. In most cases, the chemical scales are much smaller than the smallest scales of turbulence, which are the Kolmogorov scales. This is why turbulent combustion presents an even a bigger challenge than non-reactive turbulent flow. DNS studies of turbulent flames can be extremely difficult, depending on the degree of complexity that needs to be imposed on the simulations. If detailed chemistry is to be incorporated, then the computational cost can increase dramatically, in some circumstances to epic proportions. Imagine that even for the simple combustion process of methane, more than 200 reaction steps are required to adequately represent the chemistry process. It is therefore not surprising that DNS of turbulent combustion is still restricted to small-scale laboratory flames.

To realize the feasibility of resolving complex reacting flows in practical combusting systems, simplifications of the combustion process have resulted in a number of widely applied models. A short description of some of these models for non-premixed and premixed combustion is given below.

By representing the combustion of fuel as a global, one-step, infinitely fast chemical reaction, the simple chemical reacting system (SCRS) assumes the oxidant reacts with the fuel to form products at stoichiometric proportions. The intermediate reactions are ignored since we are only concerned with the global nature of the combustion process. With this model, the mass fractions of the reactants and products accompanied by the inert species can be expressed as fixed algebraic state relationships in terms of a passive scalar called the mixture fraction f. As a consequence, it is only necessary to solve one extra transport equation for f, rather than individual transport equations for each mass fraction. To account for the fluctuations of mass fractions due to turbulence, the average scalars of these variables can be obtained by weighting the instantaneous value with a probability density function for the mixture fraction f. Clipped Gaussian and beta functions are typical probability density functions that have been applied to provide the best results. Interested readers are encouraged to refer to Lockwood and Naguib (1975), Pope (1994), and Jones (1979) for further details.

The eddy break-up concepts introduced by Spalding (1971) and Magnussen and Hjertager (1976) present an alternative approach to the SCRS, where the rate of consumption of fuel is solved as a function of local flow properties. Here, the mixing-controlled rate of reaction is expressed in terms of the turbulence time scale. The model considers the slowest rate as the reaction rate of fuel, depending on the minimum dissipation rates of fuel, oxygen, and products. In this model, it is also possible to accommodate kinetically controlled reaction terms from the Arrhenius kinetic rate expression to govern the reaction rate of fuel in addition to the mixing-controlled rate of reaction. The implementation is straightforward, and it has been shown to yield reasonably good predictions, but the quality of the predictions depends greatly on the turbulence models used.

In addition to the SCRS and eddy break-up models, another popular combustion model is the consideration of laminar flamelets. This approach is based on the assumption that these flamelets are reaction–diffusion layers in a quasi-steady state that are continuously displaced and stretched within the turbulent medium. The layers are assumed to be thinner than all the turbulent scales, so that their internal structures have the compositional structure of laminar flames. As in the SCRS, a transport equation for the mixture fraction is solved.

However, the instantaneous species mass fractions are now deduced from the laminar state relationships, which can be taken from experimental measurements. The species fluctuations can also be accounted for by the probability density function described above to obtain the average variables. This model allows the inclusion of detailed chemistry. Application of the eddy-break-up and laminar flamelet models for the prediction of velocity and temperature fields in a compartment fire has been reported in Yeoh et al. (2002); the reader may wish to refer to this article for a better understanding of simulating combusting flames in the context of fire dynamics.

Many of the traditional combustion models developed above have been derived on the basis that the flames are under near-equilibrium conditions. To predict highly non-equilibrium flame events, such as ignition, lift-off, or extinction, it would be possible to modify the state relationships to include the scalar dissipation rate dependence and to distinguish between the burning and extinguishing flamelets. Peters (2000) proposed a strained laminar flamelet model to accommodate such effects. Another possible approach is to use the compositional probability density function transport model (Dopazo, 1993) that particularly simulates finite-rate chemical kinetic effects in turbulent reacting flows. This transport equation can be solved either through a Lagrangian approach using stochastic methods, as suggested by Pope (1994), or through a Eulerian framework using stochastic fields developed by Valino (1998). It is clear that combustion modeling is still very much an area of active research. With the advancements in computer speed and parallel architectures, time-accurate LES of combusting flames is becoming ever more feasible and prevalent. The flow unsteadiness within the flame has been found to be better captured using the LES approach and this has been succinctly demonstrated by the recent study of a free-standing buoyant flame, as described by the CFD application example in Chapter 7, and investigation of a strongly radiating non-premixed turbulent jet flame by Desjardin and Frankel (1999).

8.3.6   Fluid–Structure Interaction

Interaction between a flexible structure and the surrounding fluid promotes a variety of phenomena with applications like stability analysis of airplane wings, design of bridges, and flow of blood through arteries (to be discussed in more detail later), etc. In this section, we discuss the various techniques behind the emerging area of computational aeroelasticity (CAE) for performing analyses on a range of aeroelastic applications, where the basic approaches are equally applicable to other applications related to fluid–structure interaction problems.

Despite the many advancements in CFD methods and structural dynamics tools that have been developed, many approaches to fluid–structure studies still seek to synthesize independent computational approaches with fluid dynamics and structural dynamic sub-systems. Such approaches are known to be fraught with complications associated with the interaction between the two simulation modules. As reviewed by Kamakoti and Shyy (2004), the task is therefore to choose appropriate models for fluid and structure based on the application, and to develop an efficient interface for coupling the two models. CAE can be broadly categorized into three approaches: *fully coupled*, *closely coupled*, and *loosely coupled*.

In the fully coupled model, the governing equations are reformulated by combining the fluid and structural equations of motion, which are solved and integrated in time simultaneously. This method has severe limitations because of the need to solve the equations in two different reference systems: fluid equations as Eulerian and structural equations as Lagrangian. The stiff set of equations for the structural system makes it virtually impossible to solve the equations using monolithic computational schemes for large-scale problems. For the class of methodologies that belong to the loosely coupled model, unlike the fully coupled analysis, the structural and fluid equations are solved separately. This results in two different computational grids that may not be coincident at the interface or boundary. To establish a link between the respective regions, an interfacing technique is developed to exchange information back and forth between the modules. Figure 8.8 illustrates a sample of the fluid and structure solvers, along with selective interfacing methodologies for aeroelastic simulation. This method provides the flexibility for choosing different solvers but the coupling procedure leads to a loss in accuracy because the modules are updated only after sufficient convergence is achieved in each of the respective regions. It is usually limited to small perturbations and problems with

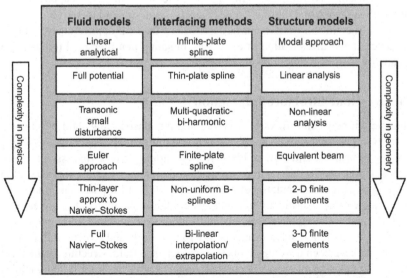

FIGURE 8.8 Sample fluid and structure solvers along with select interfacing methodologies for aeroelastic simulation, taken from Kamakoti and Shyy (2004).

moderate non-linearity. In the *closely coupled* model, the approach not only paves the way for using different solvers for the fluid and structure modules but also couples the solvers in a tight fashion through *one single module* with exchange of information taking place at the boundary. The information exchanged here includes the surface loads, which are mapped from the CFD surface grid onto the structure dynamics grid, and the displacement field, which is mapped from the structure dynamics grid onto the CFD surface grid. This implies that a moving-boundary technique is required for the CFD surface mesh so that the moving-mesh algorithms, such as the spring analogy and transfinite interpolation-based method by Hartwich and Agrawal (1997) or moving mesh partial differential equation by Huang and his co-workers (1994, 1999, 2001), are applied to enable re-meshing of the entire CFD domain as the solution marches in time.

In this short description of advances achieved in fluid–structure interactions, we have aimed to provide the reader with the culmination of some suitable solvers and interfacing methodologies for performing fluid–structure interaction calculations. The materials presented in this section are by no means exhaustive. Numerous research issues remain opened and unresolved. Particularly, future challenges lie in the development of higher-order methods to better capture the highly complex flow physics in the fluid flow and of detailed structure modeling, including non-linear effects. Ultimately, the robust application of this methodology for accurately modeling large systems still lingers in the wake of more efficient procedures.

8.3.7 Physiological Fluid Dynamics

There has been enormous interest in biomedical or bio-fluid engineering, particularly for understanding physiological applications in the human respiratory and cardiovascular systems. Owing to significant progress made in the development and application of computational methods for modeling air flow and blood flow in the respiratory and cardiovascular systems, respectively, the discoveries achieved are transforming our fundamental understanding of the fluid-mechanical behaviors within us. With medical imaging techniques, such as magnetic resonance imaging (MRI), we have the ability to construct anatomically accurate geometric models that provide new insights into the air flow and blood-flow velocities and pressure fields in normal subjects and patients. Some examples of biomedical CFD were demonstrated in Chapters 1 and 7. Despite the wealth of data that has been attained through three-dimensional computational models, significant work still remains (Grotberg and Jensen, 2004; Taylor and Draney, 2004).

In the cardiovascular system, one possible cause of heart attack or stroke could be the dangerous self-excited oscillation that arises in arterial stenoses, according to Ku (1997). The resulting static and dynamic loading on the diseased arterial wall may well be sufficiently vigorous or sustained to fracture the atherosclerotic plaque in the stenosis, thereby causing fragments to be swept downstream in the arteries, where the plaque embolism can serve as a risk factor for heart attack or stroke. In an attempt to quantify phenomena difficult to describe by in vitro techniques, computational models based on CFD methodologies have been used. Perktold et al. (1991) have adopted the finite-element method to simulate the pulsatile flow of a Newtonian fluid in a model representing the carotid bifurcation using rigid-wall approximations. This assumption, although justifiable for arteries where the wall motion is small (as may be expected in steady flow), may not be appropriate for arteries experiencing large deformation (for example, pulsatile flow in younger subjects). Rappitsch and Perktold (1996) have described the transport of albumin in a stenotic model, while Lei et al. (1997) have focused on the application of computational models in the design of end-to-end anastomoses.

Although detailed description of blood-flow regions, such as the complex three-dimensional flow in an aneurysm, can be handled by current CFD models, the future challenge is the development of closed-loop circulatory models that not only combine the *local* models (as described above) but also contain *global* models that can accurately describe the pressure-flow relations in a vascular network comprising millions of blood vessels or pressure-volume models of the heart. Because of the availability of MRI for constructing subject-specific geometric models, cardiovascular surgical planning using computational methods to predict changes in blood flow resulting from possible therapeutic interventions in individual patients is becoming a reality (Taylor et al., 1999). Recent development of cardiovascular surgical planning, as demonstrated by Wilson (2002), has led to a plan's being specifically constructed for a patient with atherosclerosis in the lower extremities. Here, three-dimensional blood-flow simulations were carried out to compute the flow velocities in simulating the scenario for the clearance of a dye in the blood stream. Such simulations have been deemed necessary because they provide useful information for identifying sites of relative flow stasis that may be prone to clot formation or sites of thickening of the inner layer of blood vessels; this information allows targeted treatment tailored specifically for the patient with cardiovascular disease.

Another challenge in cardiovascular-fluid investigations is the development of models that combine molecular- and cellular-level phenomena with macro-scale fluid mechanics. These models can allow the interaction between atherogenic agents and the artery wall and the response of the artery wall to hemodynamic forces from cellular to tissue scales to be studied. For such flows, the explicit inclusion for the particle nature of matter is required through specially developed spatial scale–multi-scale models, which are discussed in the next section.

In the respiratory system, the airways experience deformation to some degree and flow–structure interactions underlie a number of important pulmonary conditions. Of particular significance are expiratory and inspiratory flow limitations. The latter can lead to flow-induced instabilities that can generate snoring and the upper-airway obstruction that contributes to sleep apnea. The airways of the lung also are a source of pertinent problems where multi-phase fluid mechanics have important biological applications, involving flexible tubes with a liquid lining or liquid occlusion. The consideration of surface tension, elastic forces, and air flow, acting together in controlling the configuration of the deformable airway and its internal liquid lining, is critical in determining the conditions leading to airway closure; the liquid lining forms a plug, occluding (and collapsing) the airway and inhibiting gas exchange, a risk factor for an asthma attack.

Many three-dimensional models have been developed to describe the three-dimensional flow-structure interaction of an internal three-dimensional Navier–Stokes flow. It has been shown that non-axisymmetric buckling of a tube, which contributed to non-linear pressure-flow relations, can exhibit flow limitations through pure viscous mechanisms (Heil, 1997; Heil and Pedley, 1996). This work was extended to computations that described three-dimensional flows in non-uniformly buckled tubes at Reynolds numbers of a few hundred (Hazel and Heil, 2003). The development of models for bio-fluid mechanics in flexible structures is at the crossroads of intense research and the authors have just barely scratched the surface in describing some applications to respiratory systems. Understanding the physical origin and nature of phenomena contributing to a vessel's biological function or dysfunction remains computationally, experimentally, and analytically challenging.

Most physiological fluid investigations have thus far concentrated on the assumption of rigid-wall approximations in the geometrical model. Realistically speaking, pulsatile flows in the cardiovascular system and air flow through the airways in the respiratory system require flexible wall structures to appropriately represent the actual flow characteristics. In the future, fluid–structure analyses will become a required feature in studies of the respiratory system and/or cardiovascular system.

8.4 OTHER NUMERICAL APPROACHES FOR CFD

Unlike conventional numerical schemes based on the discretization of macroscopic continuum equations commonly used in many engineering applications, alternative numerical approaches have been developed lately for simulating fluid flows and modeling physics in fluids. We will review three promising methods: the lattice Boltzmann, Monte-Carlo, and particle and discrete element methods. For the particle method, an overview of the vortex methods and smooth particle hydrodynamics for the simulation of the continuum phenomena are discussed and described.

8.4.1 Lattice Boltzmann Method

The lattice Boltzmann method (LBM) is a methodology based on microscopic particle models and mesoscopic kinetic equations. According to Kadanoff (1986), the macroscopic behavior of a fluid system is generally not very sensitive to the underlying microscopic particle behavior if only collective macroscopic flow behavior is of interest. The fundamental idea behind the LBM is to construct simplified kinetic models that incorporate only the essential physics of microscopic or mesoscopic processes so that the macroscopic averaged properties obey the desired macroscopic equations. This avoids the use of the full Boltzman equation and also avoids following each particle, as in molecular dynamics simulations.

It is worth noting that even though the LBM is based on a particle representation, the principal focus remains the averaged macroscopic behavior. The kinetic nature of the LBM introduces three important features that distinguish this methodology from other numerical methods. First, the convection operator of the LBM in the velocity phase is linear. The inherent simple convection, when combined with the collision operator, allows the recovery of the non-linear macroscopic advection through multi-scale expansions. Second, the incompressible Navier–Stokes equations can be obtained in the nearly incompressible limit of the LBM. The pressure is calculated directly from the equation of state, in contrast to satisfying a Poisson equation with velocity strains acting as sources. Third, the LBM utilizes the minimum set of velocities in the phase space. Because only one or two speeds and a few moving directions are required, the transformation relating the microscopic distribution function and macroscopic quantities is greatly simplified and consists of simple arithmetic calculations.

For the LBM, the lattice Boltzmann equation (LBE) based on the lattice gas automata can be constructed via simplified, fictitious molecular dynamics in which space, time, and particle velocities are all discrete. Using the commonly adopted approach of the single relaxation time collision model of Bhatnagar, Gross, and Krook (1954), the LBE can be expressed as

$$\frac{\partial f}{\partial t} + e \cdot \nabla f = \frac{f - f^{eq}}{\lambda} \qquad (8.18)$$

where f is the particle distribution function, e is the particle velocity, λ is the relaxation time due to particle collision, and f^{eq} is the equilibrium Boltzmann-Maxwell distribution function. It can be demonstrated that the particle velocity space e can be discretized into a small set of discrete vectors e_i such that the macroscopic conservation laws are satisfied (He and Luo, 1997). Equation (8.18) can thus be discretized along each velocity direction e_i at each lattice by the following:

$$f_i(x + e_i \delta t, t + \delta t) - f_i(x, t) = -\frac{1}{\tau} \left[f_i(x, t) - f_i^{eq}(x, t) \right] \qquad (8.19)$$

Here, $f_i(x, t)$ denotes the density function along velocity e_i at lattice position x and time t, $\tau = \lambda/\delta t$ represents the relaxation parameter, and δt is the time step. With the LBM, the macroscopic density ρ and flow velocity u can be defined in terms of the particle distribution function as

$$\rho = \sum_i f_i, \quad \rho u = \sum_i f_i e_i \qquad (8.20)$$

It is shown from the above that the numerical algorithm of the LBM is relatively simple when compared with conventional Navier–Stokes methods. In terms of computational effort, the LBM consists of mainly two operations: collision on the right-hand side of Eq. (8.19) and streaming on the left-hand side. The collision operation is completely local and since streaming can be easily achieved by a simple *shift* operation, which is offered as an intrinsic function by most compilers, the LBM is well suited for parallelism. Because of the availability of very fast and massively parallel machines, LBM fulfills the requirement in a straightforward manner. However, Eq. (8.19) can also be interpreted as a special finite-difference form of the continuous form of Eq. (8.18), which means that the time step is limited by the lattice size due to the explicit nature of the lattice. Hence, the computational efficiency of LBM remains an important issue to be evaluated, particularly for steady-state flows.

Lattice Boltzmann simulations of fluid flows have been performed for flows with simple geometries, including driven-cavity flows, flow over a backward-facing step, flow around a circular cylinder, and flows with complex geometries. For the simulation of fluid turbulence, the development of a sub-grid-scale LBE turbulent model has provided the means for possibly employing LBM for the investigation of turbulent flows in industrial applications of practical interest. LBM also provides an alternative for simulating complicated multi-phase and multi-component fluid flows. The method has proven to overcome difficulties associated with the conventional macroscopic approaches in modeling interface dynamics and important related engineering applications, including flow through porous media, boiling dynamics, and dendrite formation. Interested readers are encouraged to refer to the following references for the range of applicability and future innovative developments and applications of the LBM: Hou et al. (1995, 1996), Benzi et al. (1996), Luo (1997), He and Doolen (1997), Shan (1997), Spaid and Phelan (1997), Chen and Doolen (1998), and Yang et al. (2000), among others.

8.4.2 Monte-Carlo Method

Lately, the direct simulation Monte Carlo (DSMC) has been proposed in addition to the LBM for the computation of fluid dynamics because of the practical scientific and engineering importance of solving high-Knudsen-number (Kn) flows. The dimensionless parameter Kn, by definition, generally characterizes the ratio of the molecular mean free path to the characteristic length. As the mean free path becomes comparable with, or even larger than, the characteristic length, especially for high Kn, the particle nature of matter must be explicitly accounted for, since the continuum fluid approximation breaks down, particularly in the area of micro-scale or nano-scale fluid systems. Table 8.1 shows the hierarchy of mathematical models used to describe the interactions of atoms, ions, and molecules. These models can range from very fundamental solutions of sets of elementary interactions of particles (such as molecular-dynamics

TABLE 8.1 Levels of models of many-body interactions (taken from Oran et al., 1998)

Equation	Solution Method
Newton's Law $f = ma$	Molecular dynamics (deterministic, particle-based, prescribed inter-particle forces)
Liouville equation $F(x_i, v_i, t), i = 1, N_p$	Monte-Carlo methods (statistical, particle-based methods) —DMSC
Boltzmann equation $F(x, v, t)$ binary collision (low density)	Direct solution—LBM
Navier–Stokes equation $\rho(x, t), u(x, t)$ short mean-free-path	Direct solution: finite-difference, finite-volume, spectral methods, and so on (continuum flow methods)

models) to approximation of systems in which the individual particles are replaced by continuum fluid elements (such as the Navier–Stokes equations). Molecular dynamics is the most fundamental level of this hierarchy; the LBM and DSMC lie somewhere in between the molecular dynamics and continuum fluid consideration.

DSMC is a direct particle simulation method based on kinetic theory. The fundamental idea behind the method is to track a large number of statistically representative particles. The particles' motion is later used to modify their positions, velocities, or even chemical reactions in reacting flows. Conservation of mass, momentum, and energy are enforced based on the requirements of the machine's accuracy. The primary approximation of the DSMC method is to uncouple the molecular motions and the inter-molecular collisions over small time intervals. Here, particle motions are modeled deterministically, while the collisions are treated statistically.

The core of the DSMC procedure consists of four primary processes. First, the simulated particles are moved within a time step. Boundary conditions are enforced through modeling molecule–surface interactions, which may include physical effects, such as chemical reactions, three-body collisions, and ionized flows. Second, indexing and cross-referencing of the particles are performed. This is a prerequisite for the next two steps: simulating collisions and sampling the flow field. The key to practical DSMC for large-scale processing is the accurate and fast indexing and tracking of the particles. Third, the step of simulating collisions sets DSMC apart from other deterministic simulation methods, such as molecular dynamics. The currently preferred model is the *no-time counter technique* by Bird (1994), used in conjunction with the sub-cell technique of Bird (1986). The sub-cell method ensures that collisions occur only between near-neighboring particles by calculating local collision rates based on individual cells but it restricts possible collision pairs to sub-cells. Fourth, sampling of the particles provides the macroscopic flow properties. The spatial coordinates and velocity components of molecules in a particular cell are used to calculate macroscopic quantities at the geometric center of the cell. Interested readers may wish to refer to Muntz (1989), Cheng (1993), Cheng and Emmanual (1995), and Oran et al. (1998) for more details pertaining to the recent advances and developments of DSMC.

DSMC is a time-accurate explicit procedure. The method has shown to be a good candidate for unsteady flow applications in computing the non-equilibrium structure of shocks and boundary layers as well as hypersonic viscous flows and high-temperature rarefied-gas dynamics. The latter application has sparked strong interest, especially in the production of high-speed aerospace plane and space transportation systems. In the field of material processing, the use of DSMC in handling the growth of thin films for vapor-phase processing and plasma etching is also steadily growing. To investigate the potential of DSMC as a predictive tool, results obtained by DSMC have also been compared against experiments with Navier–Stokes calculations in the low-Kn regime. Computational studies have demonstrated that DSMC solutions approach the Navier–Stokes solutions at this limit. Limitations of current computer technology, although gradually diminishing, may still inhibit DSMC from being extensively applied to the computation of fluid dynamics. DSMC is, however, being seriously considered for resolving more specialized flows, such as flows of low speed and high Knudsen number.

8.4.3   Particle Methods

The key characteristic of particle methods, especially for the simulation of the continuum phenomena. Although being deceptively simple, the key characteristic of particle methods, especially for the simulation of the continuum phenomena, can be obtained through the solution of ordinary differential equations (ODEs) that determine the trajectories and the evolution of properties carried by the particles. These methods amount to the solution of a system of ODEs in the general form given by

$$\frac{dx_p}{dt} = u_p(x_p, t) = \sum_{q=1}^{N} K(x_p, x_q; \omega_p, \omega_q) \qquad (8.21)$$

$$\frac{d\omega_p}{dt} = \sum_{q=1}^{N} F(x_p, x_q; \omega_p, \omega_q) \qquad (8.22)$$

where x_p and u_p denote the locations and velocities for the N particles; ω_p denotes the particle properties, such as density, temperature, and vorticity; and K and F represent the dynamics governing the simulated physical system. Particles are implemented with a Lagrangian formulation, and the key common characteristic of the two popular particle methods, such as vortex methods (VMs) and smooth particle hydrodynamics (SPH), involves the approximation of the Lagrangian form of the Navier–Stokes equations by replacing the derivative operators with equivalent integral operators, which are in turn discretized on the particle locations.

Particle methods are often defined as *grid-free* methods, making them an attractive alternative to *mesh-based* methods. When applied to the Lagrangian formulation of the convection-diffusion equations, particle methods enjoy an automatic adaptivity of the computational elements as dictated by the flow map. Nonetheless, truncation errors introduced in the methodology can result in the creation and evolution of spurious vortical structures through particle distortion; these methods have to be conjoined with a grid to provide consistent, efficient, and accurate simulations. The grid aims to restore regularity in the particle location via re-meshing but does not detract from the inherent adaptive character of these methods.

Vortex particle methods have been employed since they were first introduced by Rosenhead (1930) to describe the evolution of vortical structures in incompressible flows. Based on the vortex-blob approximation of Krasny (1986), the field is recovered at every location of the domain by considering the collective behavior of all computational elements. When particles do overlap, the scales of the physical quantities that are resolved are determined by the finite particle core size rather than the inter-particle distance. Viscous effects are simulated using the method of particle-strength exchange (PSE) (Koumoutsakos, 2005). The kinematic boundary conditions, such as no through-flow at solid boundaries, are enforced by boundary integral methods, whereas viscous boundary conditions, such as no-slip, are imposed by translating into a vorticity fluid boundary conditions (Cottet and Poncet, 2004) complementing the viscous part of the equations.

VMs have certainly come of age with their extensive range of engineering applications. For the DNS of flow past an impulsively started cylinder for a range of Reynolds numbers, VMs have demonstrated the capability of automatically adapting computational elements in regions of the flow where increased resolution was found to be necessary to capture the unsteady separation phenomena (Koumoutsakos and Leonard, 1995). Because these parts are usually not known *a priori*, a suitable criterion needs to be devised to add computational elements in the critical regions for mesh-based methods. In VMs, the computational elements are inherently linked to the flow physics they represent and no such additional criterion is necessary. The adaptivity and robustness of VMs have also enabled simulations of reacting flows to be carried out (Knio and Ghoniem, 1992). In comparison to mesh-based methods, the capture of highly anisotropic diffusion phenomena was better attained through the Lagrangian particle methods while accurately transporting the scalar fields. Recently, a novel particle-level-set method for capturing interfaces has been proposed by Hieber and Koumoustsakos (2005). It has been well known since the pioneering work of Krasny in the 1980s (Krasny, 1986) that particle methods are well suited for interface capturing. Comparisons of a set of benchmark problems with existing level-set formulations have demonstrated the promise of this proposed method for achieving superior results with a reduced number of computational elements required for interface capturing.

The method of SPH was first introduced by Lucy in the late 1970s and was further developed by Monaghan (1988) for grid-free astrophysics simulations. In SPH, the inter-particle distance is taken to identify the core size. Many simulations using SPH have been conducted by extending its application range from gas dynamics in astrophysics to Newtonian and viscoelastic flows, such as those found in the important works of Monaghan (1988) and Ellero et al. (2002). Like VMs, SPH also experiences the problem of particle distortion. Several techniques have been proposed to compensate for this problem (for example, dynamic conditions by Ellero et al., 2002). Inspired by techniques in VMs, the introduction of regularization of particle distortion in SPH via re-meshing has managed to improve the method to second-order accuracy; however, this detracts from the characterization of the method as being totally grid-free. Recently, the so-called meshless methods based on Galerkin-type methods have been explored for increasing the capabilities of SPH for flow simulations. Some works by Duarte and Oden (1996) and Belytschko et al. (1996) on the unifying methods of moving least squares, reproducing kernal particle methods, and element-free Galerkin in the context of SPH are noted. For an extended understanding of the background theory and formulation, numerical implementation, and challenges and extension of the particle methods to simulate molecular phenomena, readers are encouraged to refer to an excellent article written by Koumoutsakos (2005), where the subject of multi-scale flow simulations using particles has been thoroughly reviewed.

8.4.4   Discrete Element Method

The discrete element method (DEM) was originally developed to simulate granular flow problems, such as those prevalent in rock mechanics (Cundall and Strack, 1979). Similar to particle methods, DEM solves the motion of each individual element by solving the linear momentum equation as well as the angular momentum subject to forces and torques arising both from particle interaction and those imposed on the particles by surrounding fluid. In comparison to other particle simulation methods, such as molecular dynamics, dissipative particle dynamics, or Brownian dynamics, DEM differs primarily by the different particle interaction laws as well as by the imposition of random forcing to mimic the collisions or the interaction with molecules of the surrounding fluid.

DEM can be employed for a wide range of particle sizes. While other particle simulation methods are applicable only where the particle diameters are much less than the characteristics length scale of van der Waals forces (~ 10 nanometers), DEM has a propensity for handling collision of larger-size particles. Since it is now necessary to model the detailed mechanics by which particles interact during collision, particle contact models for DEM are considerably more complex than any of the other particle simulation methods. Two different approaches are adopted to characterize the interaction of two particles during collision; the approaches can be referred to as the hard-sphere and soft-sphere models.

The soft-sphere model considers inter-particle forces, namely, the normal, tangential, damping, and sliding forces, which can be modeled using springs, dashpots, and sliders. As reviewed in Kruggel-Emden et al. (2009), the treatment of the normal force acting between two particles can be based on the degree of overlap and the displacement rate. Four main groups for the evaluation of the normal force are continuous potential models, linear viscoelastic models, non-linear viscoelastic models, and hysteretic models. For molecular dynamics simulations on the atomic or molecular level, continuous potentials between particles are widely used, such as those of the Lennard-Jones potential (Verlet, 1967). The most frequently employed models in DEM are nonetheless linear (Tsuji et al., 1993). In this model, a mean coefficient of restitution and a mean collision time are defined; the related spring stiffness and the damping coefficient are then computed. Limitations of the constant coefficient of restitution and the constant duration of contact of the linear model can be overcome by applying non-linear spring damper models. Several force laws proposed by Kuwabara and Kono (1987), Tsuji et al. (1992), and Lee and Herrmann (1993) have been developed by extending the original approach of Hertz (1882). In order to include the effect of plasticity and to avoid use of the velocity-dependent damping, hysteretic models, which may be linear or non-linear, have been proposed. Notable contributions to the development of these models are Walton and Braun (1986), Sadd et al. (1993), Thornton (1997), Vu-Quoc and Zhang (1999), and Tomas (2003). In all hysteretic models, the materials in contact result in permanent deformation.

For the evaluation of the tangential force for the soft-sphere model, a wide variety of complex models are also available to describe the force-displacement behavior. Two main groups can be defined: linear and non-linear models. For linear tangential models, the earliest approach that has found wide application is the model proposed by Cundall and Strack (1979), which determines the tangential force according to a linear elastic spring unless the Coulomb force is exceeded. Since then, several versions of linear tangential displacement models have been developed. Besides the linear models, a number of non-linear tangential models have also been proposed, especially for elastic materials following the Hertz (1882) theory (Maw et al., 1976; Vu-Quoc and Zhang, 1999; Di Renzo and Di Maio, 2005). In addition to the complex models for determining normal and tangential displacements during collisions, one important feature of the time-driven, soft-sphere model is the ability to handle problems wherein particles can remain in contact for a prolonged time period to simulate particle agglomeration.

The hard-sphere model, which is based on the proposal by Hoomans et al. (1996), assumes that interactions between particles are modeled by binary collision dynamics. The main assumptions concerning the particle shape, deformation history during collision, and nature of collisions are as follows: particles are quasi-rigid and particle shape is retained after impact; collisions are quasi-instantaneous; particle contact occurs at a point (which is in contrast to the soft-sphere model, where an overlap displacement is allowed); particles are in free flight between collisions; and interaction forces are impulsive and all other finite forces are negligible during collisions. One characteristic feature of the hard-sphere model is that a sequence of collisions, is processed one collision at a time. Another important feature is that the simulations are performed with realistic values of the key parameters—restitution and friction coefficients. Particle collisions are solved using particle impulse equations. Because of the limitation of collision dynamics that has been assumed, it is not surprising that the model is much more computationally efficient than the soft-sphere model, since the particle-collision time scale does not need to be resolved. One characteristic feature of the soft-sphere model is the very small time required (often $< 10^{-6}$ s).

Fundamentally, DEM can be employed to model simple particle systems of practical interest, such as those occurring in micro-fluidics, particle filtration, or blood-flow problems in arteries and arterioles. DEM can also be useful in the development of constitutive models for the kinetic theory approach and population balance models, and to better understand the micro-structure of a large-scale system. Owing to the exponential growth of computer power and speed as well as more refined collision models, applications of DEM have increased dramatically, for example, in milling processes (Mishra, 2003a,b), environmental particulate flows (Richards et al., 2004), granular mixing (Bertrand et al., 2005), and fluidized-bed processes (Deen et al., 2007). There have also been recent attempts to extend DEM to systems containing small particle sizes in the sub-micron or even nano-scale range. Fanelli et al. (2006) have developed a model for contact forces of nano-clusters by direct consideration of the van der Waals attractive and Born repulsive forces, and they have applied DEM to investigate nano-particle dispersion. Luan and Robbins (2005) have compared the predictions of contact models with molecular dynamics simulations for nanoscale particles and found that good agreement was achieved for the normal force predictions, but they noted that atomic-scale roughness leads to significant variation in sliding resistance force between the particles.

8.5 SUMMARY

New areas of applications in fluid mechanics have brought about the development of many novel techniques. The advanced topics presented in this chapter explore the latest trends in CFD research and development. It isn't the authors' intention to provide a critical and thorough review of each of the topics in this chapter, but rather to provide a broad overview of the research methodologies adopted and key issues that still need to be addressed and resolved. Nevertheless, the significant advances achieved in numerical methods and computational models, and successful applications of some unconventional approaches, particularly the lattice Boltzmann method, direct simulation of Monte Carlo, vortex method, and smooth particle hydrodynamics, in simulating fluid flows, and the discrete element method in simulating flows of a wide range of particle sizes, should leave us awed by the sheer magnitude of CFD in the many purposeful investigations of fluid-related problems that have been covered in this chapter. Flow systems involving migration of strong shocks in transonic and hypersonic flows, complex bubble mechanistic behavior in multi-phase flow structures, or even complicated flows that exist in our body's systems, which are comprised of many cells and blood vessels capable of dramatic redistribution of blood over numerous cardiac cycles, are just some of the challenging fluid-dynamics problems that further accentuate the demands of CFD research. Like our predecessors, Newton, Euler, Bernoulli, Poiseuille, Young, Lighthill, and many others, we have found decades of intense fascination in the study of fluid mechanics. As the future unfolds, CFD methods and models will continue to remain at the forefront of intense research and development so long as the vast majority of fluid flows and processes remain unresolved.

REVIEW QUESTIONS

8.1 Simplify the general continuity equation below to a steady incompressible flow equation:

$$\frac{\partial \rho}{\partial t} + \frac{\partial(\rho u)}{\partial x} + \frac{\partial(\rho v)}{\partial y} + \frac{\partial(\rho w)}{\partial z} = 0$$

8.2 How is the pressure term used to satisfy the continuity equation in the marker-and-cell (MAC) method?

8.3 Discuss briefly the idea behind the fractional-step procedure.

8.4 What types of applications and situations involve compressible flows?

8.5 What difficulties arise from modeling a transient supersonic flow around an airfoil?

8.6 What is the biggest difficulty that has to be overcome with compressible flows?

8.7 What techniques can be used to minimize oscillations that occur in compressible flow due to discontinuities at the shock front?

8.8 How does a higher-order scheme, such as a fifth-order scheme, deal with discontinuities?

8.9 Under what circumstances would adaptive meshing be used? What would happen if a fixed mesh were employed instead?

8.10 Discuss briefly the concept behind the *r-refinement* grid-adaptive technique.

8.11 What kinds of applications commonly use moving grids?

8.12 Explain how a moving grid can be applied to simulate the screw super-charger shown below. What parts would need to remain stationary and what parts would be allowed to move?

8.13 What is the main advantage in the numerical solution of using multi-grid methods in terms of the handling of *high-* and *low-frequency* errors?

8.14 Explain what *domain decomposition* and *load balancing* are in parallel computing.

8.15 What is the immersed boundary method and how is it different from using a boundary-fitted grid?

8.16 What is a direct numerical simulation (DNS)? How does it differ from a Reynolds-averaged Navier–Stokes (RANS) approach in terms of its handling of turbulence?

8.17 What are the Kolmogorov micro-scales? How do these scales impact the mesh design?

8.18 Why can't DNS be used to solve high-Reynolds-number flows at the moment?

8.19 What is the main concept behind LES in turbulent modeling?

8.20 What are the sub-grid-scale (SGS) models in LES? How are they used to define small scales of turbulence?

8.21 In the RANS-LES coupling approach, in which region of the mesh would you apply the RANS model and in which region would you apply the LES model?

8.22 Why would you use a RANS-LES coupling approach to model high Reynolds number turbulent flows?

8.23 What is the difference between *one-way coupling* and *two-way coupling* in multi-phase flows?

8.24 Would you use a *Eulerian–Eulerian* or *Eulerian–Lagrangian* for a multi-phase flow that had a high mass loading for the secondary phase (i.e., not the continuous phase)?

8.25 Combustion is a complex phenomenon to model. What types of considerations must be made in modeling combustion?

8.26 Explain fluid–structure interaction (FSI) modeling. In what applications can it be used?

8.27 What is the key requirement (and the most complex) that enables the interaction between the fluid and structure in FSI?

8.28 What advanced techniques would be required to simulate air flow through the respiratory system into the lungs? What about pulsating blood flow through veins and arteries?

8.29 Briefly discuss the concept of the lattice Boltzmann method.

8.30 Briefly discuss the concept of the Monte-Carlo method.

8.31 Briefly discuss the concept of the particle method.

8.32 Briefly discuss the concept of the discrete element method.

Full Derivation of Conservation Equations

The full derivation of the conservation equations for momentum and energy is presented here.

The concept of *substantial derivative* is described herein. It is conveniently acceptable to collect all the density terms together by expanding Eq. (3.10) by the chain rule. This gives

$$\frac{\partial \rho}{\partial t} + u\frac{\partial \rho}{\partial x} + v\frac{\partial \rho}{\partial y} + w\frac{\partial \rho}{\partial z} + \rho\left(\frac{\partial u}{\partial x} + \frac{\partial v}{\partial y} + \frac{\partial w}{\partial z}\right) = 0 \qquad (A.1)$$

or

$$\frac{D\rho}{Dt} + \rho\left(\frac{\partial u}{\partial x} + \frac{\partial v}{\partial y} + \frac{\partial w}{\partial z}\right) = 0 \qquad (A.2)$$

where D/Dt is the substantial derivative in Cartesian coordinates. The time derivatives of $D\rho/Dt$ and $\partial\rho/\partial t$ are physically and numerically different quantities. The reader should note that the former is the time rate of change following a moving fluid element while the latter is the time rate of change at a fixed location. Also, if we consider the general variable property per unit mass denoted as ϕ, the substantial derivative of ϕ with respect to time, written as $D\phi/Dt$, is

$$\frac{D\phi}{Dt} = \frac{\partial \phi}{\partial t} + u\frac{\partial \phi}{\partial x} + v\frac{\partial \phi}{\partial y} + w\frac{\partial \phi}{\partial z} \qquad (A.3)$$

The above equation defines the rate of change of the variable property ϕ per unit mass. As in the case of mass conservation, we are interested in developing equations for rates of change per unit volume. The rate of change of the variable property ϕ per unit volume can be obtained by multiplying the density ρ with the substantial derivative of ϕ that is given by

$$\rho\frac{D\phi}{Dt} = \rho\frac{\partial \phi}{\partial t} + \rho u\frac{\partial \phi}{\partial x} + \rho v\frac{\partial \phi}{\partial y} + \rho w\frac{\partial \phi}{\partial z} \qquad (A.4)$$

It is recognized that Eq. (A.4) represents the *non-conservation form* of the rate of change of the variable property ϕ per unit volume.

The mass conservation equation derived in Section 3.2.1 defines the sum of the rate change of density and is called the *advection* term, which is

$$\frac{\partial \rho}{\partial t} + \frac{\partial(\rho u)}{\partial x} + \frac{\partial(\rho v)}{\partial y} + \frac{\partial(\rho w)}{\partial z} = 0$$

It is conceivable that the generalization of these terms for the variable property ϕ in *conservation form* can be expressed as

$$\frac{\partial(\rho\phi)}{\partial t} + \frac{\partial(\rho u\phi)}{\partial x} + \frac{\partial(\rho v\phi)}{\partial y} + \frac{\partial(\rho w\phi)}{\partial z} \tag{A.5}$$

The above formula expresses the rate of change of ϕ per unit volume with the addition of the net flow of ϕ out of the fluid element per unit volume. It is now re-written to illustrate the relationship between the conservative form of Eq. (3.16) and the non-conservative form of Eq. (3.15):

$$\frac{\partial(\rho\phi)}{\partial t} + \frac{\partial(\rho u\phi)}{\partial x} + \frac{\partial(\rho v\phi)}{\partial y} + \frac{\partial(\rho w\phi)}{\partial z} =$$

Invoking the continuity equation

$$\rho\frac{\partial\phi}{\partial t} + \rho u\frac{\partial\phi}{\partial x} + \rho v\frac{\partial\phi}{\partial y} + \rho w\frac{\partial\phi}{\partial z} + \phi\underbrace{\left[\frac{\partial\rho}{\partial t} + \frac{\partial(\rho u)}{\partial x} + \frac{\partial(\rho v)}{\partial y} + \frac{\partial(\rho w)}{\partial z}\right]}_{=0} = \rho\frac{D\phi}{Dt}$$

$$\tag{A.6}$$

Both of these forms can be used to express the conservation of a physical quantity. For brevity, the non-conservative form is adopted to derive the conservation equations for momentum and energy.

For the conservation of momentum, the net force in the x direction is the sum of the force components acting on the fluid element. Considering the velocity component u as seen in Figure 3.4, the surface forces are due to the normal stress σ_{xx} and the tangential stresses τ_{yx} and τ_{zx} acting on the surfaces of the fluid element. The net force in the normal x direction is

$$\left[\sigma_{xx} + \frac{\partial\sigma_{xx}}{\partial x}\Delta x\right]\Delta y\Delta z - \sigma_{xx}\Delta y\Delta z \tag{A.7}$$

while the net tangential forces acting along the x direction are given by

$$\left[\tau_{yx} + \frac{\partial\tau_{yx}}{\partial y}\Delta y\right]\Delta x\Delta z - \tau_{yx}\Delta x\Delta z \tag{A.8}$$

and

$$\left[\tau_{zx} + \frac{\partial\tau_{zx}}{\partial z}\Delta z\right]\Delta x\Delta y - \tau_{zx}\Delta x\Delta y \tag{A.9}$$

The total net force per unit volume on the fluid due to these surface stresses should be equal to the sum of Eqs. (A.7), (A.8), and (A.9) divided by the control volume $\Delta x\Delta y\Delta z$:

$$\frac{\partial\sigma_{xx}}{\partial x} + \frac{\partial\tau_{yx}}{\partial y} + \frac{\partial\tau_{zx}}{\partial z} \tag{A.10}$$

It is not too difficult to verify that the total net forces per unit volume on the rest of the control volume surfaces in the y direction and z direction are given by

$$\frac{\partial\tau_{xy}}{\partial x} + \frac{\partial\sigma_{yy}}{\partial y} + \frac{\partial\tau_{zy}}{\partial z} \tag{A.11}$$

and

$$\frac{\partial \tau_{xz}}{\partial x} + \frac{\partial \tau_{yz}}{\partial y} + \frac{\partial \sigma_{zz}}{\partial z} \tag{A.12}$$

Combining Eq. (A.10) with the substantial derivative of the horizontal velocity component u and body forces, the x-momentum equation becomes

$$\rho \frac{Du}{Dt} = \frac{\partial \sigma_{xx}}{\partial x} + \frac{\partial \tau_{yx}}{\partial y} + \frac{\partial \tau_{zx}}{\partial z} + \sum F_x^{body\ forces} \tag{A.13}$$

In a similar fashion, the y-momentum and z-momentum equations, using Eqs. (A.11) and (A.12), can be obtained through

$$\rho \frac{Dv}{Dt} = \frac{\partial \tau_{xy}}{\partial x} + \frac{\partial \sigma_{yy}}{\partial y} + \frac{\partial \tau_{zy}}{\partial z} + \sum F_y^{body\ forces} \tag{A.14}$$

and

$$\rho \frac{Dw}{Dt} = \frac{\partial \tau_{xz}}{\partial x} + \frac{\partial \tau_{yz}}{\partial y} + \frac{\partial \sigma_{zz}}{\partial z} + \sum F_z^{body\ forces} \tag{A.15}$$

If the fluid is taken to be Newtonian and isotropic—since all gases and the majority of liquids are isotropic—the normal stresses σ_{xx}, σ_{yy}, and σ_{zz} appearing in Eqs. (A.13)–(A.15) can be formulated in terms of pressure p and normal viscous stress components τ_{xx}, τ_{yy}, and τ_{zz} acting perpendicular on the control volume. The remaining terms contain the tangential viscous stress components as also described in Eqs. (A.13)–(A.15). In many fluid flows, a suitable model for the viscous stresses is introduced, which can be expressed as a function of the local deformation rate (or strain rate). Assuming that the fluid is Newtonian and isotropic, the rate of linear deformation on the control volume $\Delta x \Delta y \Delta z$ caused by the motion of fluid can usually be expressed in terms of the velocity gradients. The normal stress relationships can be expressed as

$$\sigma_{xx} = -p + \tau_{xx} \qquad \sigma_{yy} = -p + \tau_{yy} \qquad \sigma_{zz} = -p + \tau_{zz} \tag{A.16}$$

According to *Newton's law of viscosity*, the normal and tangential viscous stress components are given by

$$\tau_{xx} = 2\mu \frac{\partial u}{\partial x} + \lambda \left[\frac{\partial u}{\partial x} + \frac{\partial v}{\partial y} + \frac{\partial w}{\partial z} \right] \qquad \tau_{yy} = 2\mu \frac{\partial v}{\partial y} + \lambda \left[\frac{\partial u}{\partial x} + \frac{\partial v}{\partial y} + \frac{\partial w}{\partial z} \right]$$

$$\tau_{zz} = 2\mu \frac{\partial w}{\partial z} + \lambda \left[\frac{\partial u}{\partial x} + \frac{\partial v}{\partial y} + \frac{\partial w}{\partial z} \right]$$

$$\tau_{xy} = \tau_{yx} = \mu \left(\frac{\partial v}{\partial x} + \frac{\partial u}{\partial y} \right) \qquad \tau_{xz} = \tau_{zx} = \mu \left(\frac{\partial w}{\partial x} + \frac{\partial u}{\partial z} \right)$$

$$\tau_{yz} = \tau_{zy} = \mu \left(\frac{\partial w}{\partial y} + \frac{\partial v}{\partial z} \right) \tag{A.17}$$

The proportionality constants of μ and λ are the (first) dynamic viscosity that relates stresses to linear deformation and the second viscosity that relates stresses to the volumetric deformation, respectively. To the present day, not much is known about the second viscosity. Nevertheless, the Stokes hypothesis that $\lambda = -2/3\mu$ is frequently used and it has been found for gases to be a good working approximation.

When we combine Eqs. (A.16) and (A.17) with Eqs. (A.13)–(A.15), the equations for the velocity components u, v, and w in three dimensions can be rewritten as

$$\rho \frac{Du}{Dt} = -\frac{\partial p}{\partial x} + \frac{\partial}{\partial x}\left[2\mu\frac{\partial u}{\partial x} + \lambda\left(\frac{\partial u}{\partial x} + \frac{\partial v}{\partial y} + \frac{\partial w}{\partial z}\right)\right]$$
$$+ \frac{\partial}{\partial y}\left[\mu\left(\frac{\partial u}{\partial y} + \frac{\partial v}{\partial x}\right)\right] + \frac{\partial}{\partial z}\left[\mu\left(\frac{\partial u}{\partial z} + \frac{\partial w}{\partial x}\right)\right] + \sum F_x^{body\ forces} \tag{A.18}$$

$$\rho \frac{Dv}{Dt} = -\frac{\partial p}{\partial y} + \frac{\partial}{\partial y}\left[2\mu\frac{\partial v}{\partial y} + \lambda\left(\frac{\partial u}{\partial x} + \frac{\partial v}{\partial y} + \frac{\partial w}{\partial z}\right)\right]$$
$$+ \frac{\partial}{\partial x}\left[\mu\left(\frac{\partial u}{\partial y} + \frac{\partial v}{\partial x}\right)\right] + \frac{\partial}{\partial z}\left[\mu\left(\frac{\partial v}{\partial z} + \frac{\partial w}{\partial y}\right)\right] + \sum F_y^{body\ forces} \tag{A.19}$$

$$\rho \frac{Dw}{Dt} = -\frac{\partial p}{\partial z} + \frac{\partial}{\partial z}\left[2\mu\frac{\partial w}{\partial z} + \lambda\left(\frac{\partial u}{\partial x} + \frac{\partial v}{\partial y} + \frac{\partial w}{\partial z}\right)\right]$$
$$+ \frac{\partial}{\partial x}\left[\mu\left(\frac{\partial u}{\partial z} + \frac{\partial w}{\partial x}\right)\right] + \frac{\partial}{\partial y}\left[\mu\left(\frac{\partial v}{\partial z} + \frac{\partial w}{\partial y}\right)\right] + \sum F_z^{body\ forces} \tag{A.20}$$

For the conservation of energy, the rate of work done on the control volume $\Delta x \Delta y \Delta z$ is equivalent to the product of the force and velocity component, which in the x direction is the velocity component u. From Figure 3.5, the work done by the normal force in the x direction is

$$\left[u\sigma_{xx} + \frac{\partial(u\sigma_{xx})}{\partial x}\Delta x\right]\Delta y\Delta z - u\sigma_{xx}\Delta y\Delta z \tag{A.21}$$

while the work done by the tangential forces in the x direction is given by

$$\left[u\tau_{yx} + \frac{\partial(u\tau_{yx})}{\partial y}\Delta y\right]\Delta x\Delta z - u\tau_{yx}\Delta x\Delta z \tag{A.22}$$

and

$$\left[u\tau_{zx} + \frac{\partial(u\tau_{zx})}{\partial z}\Delta z\right]\Delta x\Delta y - u\tau_{zx}\Delta x\Delta y \tag{A.23}$$

The net rate of work done by these surface forces acting in the x direction divided by the control volume $\Delta x \Delta y \Delta z$ is given by

$$\frac{\partial(u\sigma_{xx})}{\partial x} + \frac{\partial(u\tau_{yx})}{\partial y} + \frac{\partial(u\tau_{zx})}{\partial z} \tag{A.24}$$

Work done due to surface stress components in the y direction and z direction can also be similarly derived, and these additional rates of work done on the fluid are

$$\frac{\partial(v\tau_{xy})}{\partial x} + \frac{\partial(v\sigma_{yy})}{\partial y} + \frac{\partial(v\tau_{zy})}{\partial z} \tag{A.25}$$

and

$$\frac{\partial(w\tau_{xz})}{\partial x} + \frac{\partial(w\tau_{yz})}{\partial y} + \frac{\partial(w\sigma_{zz})}{\partial z} \tag{A.26}$$

For heat added, the net rate of heat transfer to the fluid due to the heat flow in the x direction is given by the difference between the heat input and the heat loss at surface $x + \Delta x$, as depicted in Figure 3.5:

$$\left[q_x + \frac{\partial q_x}{\partial x}\Delta x \right]\Delta y\Delta z - q_x\Delta y\Delta z \tag{A.27}$$

Similarly, the net rates of heat transfer in the y direction and z direction may also be expressed as

$$\left[q_y + \frac{\partial q_y}{\partial y}\Delta y \right]\Delta x\Delta z - q_y\Delta x\Delta z \tag{A.28}$$

and

$$\left[q_z + \frac{\partial q_z}{\partial z}\Delta z \right]\Delta x\Delta y - q_z\Delta x\Delta y \tag{A.29}$$

The total rate of heat added to the fluid divided by the control volume $\Delta x\Delta y\Delta z$ results in

$$\frac{\partial q_x}{\partial x} + \frac{\partial q_y}{\partial y} + \frac{\partial q_z}{\partial z} \tag{A.30}$$

Combining Eqs. (A.24), (A.25), (A.26), and (A.30) with the substantial derivative for a given specific energy E of a fluid, the equation for the conservation of energy becomes

$$\begin{aligned}
\rho\frac{DE}{Dt} &= \frac{\partial(u\sigma_{xx})}{\partial x} + \frac{\partial(v\sigma_{yy})}{\partial y} + \frac{\partial(w\sigma_{zz})}{\partial z} \\
&\quad + \frac{\partial(u\tau_{yx})}{\partial y} + \frac{\partial(u\tau_{zx})}{\partial z} + \frac{\partial(v\tau_{xy})}{\partial x} + \frac{\partial(v\tau_{zy})}{\partial z} + \frac{\partial(w\tau_{xz})}{\partial x} + \frac{\partial(w\tau_{yz})}{\partial y} \\
&\quad - \frac{\partial q_x}{\partial x} - \frac{\partial q_y}{\partial y} - \frac{\partial q_z}{\partial z}
\end{aligned} \tag{A.31}$$

Thus far, we have not defined the specific energy E of a fluid. Often the energy of a fluid is defined as the sum of the internal energy, kinetic energy, and gravitational potential energy. We shall regard the gravitational force as a body force and include the effects of potential energy changes as a source term. In three dimensions, the specific energy E can be defined as

$$E = \underbrace{e}_{\text{internal energy}} + \underbrace{\frac{1}{2}\left(u^2 + v^2 + w^2\right)}_{\text{kinetic energy}} \tag{A.32}$$

For compressible flows, Eq. (A.32) is often re-arranged to give an equation for the *enthalpy*. The specific enthalpy h_{sp} and the specific (total) enthalpy h of a fluid are defined as

$$h_{sp} = e + \frac{p}{\rho}$$

and

$$h = h_{sp} + \frac{1}{2}\left(u^2 + v^2 + w^2\right)$$

Combining these two definitions with the specific energy E, we obtain

$$h = e + \frac{p}{\rho} + \frac{1}{2}\left(u^2 + v^2 + w^2\right) = E + \frac{p}{\rho} \qquad (A.33)$$

Upwind Schemes

The first-order upwind scheme is described in Chapter 4, Section 4.2.3. Here, we concentrate on the formulation of the second-order upwind and third-order QUICK schemes, as illustrated below. As an improvement to the first-order upwind scheme, the idea is to incorporate additional variables located at the neighboring grid nodal points indicated by the properties at points WW and EE, as shown in Figure B.1, in order to evaluate the interface values at the cell faces of w and e.

For the second-order upwind scheme, assuming uniform distribution of the grid nodal points, additional information on the fluid flow is introduced into the approximation by the consideration of an extra upstream variable point; that is,

$$\phi_w = \frac{3}{2}\phi_W - \frac{1}{2}\phi_{WW}$$
$$\phi_e = \frac{3}{2}\phi_P - \frac{1}{2}\phi_W \qquad \text{if } u_w > 0 \text{ and } u_e > 0 \qquad \text{(B.1)}$$

$$\phi_w = \frac{3}{2}\phi_P - \frac{1}{2}\phi_E$$
$$\phi_e = \frac{3}{2}\phi_E - \frac{1}{2}\phi_{EE} \qquad \text{if } u_w < 0 \text{ and } u_e < 0 \qquad \text{(B.2)}$$

For the third-order QUICK scheme, a quadratic approximation is introduced across two variable points at the upstream and one at the downstream, depending on the flow direction. The unequal weighting influence of this particular scheme still hinges on the knowledge

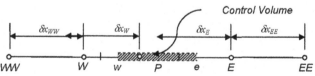

FIGURE B.1 A schematic representation of a control volume around a node P in a one-dimensional domain with surrounding grid nodal points of WW, W, E, and EE.

biased toward the upstream flow information. The interface values ϕ_w and ϕ_e based on a uniform grid nodal point distribution can be evaluated as

$$\phi_w = -\frac{1}{8}\phi_{WW} + \frac{6}{8}\phi_W + \frac{3}{8}\phi_P$$

$$\phi_e = -\frac{1}{8}\phi_W + \frac{6}{8}\phi_P + \frac{3}{8}\phi_E$$
$$\quad \text{if } u_w > 0 \text{ and } u_e > 0 \qquad \text{(B.3)}$$

$$\phi_w = -\frac{1}{8}\phi_E + \frac{6}{8}\phi_P + \frac{3}{8}\phi_W$$

$$\phi_e = -\frac{1}{8}\phi_{EE} + \frac{6}{8}\phi_E + \frac{3}{8}\phi_P$$
$$\quad \text{if } u_w < 0 \text{ and } u_e < 0 \qquad \text{(B.4)}$$

Explicit and Implicit Methods

The first-order explicit and implicit methods are described in Chapter 4, section 4.2.3. Here, we further concentrate on the formulation of the second-order *explicit* Adams-Bashford and *semi-implicit* Crank-Nicolson methods, as illustrated below.

As illustrated in Figure C.1, the extension of the first-order explicit method to the second-order explicit Adams-Bashford method requires the values not only at time level n but also at time level $n - 1$. The unsteady one-dimensional convection-diffusion of Eq. (4.43) can be recast in the form of

$$\frac{\phi_P^{n+1} - \phi_P^n}{\Delta t} = \frac{3}{2} \frac{\partial \phi}{\partial t}\bigg|^n - \frac{1}{2} \frac{\partial \phi}{\partial t}\bigg|^{n-1}$$

$$= \frac{3}{2} \left[-u_e A_E \frac{1}{2}(\phi_W + \phi_P) + u_w A_W \frac{1}{2}(\phi_P + \phi_E) \right.$$

$$\left. + \Gamma_e A_E \frac{1}{\rho}\left(\frac{\phi_E - \phi_P}{\delta x_E}\right) - \Gamma_w A_W \frac{1}{\rho}\left(\frac{\phi_P - \phi_W}{\delta x_W}\right) + \frac{S_\phi}{\rho}\Delta V \right]^n$$

$$- \frac{1}{2} \left[-u_e A_E \frac{1}{2}(\phi_W + \phi_P) + u_w A_W \frac{1}{2}(\phi_P + \phi_E) \right.$$

$$\left. + \Gamma_e A_E \frac{1}{\rho}\left(\frac{\phi_E - \phi_P}{\delta x_E}\right) - \Gamma_w A_W \frac{1}{\rho}\left(\frac{\phi_P - \phi_W}{\delta x_W}\right) + \frac{S_\phi}{\rho}\Delta V \right]^{n-1} \quad \text{(C.1)}$$

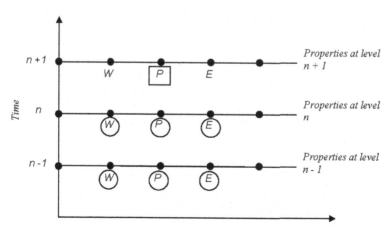

FIGURE C.1 An illustration of the second-order Adams-Bashford.

For the second-order Crank-Nicolson method, this special type of differencing in time requires the solution of ϕ_P^{n+1} to be obtained through averaging the properties between time levels n and $n+1$. There are many versions of the Crank-Nicolson form in CFD. Some may only include the implicit evaluation of the diffusive term, while others may even choose to consider the convective term in addition to the diffusive term to be implicitly determined, but the source term remains in the previous time level n regardless. The authors present the latter form as depicted below:

$$
\begin{aligned}
\frac{\phi_P^{n+1} - \phi_P^n}{\Delta t} = & -\frac{u_e^n A_E}{2}\left[\frac{1}{2}\left(\phi_W^{n+1} + \phi_W^n\right) + \frac{1}{2}\left(\phi_P^{n+1} + \phi_P^n\right)\right] \\
& +\frac{u_w^n A_W}{2}\left[\frac{1}{2}\left(\phi_P^{n+1} + \phi_P^n\right) + \frac{1}{2}\left(\phi_E^{n+1} + \phi_E^n\right)\right] \\
& +\frac{\Gamma_e^n A_E}{\rho}\left[\frac{\frac{1}{2}\left(\phi_E^{n+1} + \phi_E^n\right) - \frac{1}{2}\left(\phi_P^{n+1} + \phi_P^n\right)}{\delta x_E}\right] \\
& -\frac{\Gamma_w^n A_W}{\rho}\left[\frac{\frac{1}{2}\left(\phi_P^{n+1} + \phi_P^n\right) - \frac{1}{2}\left(\phi_W^{n+1} + \phi_W^n\right)}{\delta x_W}\right] + \frac{S_\phi^n}{\rho}\Delta V \quad \text{(C.2)}
\end{aligned}
$$

The above equation represents an example of a *semi-implicit* approach. Like the *fully implicit* approach, it also requires a simultaneous solution of the unknowns at *all* grid nodal points in the respective difference equations for a given time level of $n+1$.

Learning Program

The materials presented in this book have been partially designed from teaching *Introduction to Computational Fluid Dynamics,* a course for senior undergraduate students in the School of Aerospace, Mechanical and Manufacturing Engineering at RMIT University, Australia. This learning program can be adopted by an instructor to conduct either a 1-semester (6–8 hours/week) or 2-semester (3–4 hours/week) CFD course in any engineering department. For example, the instructor may wish to assign 3 or 4 hours per week of lectures and 3 or 4 hours per week of CFD labs for a 1-semester course. The appropriate allocation of hours for lectures or CFD labs is entirely up to the instructor. He/she may reduce the number of hours for lectures and concentrate more on CFD labs to allow students to attain more practical experience in handling real fluid-flow problems through CFD methods.

The program consists of the student's own reading of the relevant chapters described below, working out the assignments, and completing a final CFD project (see Appendix E). This teaching approach has worked very well in helping students to better engage in real problem-based assignments and projects. From the students' perspective, an air of excitement pervades from the beginning of the course through working relatively easier problems in early assignments toward solving a real-life fluid-dynamics problem chosen by the student at the final stage of the course. Details about mathematical formulations are kept to a minimum and computer programming is avoided during the teaching of the course. Rather, basic and practical knowledge about analyzing a CFD solution are emphasized. The lectures are thus directed primarily toward presenting the major theories and methodologies used in CFD and toward guiding students to where their learning efforts should be concentrated. The CFD labs are aimed at facilitating learning of basic theories and developing analysis capability and the ability to resolve practical engineering problems through the use of CFD software. During the CFD labs, tutorials are also used to introduce and to discuss the problems in assignments, as well as to provide hands-on assistance and feedback to the students. The final CFD project allows the student to attain experience in the application of CFD methods and analysis to real-world engineering problems.

LEARNING PROGRAM FOR A ONE-SEMESTER CFD COURSE

Week 1	Lecture:	Introduction to CFD and CFD Procedure
	Reading:	Chapters 1 and 2
	CFD Lab:	Introduction to CFD Software Assignment 1 (Introduction)
Week 2	Lecture:	Basic CFD Equations
	Reading:	Chapter 3
	CFD Lab:	Assignment 1 (Discussion)

Week 3	Lecture:	Mesh Generation and Boundary Conditions
	Reading:	Chapters 3 and 6
	CFD Lab:	Assignment 1 (Finalization)
Week 4	Lecture:	Basic Numerical Methods
	Reading:	Chapter 4
	CFD Lab:	Assignment 2 (Introduction)
Week 5	Lecture:	Basic Numerical Techniques
	Reading:	Chapter 4
	CFD Lab:	Assignment 2 (Discussion)
Week 6	Lecture:	CFD Solution Analysis
	Reading:	Chapter 5
	CFD Lab:	Assignment 2 (Finalization)
Week 7	Lecture:	Turbulence Modeling
	Reading:	Chapter 6
	CFD Lab:	Assignment 3 (Introduction)
Week 8	Lecture:	Practical Guidelines and Case Study
	Reading:	Chapter 6
	CFD Lab:	Assignment 3 (Discussion)
Week 9	Lecture:	CFD Applications
	Reading:	Chapter 7
	CFD Lab:	Assignment 3 (Finalization)
Week 10	Lecture:	Invited Seminar: Engineering Design and Optimization Using CFD
	Reading:	Chapter 7
	CFD Lab:	Introduction to CFD Project
Week 11	Lecture:	Advanced CFD topics
	Reading:	Chapter 8
	CFD Lab:	CFD Project (Discussion)
Week 12		CFD Project (Feedback)
Week 13		CFD Project (Finalization)
Week 14		Revision and Final Examination

CFD Assignments and Guideline for CFD Project

A sample of three assignments and a guideline for a CFD project clarifying the aim and objectives are described in this appendix. Students who do not have their own project topics are welcome to select one from the project topics (CFD Projects A–C) exemplified herein.

ASSIGNMENT 1

Background and Aim

The backward-facing step is commonly used as a benchmark for validating numerous flow characteristics, including flow recirculation and reattachment, as well as for testing of numerical models and methods. This problem has numerous applications in industry, such as for HVAC, a combustion chamber, etc. The aims of this problem are

1. To learn the process of creating and exporting a mesh by using any available mesh-generation software packages. For this assignment, the mesh generator GAMBIT is employed.
2. To learn how to set suitable boundary conditions and numerical models using any available CFD software packages. ANSYS-FLUENT is used to solve the flow problem in this assignment.
3. To explore the post-processing facilities of the CFD code to analyze the numerical results.
4. To formulate concise professional reports.

Problem Description

The student is required to compute laminar flows through a backward-facing step as detailed below. The coordinates given for the geometry are normalized against the characteristic length scale (see Figure E.1). For the case of the backward-facing step, the characteristic length scale is the **step height** (in this case, a length of unity is assumed). The normalized fluid properties at the **velocity inlet** are given as

- Inlet velocity: $u_x = 1$ and $u_y = 0$
- Fluid properties: density, $\rho = 1$, dynamic viscosity, $\mu = 1/Re$, where Re is the Reynolds number

The outlet boundary is defined as an outflow condition, while the no-slip condition is invoked for the rest of the computational walls. Turbulence is ignored and no heat transfer exists within the system. It is noted that the dimensions of the computational domain may need to be altered to ensure that the flow is sufficiently developed at the outlet boundary.

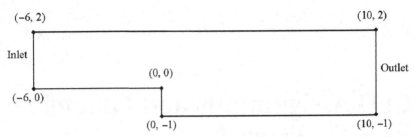

FIGURE E.1 A schematic illustration of two-dimensional geometry of a backward-facing step.

Instructions

1. Initiate GAMBIT to create a mesh for the backward-facing step. Assign appropriate boundary conditions to the computational domain. Structured mesh is preferred, but the user may alternatively generate the geometry with an unstructured mesh. Ensure that proper mesh quality is achieved. Provide explanations for areas that require further mesh refinement. Export the two-dimensional mesh to ANSYS-FLUENT, Version 6.

2. Using ANSYS-FLUENT, solve the simulation to obtain the velocity and pressure contours and the velocity vectors for $Re = 100$. Ensure that the flow is fully developed (ensuring no flow reversal) or is close to developed at the outlet boundary. Discuss any observed flow characteristics using the physical parameters.

3. Repeat Steps 2 and 3 for other meshes of varying densities. Use the same flow settings for $Re = 100$ to determine the sensitivity of the mesh to the reattachment point of the recirculation zone. Plot a graph relating the mesh size against the reattachment location, highlighting the most economical mesh for numerical computations. (Hint: When grid independence is achieved, the reattachment point will not vary with increasing mesh density.)

4. Using the mesh determined from Step 3, perform simulations for Reynolds numbers of 50, 150, and 200. Compile your results and create a graph illustrating the relationship between the Reynolds numbers and reattachment points. Explain the phenomenon and provide your own conclusions.

ASSIGNMENT 2

Background and Aim

One common CFD application is the study of flows over external structures. In the automotive industry, it is important to determine the aerodynamic effects of the spacing between adjacent motor vehicles. The Ahmed model is often used in experiments as a representation of the motor vehicle due to its simple geometry and the ease of varying a number of important parameters. The aims of simulating these models are

1. To create a CFD simulation of a single Ahmed model and extract meaningful data.
2. To obtain CFD simulations as well as study the effects of spacing between two Ahmed models.

3. To gain an understanding on the model requirements for turbulent flow and the importance of the distribution of y^+ values.
4. To better understand boundary-layer flows.
5. To assess the computational results against available published experimental data.

Problem Description

Single-Car Configuration

The geometry of a typical Ahmed configuration is shown in Figure E.2. Students are required to develop a model simulation of a single two-dimensional Ahmed configuration model.

The coordinates for the Ahmed model are given as

x	y
−0.94400	0.00000
0.00000	0.00000
−1.04400	0.10000
0.00000	0.17700
−1.04400	0.18800
−0.94400	0.28800
−0.19226	0.28800

Note that the characteristic length (taken as the length of the vehicle) of the Ahmed configuration is not equivalent to unity. Consideration must be taken into account when normalizing the Reynolds number and the spacing between adjacent vehicles.

Simulations are to be performed in air. The outer domain for the **single-model case** should be constructed according to the Cartesian coordinates below, which should allow any wakes and vortices to be properly resolved within the computational domain.

x	y
−9.39600	−0.05000
−9.39600	4.12600
19.83600	−0.05000
19.83600	4.12600

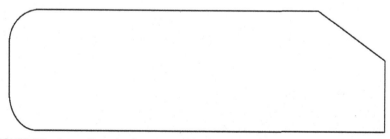

FIGURE E.2 A two-dimensional geometry of an Ahmed model.

At the inlet, the velocity should be set according to the **Reynolds number** (with respect to the car length) of 2.3×10^6. The flow **turbulence intensity** is assumed to be **1.8%**.

Instructions

1. Students are required to generate another mesh for a second **vehicle trailing the one created above.** (Hint: The domain may need to be purposefully extended to accommodate the additional vehicle.)
2. Appropriate meshing should be employed, preferably similar to above. The same meshing methods and boundary conditions used in the previous section should be used herein.
3. **Vary the distance** between the trailing and leading vehicles. Formulate at least three additional cases. Discuss the flow characteristics and compare the drag and lift coefficients for both the lead and the rear car models (Figures E.3 and E.4).
4. Compare the drag and lift coefficients against experimental results (Figures E.3 and E.4).
5. Discuss the flow physics obtained from the predicted drag and lift coefficients.

Drafting Configuration

Instructions

1. Generate a mesh of sufficient quality (keeping in mind the numerical considerations of aspect ratios and grid skewness). Ensure **mesh independence** is reached. (Hint: Use two or three different types of mesh densities.)
2. Students must apply the knowledge gained from tutorials and the previous assignment to determine **suitable boundary conditions.** (Note: Simulations can be performed by allowing the **floor to travel in the same speed as the air.**)
3. Discuss the importance of y^+ values in turbulent flows. Students must ensure that acceptable values are achieved. If otherwise, provide an explanation.

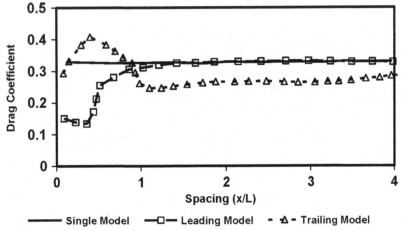

FIGURE E.3 Drag coefficient for a single Ahmed model and two Ahmed models at different vehicle spacing. *(From Watkins and Vino, 2004)*

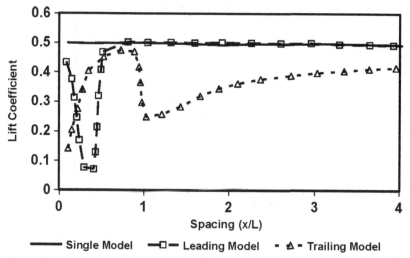

FIGURE E.4   Lift coefficient for a single Ahmed model and two Ahmed models at different vehicle spacing. *(From Watkins and Vino, 2004)*

ASSIGNMENT 3

Background and Aim

CFD has the ability to model fluid flows coupled with heat transfer. The thermal and hydrodynamic behavior of fluid within a channel has long been an established area of research. CFD simulations can provide important insights into flow behavior and heat transfer to improve the heat transfer within a complex channel geometry having a wave-shaped wall (see Figure E.5), which is increasingly being explored in industrial heat exchangers. The aims of this assignment are

1. To create a wavy channel consisting of a sufficient number of complete waves to provide developed flow conditions.
2. To create a wavy channel segment using the periodic boundary conditions.
3. To investigate the effect of different turbulence models and wall functions on the solution.
4. To implement a constant surface temperature as well as a constant surface heat flux on the wavy wall and to determine the relationship between the Reynolds number and the thermal properties.
5. To compare and discuss simulation results with experimental results.

Problem Description

The channel consists of a **repeated section** consisting of a straight wall at the top and a sine-wave-shaped wall at the bottom.

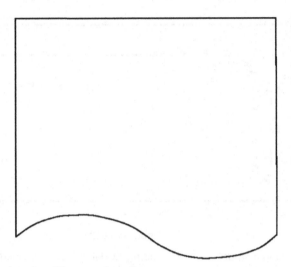

FIGURE E.5   A section of the wavy channel.

The **coordinates** for the geometry are given as

0	0
0.25	0.1
0.5	0
0.75	−0.1
1	0
1	0
0	1
1	1

The **flow properties** of the air are

- Mass flow rate: $\dot{m} = 0.816$ kg/s
- Density: $\rho = 1$ kg/m^3
- Dynamic viscosity: $\mu = 0.0001$ kg/ms
- Bulk fluid temperature: $T_b = 300$ K

Other thermal fluid properties are set as default.
   The flow is **initialized** according to

- X velocity $= 0.816$ m/s
- Turbulence kinetic energy $= 1$ m^2/s^2
- Turbulence dissipation rate $= 1 \times 10^5$ m^2/s^3

   Enhanced wall treatment is applied for all turbulence models (suitable y^+ values are
adjusted accordingly as discussed in Chapter 6).

Instructions

1. Create a channel consisting of a sufficient number of the sections given above to provide developed flow conditions. (Hint: approximately 12 sections.)
2. Compare the **normalized** axial velocity at the crest and the trough of the full model where the flow has become fully developed with the experimental results.
3. Create a section as described above using periodic boundary conditions (see Chapter 7).
4. Compare the **normalized** axial velocity at the crest ($x=0.25$) and the trough ($x=0.75$) of the periodic model with the experimental results. Discuss the relationship between the full model, periodic model, and experimental results.
5. Using the periodic model, investigate the accuracy of using the standard k-ε, RNG, and realizable turbulence models and compare them against the experimental results. Discuss which turbulence model is the most aptly suited for such a complex configuration.
6. Implement a **constant wall temperature of 500 K** on the wavy wall surface for a full model and discuss the thermal characteristics. Vary the flow to provide Reynolds numbers of 10,000 and 5,400. Compare and discuss the effect of changing the flow rate.
7. Implement a **constant wall heat flux of 1000 W/m^2** on the wavy wall surface for a full model and discuss the thermal characteristics. Use only the initial Reynolds number.
8. Generate other results of interest: velocity, turbulence parameters, temperature distribution (wall and fluid), total heat transfer, and Nusselt number.

PROJECT GUIDELINE
Aim

The aim of the CFD project is to provide an opportunity for students to demonstrate their understanding of the fundamentals and use of CFD software as well as to introduce them to the numerous applications within the software. Students are allowed to freely determine any topic of interest. It may be desirable that the intended CFD project topic coincide with the undertaking of the student's final year project, keeping in mind the many constraints and the complexity of the flow problem to be simulated. Students are therefore strongly encouraged to consult and discuss the project proposal with the lecturer/instructor before embarking on the next stage of their numerical study.

Objectives

The following abilities should be demonstrated and explained in the project report:

1. The ability to use the commercial CFD software packages for
   - Mesh generation and grid quality.
   - Defining the settings of a flow problem, i.e., boundary conditions and solver settings.
   - Selecting appropriate CFD models, i.e., turbulence models, heat transfer, or other types of simulation.
2. The ability to use CFD as a tool for engineering design in
   - Reducing drag or increasing lift for flow over geometries, e.g., car bodies or airfoils.
   - Increasing heat transfer for cooling of a car engine, or reducing heat transfer to prevent heat loss.
   - Creating desired flow control, i.e., the ability to cause flow to move within a desired region.

3. The ability to apply CFD knowledge to analyze numerical results:
   - Discussions of the flow patterns and behaviors (e.g., wake flow, flow separation, boundary-layer and convective effects).
   - Discussions of the accuracy of the CFD solution with regard to mesh quality, flow models chosen, and boundary/domain setting.
   - Assumptions made on the actual model to allow for modeling simplifications to the original geometry.

Note: Not all the items listed above are applicable to all types of CFD problems. Students should consult with lecturers or tutors concerning the above, as well as for ascertaining the suitability of the project. A brief project proposal outlining the aims and scope of the project, problem description, and objectives of the project should be prepared and submitted to the lecturers/tutors; the proposal will account for 5% of the project grade.

Examples of past CFD project topics:

- Effect of vehicle spacing on vehicles in convoy
- CFD analysis of the wing-in-ground effect
- Investigation of the effect of winglet design on lift and drag performance
- CFD simulation of flow over a bicycle helmet
- Study of the reduction of aerodynamic drag on a car-caravan combination
- Turbulent flow analysis over two two-dimensional wings of variable horizontal separation
- Investigation of the flow field in areas of different hydro power plants
- Modeling of a car air intake system and comparison of different designs
- Numerical investigation of trailing-edge flows
- Comparison of Fowler flap systems using CFD

EXAMPLE—CFD PROJECT PROPOSAL PREPARED BY THE STUDENT

Introduction

Airfoil design plays a pivotal role in wing and control surface performance in aerospace engineering. Because wind-tunnel testing can be time consuming and expensive, CFD provides an attractive alternative. Modeling of flow over an airfoil is an important CFD problem. Among the many airfoil design features, the flap (Figure E.6), which when deployed increases the camber of the wing to give increased lift (and drag), is utilized during take-off and landing in most aircraft as a lift- (and drag-) enhancing device (Figure E.7).

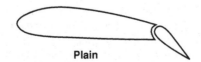

Plain

FIGURE E.6 Plain flap.

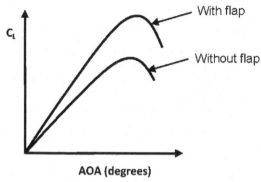

FIGURE E.7 Lift curve slope.

Scope

The project will focus on the aerodynamic characteristics of a two-dimensional NACA 23012 airfoil with a 20% scaled NACA 23012 plain flap. The airfoil will be modelled at 0° angle of attack with flap settings of 0°, 10°, 20°, 30°, and 40°. A Reynolds number of 1,400,000 helps in the study of the turbulent characteristics of the combo. The lift and drag coefficients and flow separation will be determined using the standard, RNG k-ε and realizable k-ε turbulence models to assess the most appropriate model to be applied.

Results will be validated against benchmark experimental data of Carl Wenzinger (Wenzinger, 1937, *Pressure distribution over an NACA 23012 airfoil with an NACA 23012 external-airfoil flap*. NACA Report No. 614).

Objectives

- Create a quality mesh around the airfoil/flap.
- Model the air flow around the airfoil/flap at $Re = 1,400,000$ and SSL conditions with flap settings varying from 0° to 40° using a standard k-ε turbulence model.
- Measure the lift and drag coefficient and pressure and velocity distribution of the airfoil/flap.
- Repeat with RNG k-ε and realizable k-ε turbulence models.
- Compare and discuss the results of the three turbulence models.
- Evaluate the accuracy of the results against experimental data.

OTHER TOPICS FOR CFD PROJECTS

CFD Project A: CFD Simulation of Turbulent Flow over a Backward-Facing Step

Background

The backward-facing step is commonly used as a benchmark for validation of numerous flow characteristics, including turbulence model, multi-phase flows, and fundamental numerical methods. This model also has applications in industry, such as for combustion and HVAC.

Objectives

The aims of this simulation are to

1. Create a backward-facing step simulation appropriate for turbulence modeling.
2. Determine the effects of different meshing schemes.
3. Simulate turbulent flow and compare it with the benchmark data.
4. Determine the effects of different turbulence models.
5. Understand the relevant flow characteristics.
6. Prepare a concise and well-written professional report.

Problem Description

Dry air at 27°C flows through a two-dimensional duct with a backward-facing step at a Reynolds number $Re_h = 5100$, based on the step height, h. The dimensions of this flow configuration are shown in Figure E.8. "No-slip" conditions are applied at the walls, and it can be assumed they are perfectly smooth (i.e., roughness height $= 0$ m). The turbulence intensity of the incoming flow can be assumed to be 0.01% and thermal interaction between the walls and the fluid is assumed to be adiabatic (i.e., no heat transfer). For the purpose of this assignment, it is expected that no vortex shedding will occur (i.e., steady-state analysis). From extremely accurate direct numerical simulations (DNSs) it has been determined that re-attachment of the separated boundary layer occurs on the bottom wall at $x = 16.28\,h$.

Required Discussions

1. Compare the following meshing schemes: (a) uniform structured quadrilateral mesh; (b) non-uniform structured quadrilateral mesh with mesh refinement in appropriate regions (with similar number of cells); and (c) unstructured triangular mesh with refinement in appropriate regions (with similar number of cells). Discuss the differences and the necessity of mesh refinement.
2. Using the best mesh determined from (1), run the case using appropriate settings for (a) a *standard k-ε* turbulence model and (b) a *realizable k-ε* turbulence model. Determine which model gives the better prediction. Discuss the differences between the k-ε models, referring to their characteristic equations.
3. Using the best model determined from (1) and (2), replace the dry air with ethylene glycol at 37°C flowing at the same Reynolds number ($Re_h = 5100$). Discuss any similarities to, or differences from, the dry air simulation.
4. Discuss the relevant flow characteristics of the simulation.

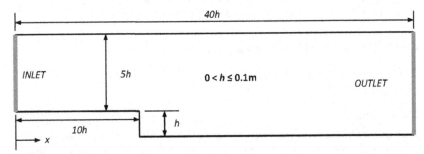

FIGURE E.8 Dry air at 27°C flow configuration.

CFD Project B: CFD Simulation of Pickup Trucks with Open/Closed Beds

Background

CFD has been a vital tool in studying vehicle aerodynamics. These simulations aim to provide a better understanding of flow behavior over complex geometries as well as give insights into methods of increasing vehicle efficiency.

Objectives

The aims of this simulation are to

1. Create two-dimensional models of a pickup truck with the bed open and closed.
2. Create a mesh of sufficient quality to achieve grid independence.
3. Simulate turbulent flow and compare with benchmark data.
4. Understand and discuss the relevant flow characteristics, such as flow velocity and pressure, to design parameters, such as drag and lift forces.
5. Create a trailing pickup truck and investigate the effect of different distances on flow characteristics.
6. Prepare a concise and well-written report.

Problem Description

A truck, as shown in Figure E.9 (dimensions provided in Addendum) is to be modeled (with tailgate closed). The moving truck is simulated traveling at a Reynolds number of 3.3×10^6 (based on the truck length). The standard k-ε turbulence model and standard wall functions are employed. Students are to create a sufficiently large domain to capture the generated wake. The inlet should have a turbulence intensity of 2% and the lower boundary (parallel to the x-axis) should be set to a non-moving smooth wall so as to simulate wind-tunnel conditions. The system is assumed to be adiabatic and time-independent. For a single pickup truck with a closed tailgate, the drag coefficient was found to be 0.44.

Required Discussions

1. Determine appropriate boundary-layer settings to provide a mesh of sufficient quality for this simulation, including calculations of the first-layer thickness for a standard wall model turbulence case.
2. Compare the simulation drag coefficient with the benchmark data to determine mesh independence.

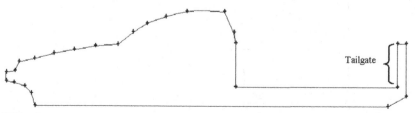

FIGURE E.9 Simulation of a moving truck traveling at a Reynolds number of 3.3×10^6.

3. Compare the flow characteristics of a single truck with the **tailgate open** with one with the **tailgate closed** to determine if tailgate position has an effect on the overall drag of the vehicle. Discuss.
4. Compare the effect of adding a trailing truck of variable distance (recommended you do not exceed 3 lengths of the truck) on both the leading and trailing vehicles. You may set either truck to have an open or closed tailgate for this investigation. Plot and compare on a graph the drag and lift coefficients for both trucks against different separation lengths. Discuss.

Note: You may need to extend the length of the domain to incorporate the extra truck.

Addendum

Coordinates for the pickup truck with closed tailgate:

X	Y
0	0.354
0	0.224
0.0215	0.246
0.0215	0.354
0.045	0.199
0.4226	0.246
0.4226	0.354
0.4264	0.38
0.4504	0.432
0.5458	0.432
0.5981	0.419
0.6438	0.402
0.6795	0.38
0.7185	0.35
0.7743	0.346
0.8284	0.337
0.8792	0.328
0.9263	0.199
0.9263	0.199
0.9267	0.315
0.935	0.229
0.9518	0.246
0.966	0.307
0.9771	0.285
0.9798	0.255
1	0.26
1	0.281

CFD Project C: Investigation of Cooling Electronic Components within a Computer

Background

CFD is a powerful tool for optimizing applications involving heat transfer. The cooling of electronic components within a computer has recently become an important issue, particularly because increased processing power tends to generate more heat, which may cause hardware damage or failure.

Objectives

The aims of this simulation are to

1. Prepare a suitable mesh of sufficient quality to analyze the thermal and flow characteristics of air cooling within a computer.
2. Understand the requirements for cooling electronic components within the system.
3. Design the system by placing the components in optimal positions to allow operation within the safety range through understanding of the thermal and flow behavior.
4. Determine the best positions for the electronic components.

Problem Description

A system case is shown in Figure E.10 with the dimensions listed in millimeters. There are two exhaust openings (50 mm wide) and one air intake (80 mm wide). Air is to enter the system at a temperature of 20°C at a rate of 0.1 m³/s with a turbulence intensity of 5%. The walls of the system are assumed to be adiabatic and smooth. Within the system, add (a) 1 processor with dimensions 25 mm × 25 mm with a heat output of 800 W/m²; (b) three RAM modules with dimensions 135 mm × 30 mm with a heat output

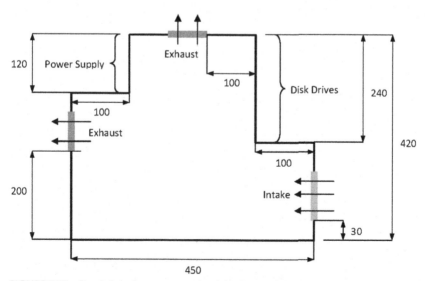

FIGURE E.10   Simulation of a system case involving heat transfer.

of 60 W/m^2; and (c) three other components of dimensions 100 mm $\times$ 50 mm with a heat output of 150 W/m^2 per module. Use the standard k-ε turbulence model to simulate the flow turbulence per component.

Required Discussions

Arrange the components listed above within the system casing with the given boundary conditions in at least three different designs. In addition, analyze the fluid flow and thermal characteristics to determine the optimal locations for your components, bearing in mind that no component surface should exceed 80°C.

References

Ahmed SR, Ramm G, Faltin G. (1984). Some salient features of the time-averaged ground vehicle wake, *SAE Technical Paper 840300*, Detroit, MI.

Aiba S, Tsuchida H, Ota T. (1982a). Heat transfer around tubes in in-line tube banks, *Bull JSME*, Series 840300, 25:919–926.

Aiba S, Tsuchida H, Ota T. (1982b). Heat transfer around tubes in staggered tube banks, *Bull JSME*, 25:927–933.

American Institute of Aeronautics and Astronautics (1998). Guide for the verification and validation of computational fluid dynamics, *AIAA G-077-1998*, Reston, VA.

Anderson W, Thomas JL, Van Leer B. (1986). Comparison of finite volume flux vector splittings for the Euler equations, *AIAA J.*, 24:1453–1460.

Andersonp JD Jr. (1995). *Computational Fluid Dynamics—The Basics with Applications*, New York, McGraw-Hill.

Apsley D, Chen W-L, Leschziner M, Lien F-S. (1997). Nonlinear eddy viscosity modelling of separated flows, *IAHR J. Hydraulic Research*, 35:723–748.

Arcilla AS, Häuser J, Eiseman PR, Thompson JF. editors. (1991). *Numerical Grid Generation in Computational Fluid Dynamics and Related Fields*. Amsterdam, North-Holland.

Baker AJ. (1983). *Finite Element Fluid Mechanics*, New York, McGraw-Hill.

Balaras E, Benocci C, Piomelli U. (1996). Two-layer approximate boundary conditions for large-eddy simulations, *AIAA J.*, 34:1111–1119.

Baldwin WS, Lomax H. (1978). Thin-layer approximation and algebraic model for separated turbulent flow, *AIAA-78-0257*, Jan.

Battan P, Goldberg U, Chakravarthy S. (2004). Interfacing statistical turbulence closures with large-eddy simulation, *AIAA J.*, 42:485–492.

Baum HR, McCaffrey BJ. (1989). Fire induced flow field—theory and experiment, *Proc. Fire Safety Science 2nd Int. Symp,* New York, Hemisphere, 129–148.

Beam RM, Warming RF. (1978). An implicit factored scheme for the compressible Navier–Stokes equations, *AIAA J.*, 16:393–402.

Belytschko T, Krongauz Y, Organ D, Fleming M, Krysl P. (1996). Meshless methods: an overview and recent developments, *Comp. Meth. Appl. Mech. Eng.*, 139:3–47.

Benson RA, McRae DS. (1991). A solution adaptive mesh algorithm for dynamic/static refinement of two and three dimensional grids, *3rd Int. Conf. Num. Grid Generation in Computational Field Simulations*, Barcelona, Spain.

Benzi R, Struglia MV, Tripiccione R. (1996). Extended self-similarity in numerical simulations of three-dimensional anisotropic turbulence, *Phys. Rev. Eng.*, 53:5565–5568.

Bertrand F, Leclaire L-A, Levecque G. (2005). DEM-based models for the mixing of granular materials, *Chem. Eng. Sci.*, 60:2517–2531.

Bhatnagar PL, Gross EP, Krook M. (1954). A model for collision processes in gases. I: small amplitude processes in charged and neutral one-component systems, *Phys. Rev.* 94:511–525.

Bird GA. (1986). Direct simulation of gas at the molecular level, *Communications in Applied Numerical Methods*, 4(2), 165–172.

Bird GA. (1994). *Molecular Gas Dynamics and the Direct Simulation of Gas Flows*. Clareton, Oxford University Press.

Bourgoyne DA, Ceccio SL, Dowling DR, Jessup S, Park J, Brewer W, Pankajakshan R. (2000). Hydrofoil turbulent boundary layer separation at high Reynolds numbers, *Proc. 23rd Symp. Naval Hydrodynamics*, Val de Reuil, France, September 17–22.

Boussinesq J. (1868). Memorie sur l'influence des frottements dans les mouvements reguliers des fluids, *J. Math Oures Appl.*, 13:377.

Bradshaw P. (1994). Turbulence: The chief outstanding difficulty of our subject, *Exp. Fluids*, 16:203–216.

Briley WR, McDonald H. (1977). Solution of the multidimensional compressible Navier–Stokes equations by a generalized implicit method, *J. Comp. Phys.*, 24:372–379.

Canuto C, Hussaini MY, Quateroni A, Zang TA. (1987). *Spectral Methods in Fluid Dynamics*, Berlin, Springer-Verlag.

Çengel YA. (2003). *Heat Transfer: A Practical Approach*, 2nd ed., New York, McGraw-Hill.

Çengel YA, Turner R. (2005). *Fundamentals of Thermal-Fluid Science*, 2nd ed., New York, McGraw-Hill.

Chang JLC, Kwak D. (1984). On the method of pseudo compressibility for numerically solving incompressible flows, *AIAA Paper 84-0252*.

Chen L, Tu JY, Yeoh GH. (2003). Numerical simulation of turbulent wake flows behind two side-by-side cylinders, *J. Fluids Structures*, 18:387–403.

Chen S, Doolen GD. (1998). Lattice Boltzmann method for fluid flows, *Ann. Rev. Fluid Mech.*, 30:329–364.

Cheng HK. (1993). Perspectives on hypersonic viscous flow research, *Ann. Rev. Fluid Mech.*, 25:455–484.

Cheng HK, Emmanual G. (1995). Perspectives on hypersonic nonequilibrium flow, *AIAA J.*, 33:385–400.

Cheung SCP, Yeoh GH, Cheung ALK, Yuen RKK. (2007). Flickering behaviour of turbulent fires using large eddy simulation, *Numer. Heat Transfer, Part A: Applications*, 52:679–712.

Chew LP. (1989). Constrained Delaunay triangulations, *Algorithmica*, 4:97–108.

Chiesa M, Mathiesen V, Melheim JA, Halvorsen B. (2005). Numerical simulation of particulate flow by the Eulerian–Lagrangian and the Eulerian–Eulerian approach with the application to a fluidized bed, *Comp. Chem. Eng.*, 29:291–304.

Chorin AJ. (1968). Numerical solution of Navier–Stokes equations, *Math. Comp.*, 22:745–762.

Choy Y-H, Merkle CL. (1993). The application of preconditioning in viscous flows, *J. Comp. Phys.*, 105:207–223.

Chung TJ. (2002). *Computational Fluid Dynamics*, Cambridge, Cambridge University Press.

Colburn AP. (1933). A method of correlating forced convection heat transfer data and a comparison with fluid friction, *Trans. American Institute of Chemical Engineers, 29*. American Institute of Electrical Engineers, New York, 174–210.

Colella P, Woodward P. (1984). The piecewise parabolic method for gas-dynamical simulations, *J. Comp. Phys.*, 54:174.

Cottet GH, Poncet P. (2004). Advances in direct numerical simulations of 3D wall-bounded flows by vortex-in-cell methods, *J. Comp. Phys.*, 193:136–158.

Cox G, Chitty R. (1980). A study of the deterministic properties of unbounded fire plumes, *Combust, Flame*, 39:191–209.

Crowe C, Schwarzkopf J, Sommerfield M, Tsuji Y. (1998). *Multiphase Flows with Droplets and Particles*, Boca Raton, CRC Press.

Crowe TC, Elger DF, Roberson JA. (2005). *Engineering Fluid Mechanics*, 8th ed. New York, Wiley.

Cundall PA, Strack ODL. (1979). A discrete numerical model for granular assemblies, *Geotechnique*, 29:47–65.

Daru V, Tenaud C. (2004). High order one-step monotonicity-preserving schemes for unsteady compressible flow calculations, *J. Comp. Phys.*, 193:563–594.

Davidson L, Peng S-H. (2003). Hybrid LES-RANS: A one-equation SGS model combined with a k-ω model for predicting recirculating flows, *Int. J. Num. Meth. Fluids*, 43: 1003–1018.

de Berg M, van Kreveld M, Overmars M, Schwarzkopf O. (2000). *Computational Geometry: Algorithms and Applications*, 2nd ed. Berlin, Springer-Verlag.

Deen NG, Van Sint Annaland M, Van Der Hoef MA, Kuipers JAM. (2007). Review of discrete particle modeling of fluidized beds, *Chem. Eng. Sci.*, 62:28–44.

de Langhe C, Merci B, Dick E. (2005a). Hybrid RANS/LES modelling with an approximate renormalization group. I. Model development, *J. Turb*, 6(13):1–18.

de Langhe C, Merci B, Lodefier K, Dick E. (2005b). Hybrid RANS/LES modelling with an approximate renormalization group. II. Applications, *J. Turb*, 6(14):1–16.

Demirdzic I, Muzaferija S, Peric M, Schreck E. (1997). Numerical method for simulation of flow problems involving moving and sliding grids, *Proc. 7th Int. Symp. Computational Fluid Dynamics*, Beijing, China.

Deng S, Jiang L, Liu C. (2007). DNS for flow separation control around airfoil by pulsed jets, *Computers & Fluids*, 36:1040–1060.

Desjardin PE, Frankel SH. (1999). Two-dimensional large eddy simulation of soot formation in the near-field of a strongly radiating nonpremixed acetylene-air turbulent jet flame, *Comb, Flame*, 119:121–132.

Di Renzo A, Di Maio FP. (2005). An improved integral non-linear model for the contact of particles in distinct element simulations, *Chem. Eng. Sci.*, 59:3461–3475.

Dopazo C. (1993). *Recent Development in PDF Methods, Turbulent Reacting Flows*, P. A. Libby and F. A. Williams, editors. New York, Academic Press.

Drikakis D. (2002). Embedded turbulence model in numerical methods for hyperbolic conservation laws, *Int. J. Num. Meth. Fluids*, 39:763–781.

Drikakis D, Rider W. (2005). *High-Resolution Methods for Incompressible and Low-Speed Flows*, Berlin, Springer-Verlag.

Duarte CA, Oden JT. (1996). AN H-P adaptive method using clouds, *Comp. Meth. Appl. Mech. Eng.*, 139:237–262.

Dufort EC, Frankel SP. (1953). Stability conditions in the numerical treatment of parabolic differential equations, *Mathematics of Computation*, 7:135–152.

Ellero M, Kröger M, Hess S. (2002). Viscoelastic flows studied by smoothed particle dynamics, *J. Non–Newton. Fluid Mech.*, 105:35–51.

Fanelli M, Feke DL, Manas–Zloczower I. (2006). Prediction of the dispersion of particle clusters in the nano-scale, Part I: Steady shearing responses, *Chem. Eng. Sci.*, 61:473–488.

Farhat C. (2005). CFD on moving grids: From theory to realistic flutter, maneuvering, and multidisciplinary optimization, *Int. J. Comp. Fluid Dyn.*, 19:595–603.

Fauci LJ, McDonald A. (1994). Sperm motility in the presence of boundaries, *Bull. Math. Biol.*, 57:679–699.

Ferziger JH, Perić M. (1999). *Computational Methods for Fluid Dynamics*, Berlin: Springer-Verlag.

Fletcher CAJ. (1984). *Computational Galerkin Methods*, Berlin, Springer-Verlag.

Fletcher CAJ. (1991). *Computational Techniques for Fluid Dynamics, Volume 1, Fundamental and General Techniques*, 2nd ed. New York, Springer-Verlag.

Fletcher CAJ. (1991). Computational Techniques for Fluid Dynamics, Volume 2, Specific Techniques for Different Flow Categories, 2nd ed. New York, Springer-Verlag.

Forum HPF. (1993). High-performance Fortran language specification, *Sci. Prog.*, 2:1–70.

Forum MPI. (1994). MPI: A message-passing interface standard, *J. Supercomputer Appl*, 8(3/4):165–414.

Fujii K. (2005). Progress and future prospects of CFD in aerospace—wind tunnel and beyond, *Prog. Aero. Sci.*, 42:455–470.

Fujii K, Obayashi S. (1987a). Navier–Stokes simulations of transonic flows over a practical wing configuration, *AIAA J.*, 25:368–370.

Fujii K, Obayashi S. (1987b). Navier–Stokes simulations of transonic flows over a wing fuselage configuration, *AIAA J.*, 25:1587–1596.

Garnier E, Mossi M, Sagaut P, Comte P, Deville M. (1999). On the use of shock-capturing schemes for large-eddy simulation, *J. Comp. Phys.*, 153:273–311.

Germano M, Piomelli U, Moin P, Cabot W. (1991). A dynamic subgrid-scale eddy viscosity model, *Phys. Fluids*, 3:1760–1765.

Ghias R, Mittal R, Lund TS. (2004). A non-body conformal grid method for simulation of compressible flows with complex immersed boundaries, *AIAA Paper 2004–0080*.

Giles MB. (1990). Nonreflecting boundary conditions for Euler equation calculation, *AIAA J.*, 28:2050–2058.

Godunov SK. (1959). A finite difference method for the numerical computation of discontinuous solutions of the equations of fluid dynamics, *Mat. Sb.*, 47:357.

Gresho PM. (1991). Incompressible fluid dynamics: Some fundamental formulation issues, *Ann. Rev. Fluid Mech.*, 23:182–188.

Gresho PM, Sani RL. (1990). On pressure boundary conditions for the incompressible Navier–Stokes equations, *Int. J. Num. Meth. Fluids*, 7:11–46.

Grotberg JB, Jensen OE. (2004). Biofluid mechanics in flexible tubes, *Ann Rev. Fluids Mech.*, 36:121–147.

Guerts BJ. (2001). *Modern Simulation Strategies for Turbulent Flow*, Philadelphia, Edwards, Inc.

Gumbert C, Lohner, R, Parikh, P, and Pirzadeh, S. (1989), A package for unstructured grid generation and finite element flow solvers, *AIAA paper*, 89–2175.

Gunzburger MD, Nicolades RA. editors. (1993). *Incompressible Computational Fluid Dynamics Trends and Advances*. London, Cambridge University Press.

Gustafsson B, Lotsedt P, Goran A. (2002). A fourth-order difference method or the incompressible Navier–Stokes equations. Numerical simulation of incompressible flows, Washington, DC, World Scientific Press.

Hafez M. (2002). *Numerical Simulation of Incompressible Flows*. Washington, DC, World Scientific Press.

Hagen GHL. (1839). Über die Bewegung des Wassers in engen cylindrischen Röhren, *Poggendorfs Annalen der Physik und Chemie*, 46:423–442.

Hahn M, Drikakis D. (2005). Large eddy simulation of compressible turbulence using high-resolution methods, *Int. J. Num. Meth. Fluids*, 47:971–977.

Hájek J, Kermes V, Stehlík P, Šikula J. (2005). Utilizing CFD as an efficient tool for improved equipment design, *Heat Transfer Engineering*, 26:15–24.

Harlow FH, Welch JE. (1965). Numerical calculation of time-dependent viscous incompressible flow with free surface, *Phys. Fluids*, 8:2182–2189.

Harten A. (1989). ENO schemes with subcell resolution, *J. Comp. Phys.*, 83:148.

Hartwich PM, Agrawal S. (1997). Method for perturbing multi-block patched grids in aeroelastic and design optimization applications, *AIAA Paper 91-2038*.

Hassan O, Probert EJ. (1999). Grid control and adaptation. In: *Handbook of Grid Generation*, Thompson JF, Soni BK, Wetherhill NP. editors. Boca Raton: CRC Press: 35–1 to 35–29.

Hazel AL, Heil M. (2003). Steady finite-Reynolds-number flows in three-dimensional collapsible tubes, *J. Fluid Mech.*, 486:79–103.

He X, Doolen GD. (1997). Lattice Boltzmann method on curvilinear coordinates system: vortex shedding behind a circular cylinder, *Phys. Rev. E.*, 56:434–440.

He X, Luo L-S. (1997). A priori derivation of the lattice Boltzmann equation, *Phys. Rev. E.*, 55:6333–6336.

Heil M. (1997). Stokes flow in collapsible tubes: computation and experiment, *J. Fluid Mech.*, 353:285–312.

Heil M, Pedley TJ. (1996). Large post-buckling deformations of cylindrical shells conveying viscous flow, *J. Fluids Struct.*, 10:565–599.

Hertz H. (1882). Über die berührung fester elatischer körper, *J. Reine Angew. Math.*, 92:156–171.

Hieber SE, Koumoutsakos P. (2005). A Lagrangian particle level set method, *J. Comp. Phys.*, 210:342–367.

Hirt CW, Nichols BD. (1981). Volume of fluid (VOF) method for dynamics of free boundaries, *J. Comp. Phys.*, 39:201–225.

Höhne T, Krepper E, Rohde U. (2010). Application of CFD codes in nuclear reactor safety analysis, *Sci Tech. Nucl. Installations*, Hindawi Publisher Article ID 198758.

Hoomans BPB, Kuipers JAM, Briels WJ, Swaaij WP. (1996). Discrete particle simulation of bubble and slug formation in a two-dimensional gas-fluidized bed: a hard-sphere approach, *Chem Eng. Sci*, 51:99–118.

Hörschler I, Meinke M, Schröder W. (2003). Numerical simulation of the flow field in a model of the nasal cavity, *Comput. Fluids*, 32:39–45.

Hou S, Sterling J, Chen S, Doolen G. (1996). A lattice Boltzmann subgrid model for high Reynolds number flows, *Fields Inst. Comm.*, 6:151–166.

Hou S, Zuo Q, Chen S, Doolen G, Cogley AC. (1995). Simulation of a cavity flow by the lattice Boltzmann method, *J. Comp. Phys.*, 118:329–347.

Hu FQ, Hussaini MY, Manthey JL. (1996). Low-dissipation and low-dispersion Runge–Kutta schemes for computational acoustics, *J. Comp. Phys.*, 124:177–191.

Huang W. (2001). Practical aspects of formulation and solution of moving mesh partial differential equations, *J. Comp. Phys.*, 171:753–775.

Huang W, Ren Y, Russel RD. (1994). Moving mesh methods based on moving mesh partial differential equations, *J. Comp. Phys.*, 113:279–290.

Huang W, Russel RD. (1999). Moving mesh strategy based on a gradient flow equation for two-dimensional problems, *SIAM J. Sci. Comp.*, 20:998–1015.

Hubbard BJ, Chen HC. (1994). A chimera scheme for incompressible viscous flows with applications to submarine hydrodynamics, *AIAA Paper 94-2210*.

Hubbard BJ, Chen HC. (1995). Calculation of unsteady flows around bodies with relative motion using a chimera RANS method, *Proc. 10th ASCE Engineering Mechanics Conference*, II:782–785, Univ. of Colorado at Boulder, May 21–24.

Hucho WH. (1996). *Road Vehicle Aerodynamic Design: An Introduction*, Edinburgh, Addison Wesley Longmann Ltd.

Humphrey JAC, Whitelaw JA, Yee G. (1981). Turbulent flow in a square duct with strong curvature, *J. Fluid Mech.*, 103:443–463.

Iaccarino G, Kalitzin G, Elkins CJ. (2003). Numerical and experimental investigation of the turbulent flow in ribbed serpentine passage, *Ann. Res. Briefs, Cent. Turb. Res.*, 379–388.

Ince NZ, Launder BE. (1995). Three-dimensional and heat-loss effects on turbulent flow in a nominally two-dimensional cavity, *Int. J. Heat Fluid Flow*, 16:171–177.

Incropera FP, DeWitt DP. (1985). *Fundamentals of Heat and Mass Transfer*, New York, Wiley.

Incropera FP, DeWitt DP. (2005). *Fundamentals of Heat and Mass Transfer*, 6th ed., New York, Wiley.

Ingram CL, McRae DS, Benson RA. (1993). Time accurate simulation of a self-excited oscillatory supersonic external flow with a multi-block solution adaptive mesh algorithm, AIAA Paper 93-3387, *Proc. 11th Computational Fluid Dynamics Conf*, Orlando.

Inthavong K, Tian ZF, Li HF, Tu JY, Yang W, Xue CL, Li CG. (2006). A numerical study of spray particles deposition in a human nasal cavity, *Aerosol Sci. Tech.*, 40:1034–1045.

Ishii M, Hibiki T. (2006). *Thermo-Fluid Dynamics of Two-Phase Flows*, Berlin, Springer-Verlag.

Issa RI. (1986). Solution of the implicitly discretised fluid flow equations by operator-splitting, *J. Comp. Phys.*, 62:40–65.

Jiang G-S, Shu C-W. (1996). Efficient implementation of weighted ENO schemes, *J. Comp. Phys.*, 126:202–228.

Jiang L, Choudari M, Chang C, Liu C. (2006). Numerical simulations of laminar-turbulent transition in supersonic boundary layers, *AIAA Paper 2006-3224.*

Johansen ST. (2003). Mathematical modeling of metallurgical processes, *Proc. 3rd Int. Conference on CFD in the Minerals and Process Industries*, CSIRO, Melbourne, Australia.

Jones WP. (1979). Models for turbulent flows with variable density. In: Kollmann W. (Ed.), *Prediction Methods for Turbulent Flows, VKI Lecture Series*. New York, Hemisphere Publishing Corporation, pp. 378–421.

Kadanoff L. (1986). On two levels, *Phys. Today*, 39:7–9.

Kader B. (1993). Temperature and concentration profiles in fully turbulent boundary layers, *Int. J. Heat Mass Transfer*, 24:1541–1544.

Kallinderis Y. editor. (2000). Adaptive methods for compressible CFD, *Comp. Meth. Appl. Sci. Eng.*, 189: Preface.

Kamakoti R, Shyy W. (2004). Fluid–structure interaction for aeroelastic applications, *Prog. Aero. Sci.*, 40:535–558.

Kandil OA, Chung HA. (1988). Unsteady vortex-dominated flows around maneuvering wings over a wide range of Mach numbers, AIAA Paper No. 88–0317, *AIAA 26th Aerospace Sciences Meeting & Exhibit*, Reno, Nevada.

Kato M, Launder BE. (1993). Three-dimensional modelling and heat-loss effects on turbulent flow in a nominally two-dimensional cavity, *Int. J. Heat and Fluid Flow*, 16:171–177.

Keyhani K, Scherer PW, Mozell MM. (1995). Numerical simulation of airflow in the human nasal cavity, *J. Biomech. Eng.*, 117:429–441.

Khan WA, Culham JR, Yovanovich MM. (2006). Analytical model for convection heat transfer from tube banks, *J, Thermophysics & Heat Transfer*, 20:720–727.

Kim SE, Choudhury D. (1995). A near-wall treatment using wall functions sensitized to pressure gradient, *ASME FED*, 217, *Separated and Complex Flows.*

Kiris C, Kwak D, Rogers S. (2002). Incompressible Navier-Stokes solvers in primitive variables and their applications to steady and unsteady flow simulations. In: *Numerical Simulation of Incompressible Flows*, Hafez M. editor. Washington, DC, World Scientific Press.

Kliafas Y, Holt M. (1987). LDV measurements of a turbulent air-solid two-phase flow in a 90° bend, *Exp, Fluids*, 5:73–85.

Knio OM, Ghoniem AF. (1992). The three-dimensional structure of periodic vorticity layers under non-symmetrical conditions, *J. Fluid Mech.*, 243:353–392.

Koelbel C, Loveman D, Schreiber R, Steele G, Zosel M. (1994). *The High Performance Fortran Handbook*, Cambridge, MIT Press.

Koobus B, Tran H, Farhat C. (2000). Computation of unsteady viscous flows around moving grids using k-ε turbulence model on unstructured dynamics grids, *Comp. Meth. Appl. Mech. Eng.*, 190:1441–1466.

Koumoutsakos P. (2005). Multiscale flow simulations using particles, *Ann. Rev. Fluid Mech.*, 37:457–487.

Koumoutsakos P, Leonard A. (1995). High resolution simulation of the flow around an impulsively started cylinder using vortex methods, *J. Fluid Mech.*, 296:1–38.

Krasny R. (1986). A study of singularity formation in a vortex sheet by the point vortex approximation, *J. Fluid Mech.*, 167:65–93.

Kruggel-Emden H, Wirtz S, Scherer V. (2009). Applicable contact force models for the discrete element model: the single particle perspective, *ASME J. Pressure Vessel Technol.*, 131:1–11.

Ku D. (1997). Blood flow in arteries, *Ann. Rev Fluids Mech.*, 29:399–434.

Kuwabara G, Kono K. (1987). Restitution coefficient in collision between two spheres, *Jpn. J. Appl. Phys., Part 1*, 26:1230–1233.

Kuzan JD. (1986). Velocity measurements for turbulent separated and near-separated flows over solid waves, Ph.D. Thesis, Dept. Chem. Eng, University of Illinois, Urbana.

Kwak D, Kiris C. (1991). Steady and unsteady solutions of the incompressible Navier–Stokes equations, *AIAA J.*, 29:603–610.

Kwak D, Kiris C, Kim CS. (2005). Computational challenges of viscous incompressible flows, *Comp. Fluids*, 34:283–299.

Labournasse E, Sagaut P. (2004). Advance in RAN-LES coupling, a review and an insight on the NLDE approach, *Arch. Comp. Meth. Eng.*, 11:199–256.

Launder BE. (1989). Second-moment closures: present and future? *Int. J. Heat Fluid Flow*, 10:282–300.

Launder BE, Reece GJ, Rodi W. (1975). Progress in the development of a Reynolds stress turbulence closure, *J. Fluid Mech.*, 68(Pt. 3):537–566.

Launder BE, Spalding DB. (1974). The numerical computation of turbulent flows, *Comp. Meth. Appl. Mech. Eng.*, 3:269–289.

Lax PD, Richtmyer RD. (1956). Survey of the stability of linear finite difference equations, *Communication on Pure Applied Mathematics*, 9:267–293.

Lee J, Herrmann HJ. (1993). Angle of repose and angle of marginal stability—molecular dynamics of granular particles, *Phys. Rev. E*, 52:3288–3291.

Lei M, Archie J, Kleimstreuer C. (1997). Computational design of a bypass graft that minimizes wall shear stress gradients in the region of the distal anastomosis, *J. Vasc. Surg.*, 25:637–646.

Leonard BP. (1991). The ULTIMATE conservative difference scheme applied to unsteady one-dimensional advection, *Comp. Meth. Appl. Mech. Eng.*, 88:17.

Li Y, Yeoh GH, Tu JY. (2004). Numerical investigation of static flow instability in a low−pressure subcooled boiling channel, 2004, *Heat and Mass Transfer*, 40:355–364.

Lilek Z, Muzaferija S, Peric M, Sedil V. (1997). Computation of unsteady flows using non-matching blocks of structured grids, *Num, Heat Transfer, Part B: Fund.*, 23:369–384.

Lilly DK. (1992). A proposed modification of the Germano subgrid-scale closure model, *Phys, Fluid*, 4:633–635.

Liseikin VD. (1999). *Grid Generation Methods*, Berlin, Springer-Verlag.

Liu X, Osher S, Chan T. (1994). Weighted essentially non-oscillatory schemes, *J. Comp. Phys.*, 115:200–212.

Lo SH. (1985). A new mesh generation scheme for arbitrary planar domains, *Int. J. Num. Meth. Eng.*, 21:1403–1426.

Lockwood FC, Naguib AS. (1975). The prediction of fluctuation in the properties of free, round jet turbulent diffusion flames, *Comb. Flame*, 24:109–124.

Loner R, Yang C, Cebral J, Soto O, Camelli F. (2002). On incompressible flow solvers. In: *Numerical Simulation of Incompressible Flows*, Hafez M, editor. *World Scientific Press, Washington, DC*.

Luan B, Robbins MO. (2005). The breakdown of continuum model for mechanical contacts, *Nature*, 435:929–932.

Luo L. (1997). Symmetry breaking of flow in 2-D symmetric channels: simulations by lattice Boltzmann method, *Int. J. Mod. Phys. C*, 8:859–867.

MacCormack RW. (1969). The effect of viscosity in hypervelocity impact cratering, *AIAA Paper 60-354*.

MacCormak RW, Paullay AJ. (1972). Computational efficiency achieved by time splitting of finite difference operators, *AIAA Paper 72-154*.

Magnussen BF, Hjertager BH. (1977). On mathematical modelling of turbulent combustion with special emphasis on soot formation and combustion, *Proc. 16th Symp. (Int.) Comb.*, The Combustion Institute, 16:719–729.

Malalasekara WMG, Versteeg HK, Gilchrist K. (1996). A review of research and an experimental study of the pulsation of buoyant diffusion flames and pool fires, *Fire Mater.*, 20:261–271.

Marcum DL, Weatherill NP. (1985). Unstructured grid generation using iterative point insertion and local reconnection, *AIAA J.*, 33:1619–1625.

MathWorks, The (1992). *The Student Edition of Matlab*, Englewood Cliffs, NJ, Prentice Hall.

Mattson T. (1995). Progamming environments for parallel and distributed computing: a comparison of p4, PVM, Linda and TCGMSG, *Int. J. Supercomputing*, 9:138–161.

Mavriplis D. (1988). Multigrid solution of the two-dimensional Euler equations on unstructured triangular meshes, *AIAA J.*, 26:824–831.

Mavriplis DJ. (1997). Unstructured grid techniques, *Annual Review of Fluid Mechanics*, 29:473–514.

Maw N, Barber JR, Fawcett JN. (1976). Oblique impact of elastic sphere, *Wear*, 38:101–114.

McCaffrey BJ. (1979). Purely buoyant diffusion flames: some experimental results, *NBSIR 79-1910*, NIST, Washington, D.C.

McCaffrey BJ. (1983). Momentum implications for buoyant diffusion flames, *Combust, Flame*, 52:149–216.

McDonald PW. (1971). The computation of transonic flow through two-dimensional gas turbine cascades, ASME Paper 71-GT-89, *Gas Turbine Conference and Products Show*, Houston.

McRae DS. (2000). R-refinement grid adaptation and issues, *Comp. Meth. Appl. Sci. Eng.*, 189:1288–1294.

Meneveau C, Lund TS, Cabot WH. (1996). A Lagrangian dynamic subgrid-scale model of turbulence, *J. Fluid Mech.*, 319:353–385.

Menter FR. (1994a). Two-equation eddy-viscosity turbulence models for engineering applications, *AIAA J.*, 32:1598–1605.

Menter FR. (1994b). Eddy-viscosity transport equations and their relation to the k-ε model, *NASA-TM-108854*.

Métais O, Lesieur M. (1992). Spectral large-eddy simulation of isotropic and stably stratified turbulence, *J. Fluid Mech.*, 256:157–194.

Mishra BK. (2003a). A review of computer simulation of tumbling mills by the discrete element method, Part I—contact mechanics, *Int. J. Miner. Process*, 71:73–93.

Mishra BK. (2003b). A review of computer simulation of tumbling mills by the discrete element method, Part II—practical applications, *Int. J. Miner. Process*, 71:95–112.

Mittal R, Iaccarino G. (2005). Immersed boundary methods, *Ann. Rev. Fluid Mech.*, 37:239–261.

Mittal R, Moin P. (1997). Suitability of upwind-biased finite difference schemes for large-eddy simulation of turbulent flows, *AIAA J.*, 35:1415.

Mittal R, Seshadri V, Udaykumar HS. (2004). Flutter, tumble and vortex induced autorotation, *Theo. Comp. Fluid Dyn.*, 17:165–170.

Mittal R, Utturkar Y, Udaykumar HS. (2002). Computational modeling and analysis of biomimetric flight mechanisms, *AIAA Paper 2002-0865*.

Monaghan JJ. (1988). An introduction to SPH, *Comp. Phys. Comm.*, 48:89–96.

Morsi SA, Alexander JA. (1972). An investigation of particle trajectories in two-phase systems, *J. Fluid Mech.*, 55:193–201.

Morsi YS, Tu JY, Yeoh GH, Yang W. (2004). Principal characteristics of turbulent gas-particulate flow in the vicinity of single tube and tube bundle structure, *Chem. Eng. Sci.*, 59:3141–3157.

Morris PJ, Long LN, Wang Q, Lockard DP. (1997). Numerical prediction of high-speed jet noise, *AIAA Paper 1598-1997*, Atlanta.

Morris PJ, Long LN, Wang Q, Pilon AR. (1998). High-speed jet noise simulations, *AIAA Paper 98-2290*.

Muntz EP. (1989). Rarefied gas dynamics, *Ann. Rev Fluid Mech.*, 21:387–417.

Murakami S. (1993). Comparison of various turbulence models applied to a bluff body, *J. Wind Eng. Ind. Aerodyn.*, 46:21–36.

Najm HH, Wyckoff PS, Knio OM. (1998). A semi-implicit numerical scheme for reacting flow: I. Stiff chemistry, *J. Comp. Phys*, 143:381–402.

Nakayama Y. (Ed.), (1998). *Visualized Flow*. Oxford, Pergamon Press.

Nikala I, Kolev. (2005). *Multiphase Flow Dynamics 1*, Berlin, Springer-Verlag,

Oberkampf WL, Trucano TG, Hirsch C. (2002). Verification, validation and predictive capability in computational engineering and physics, *Proc. 21st Century Workshop*, Johns Hopkins University Applied Physics Laboratory, Baltimore.

Oran ES, Oh CK, Cybyk BZ. (1998). Direct simulation Monte Carlo: recent advances and applications, *Ann. Rev. Fluid Mech.*, 30:403–441.

Orszag SA, Israeli N, Deville MO. (1986). Boundary conditions for incompressible flow, *J. Sci. Comp.*, 1:75–111.

Park N, Yoo JY, Choi D. (2004). Discretisation errors in large-eddy simulation on the suitability of centered and upwind-biased compact difference schemes, *J. Comp. Phys.*, 198:580–616.

Patankar SV. (1980). *Numerical Heat Transfer and Fluid Flow*, New York, Hemisphere Publishing Corporation, Taylor and Francis Group.

Patankar SV, Spalding DB. (1972). A calculation procedure for heat, mass and momentum transfer in three-dimensional parabolic flows, *Int. J. Heat Mass Transfer*, 15:1787–1806.

Patel VC, Rodi W, Scheuerer G. (1985). Turbulence model for near-wall and low Reynolds number flows: a review, *AIAA J.*, 23:1308–1319.

Peaceman DW, Rachford HH Jr. (1955). The numerical solution of parabolic and elliptic differential equations, *J. Soc. Ind. Appl. Math.*, 3(1):28–41.

Perktold K, Resch M, Peter RO. (1991). Three-dimensional numerical analysis of pulsatile flow and wall shear stress in the carotid artery bifurcation, *J. Biomech.*, 24:409–420.

Peskin CS. (1972). Flow patterns around heart valves: a digital method for solving the equations of motion, Ph.D. Thesis, Physiol., Albert Einstein College of Medicine, New York; Univ. of Michigan Microfilms, 378:72–30.

Peters N. (2000). *Turbulent Combustion*, Cambridge, Cambridge University Press.

Piomelli U, Ferziger JH, Moin P. (1987). Models for large eddy simulation of turbulent channel flows including transpiration, Technical Report, Report TF-32, Dept. Mech. Eng., Stanford University, Stanford, CA.

Piomelli U, Ferziger JH, Moin P, Kim J. (1989). New approximate boundary conditions for large eddy simulations of wall-bounded flows, *Phys, Fluids*, A1:1061–1068.

Poiseuille JLM. (1840). Recherches exp_erimentelles sur les mouvement des liquides dans les tubes de tr_es petits diam_etres, *C. R. Acad. Sci.*, 11:961–1041.

Pope SB. (1994). Lagrangian PDF methods for turbulent flows, *Ann. Rev. Fluid Mech.*, 26:23–63.

Portscht R. (1975). Studies on characteristic fluctuations of the flame radiation emitted by fires, *Combust. Sci. Tech.*, 10:73–84.

Posner JD, Buchanan CR, Dunn-Rankin D. (2003). Measurement and prediction of indoor air flow in a model room, *Energy Buildings*, 35:269–289.

Rai MM, Moin P. (1991). Direct simulation of turbulent flow using finite-difference schemes, *J. Comp. Phys.*, 96:15–53.

Rajamani GK. (2006). CFD analysis of air flow interactions in vehicle platoons, M. Eng. Thesis, RMIT University, Melbourne, Australia.

Rappitsch G, Perktold K. (1996). Pulsatile albumin transport in large arteries, *J. Biomech. Eng.*, 118:511–519.

Rhie CM, Chow WL. (1983). A numerical study of the turbulent flow past an isolated airfoil with trailing edge separation, *AIAA J.*, 21:1525–1532.

Richards K, Bithell M, Dove M, Hodge R. (2004). Discrete-element modelling: methods and applications in the environmental sciences, *Phil. Trans. R. Soc. Lond. A*, 362:1797–1861.

Rizzi AW, Inuoye M. (1973). Time split finite volume method for 3D blunt body flows, *AIAA J.*, 11:1478–1485.

Roache PJ. (1997). Quantification of uncertainty in computational fluid dynamics, *Ann. Rev. Fluid Mech.*, 29:123–160.

Roach PJ. (1998). *Fundamental of Computational Fluid Dynamics*. Albuquerque, NM, Hermosa Publishers.

Rodi W. (1991). Experience with two-layer models combining the k-ε model with a one-equation model near the wall, *AIAA Paper 91-0216*.

Rodi W. (1993). *Turbulence Models and Their Application in Hydraulics: A State of the Art Review*, Rotterdam, Balkema.

Roe PL. (2005). Computational fluid dynamics—retrospective and prospective, *Int. J. Comp. Fluid Dyn.*, 19:581–594.

Rogers SE, Kwak D, Kiris C. (1991). Steady and unsteady solutions of the incompressible Navier–Stokes equations, *AIAA J.*, 29:603–610.

Roma AM, Peskin CS, Berger MJ. (1999). An adaptive version of the immersed boundary method, *J. Comp. Phys.*, 153:509–534.

Rosenfeld M, Kwak D, Vinokur M. (1991). A fractional-step method for unsteady incompressible Navier–Stokes equations in generalized coordinate systems, *J. Comp. Phys.*, 94:102–137.

Rosenhead L. (1930). The spread of vorticity in the wake behind a cylinder, *Proc. R. Soc. A*, 127:590–612.

Ruck B, Makiola B. (1988). Particle dispersion on a single-sided backward facing step flow, *Int. J. Multiphase Flow*, 14:787–800.

Ruppert J. (1993). A new and simple algorithm for quality two-dimensional mesh generation, *Proc. 4th ACM-SIAM Symp. Discrete Algorithms*, 83–92.

Ryck B, Makiola B. (1988). Particle dispersion in a single-sided backward facing step flow, *Int. J. Multiphase Flow*, 14:787–800.

Saad Y, Schultz M. (1985). Conjugate gradient-like algorithms for solving nonsymmetric linear systems, *SIAM J.*, 44:417–424.

Sadd MH, Tai QM, Shukla A. (1993). Contact law effects on wave-propagation in particulate materials using distinct element modeling, *Int. J. Non-Linear Mech.*, 28:251–265.

Sagaut P. (2004). *Large-Eddy Simulation for Incompressible Flows—An Introduction*, Berlin, Springer-Verlag.

Sethan JA. (1996). *Level Set Methods*, London, Cambridge University Press.

Shah KB, Ferziger JH. (1997). A fluid mechanisms view of wind engineering: large eddy simulation of flow over a cubical obstacle. In: *Computer Wind Engineering*, Meroney RN, Bienkiewicsz B, editors. Amsterdam, Elsevier.

Shan X. (1997). Simulation of Rayleigh-Bénard convection using a lattice Boltzmann method, *Phys. Rev. E*, 55:2780–2788.

Shan H, Jiang L, Liu C. (2005). Direct numerical simulation of flow separation around a NACA0023 airfoil, *Computers & Fluids*, 34:1096–1114.

Shepard MS, Georges MK. (1991). Three-dimensional mesh generation by finite Octree technique, *Int. J. Num. Meth. Eng.*, 32:709–749.

Shewchuk JR. (2002). Delaunay refinement algorithms for triangular mesh generation, *Comp, Geo*, 22:21–74.

Shih T-H, Liou WW, Shabbir A, Yang Z, Zhu J. (1995). A new k-ε eddy viscosity model for high Reynolds number turbulent flows, *Computers & Fluids*, 24:227–238.

Shu C-W, Osher S. (1988). Efficient implementation of essentially non-oscillatory schemes. I., *J. Comp. Phys.*, 77:439–471.

Shu C-W, Osher S. (1989). Efficient implementation of essentially non-oscillatory schemes. II., *J. Comp. Phys.*, 83:32–78.

Simon H. editor. (1992). *Parallel Computational Fluid Dynamics*, Cambridge, MIT Press.

Smagorinsky J. (1963). General circulation experiments with the primitive equation, Part 1: The basic experiment, *Mon. Weather Rev.*, 91:99–164.

Smith RE. (1982). Algebraic grid generation. In: *Numerical Grid Generation*, Thompson, JF. editor, Amsterdam, North-Holland.

Soni B, Thompson D, Koomullil R, Thornburg H. (2001). GGTK: ATOllKit for static and dynamic geometry-grid generation, AIAA 2001-1164, *Proc. AIAA 39th Aerospace Science Meeting*, Reno, Nevada.

Spaid MAA, Phelan FR Jr. (1997). Lattice Boltzmann methods for modeling microscale flow in fibrous porous media, *Phys, Fluids*, 9:2468–2474.

Spalart PR, Jou WH, Strelets M, Allamaras SR. (1997). Comments on the feasibility of LES for wings, and on a hybrid RANS/LES approach. In: *Advances in DNS/LES*, Liu C, Liu Z, editors. Columbus, OH, Greyden Press.

Spalding DB. (1971). Mixing and chemical reaction in steady confined turbulent flames, *Proc. 13th Symp. (Int.) Comb.* The Combustion Institute, 13(1):649–657.

Spalding DB. (1980). Numerical computation of multi–phase fluid flow and heat transfer. In: *Recent Advances in Numerical Methods in Fluid*. Taylor C, Morgan K., editors, Whiting, NJ: Pineridge Press.

Speziale CG. (1998). Turbulence modeling for time-dependent RANS and VLES: a review, *AIAA J.*, 36:173–184.

Squires KD, Constantinescu GS. (2003). LES and DES investigations of turbulent flow over a sphere at Re = 10,000, *Flow, Turb. & Comb.*, 70:267–298.

Stone HL. (1968). Iterative solution of implicit approximations of multidimensional partial differential equations, *SIAM. J. Numer. Anal.*, 5(3):530–558.

Strelets M. (2001). Detached eddy simulation of massively separated flows, *AIAA Paper 2001–0879*.

Sunderam V, Geist G, Dongarra J, Manchek R. (1990). PVM: A framework for parallel distributed computing, *J. Concurrency: Prac. & Exp.*, 2:315–339.

Suresh A, Huynh HT. (1997). Accurate monotonicity-preserving schemes with Runge–Kutta time stepping, *J. Comp. Phys.*, 136:83–99.

Sweby PK. (1984). High resolution schemes using flux limiters for hyperbolic conservation laws, *SIAM J, Num. Anal.*, 21:995–1011.

Sweet RA. (1973). Direct methods for the solution of Poisson's equation on a staggered grid, *J. Comp. Phys.*, 12:422–428.

Taylor CA, Draney MT. (2004). Experimental and computational methods in cardiovascular fluid mechanics, *Ann. Rev. Fluids Mech.*, 36:197–231.

Taylor CA, Draney MT, Ku JP, Parker D, Steele BN et al. (1999). Predictive medicine: computational techniques in therapeutic decision-making, *Comp. Aided Surg.*, 4:231–247.

Temmerman L, Hadziabdic M, Leschziner MA, Hanjalic K. (2005). A hybrid two-layer URANS-LES approach for large eddy simulation at high Reynolds numbers, *Int. J. Num. Meth. Fluids*, 26:173–190.

Tennekes H, Lumley JL. (1976). *A First Course in Turbulence*, Cambridge, MIT Press.

Tezduyar TE. (2001). Finite element methods for flow problems with moving boundaries and interfaces, *Arch. Comp. Meth. Eng.*, 8:83–130.

Thomas JL, Diskin B, Brandt A. (2003). Textbook multigrid efficiency for fluid simulations, *Ann. Rev. Fluid Mech.*, 35:317–340.

Thomasset F. (1981). *Implementation of Finite Element Methods for Navier–Stokes Equations*, Berlin, Springer-Verlag.

Thompson JF, Soni BK, Weatherhill NP. editors. (1999). *Handbook of Grid Generation*. Boca Raton, CRC Press.

Thompson JF, Warsi ZUA, Mastin CW. (1982). Boundary-fitted coordinate systems for numerical solution of partial differential equations—a review, *J. Comp. Phys.*, 47:1–108.

Thompson JF, Warsi ZUA, Mastin CW. (1985). *Numerical Grid Generation—Foundations and Applications*, New York, Elsevier.

Thornton C. (1997). Force transmission in granular media, *KONA Powder Particle*, 15:81–90.

Tian ZF, Tu JY, Yeoh GH. (2005). Numerical simulation and validation of dilute gas–particle flow over back facing step, *Aerosol Sci. Tech.*, 39:319–332.

Tian ZF, Tu JY, Yeoh GH. (2006). On the numerical study of contaminant particle concentration in indoor airflow, *Build. Environ.*, 41:1504–1514.

Timmermann G. (2000). A cascadic multigrid algorithm for semilinear elliptic problems, *Numerische Mathematik*, 86:717–731.

Tomas J. (2003). Mechanics of nanoparticle adhesion—a continuum approach. In: *Particles on Surfaces 8: Detection Adhesion and Removal*, Mittal VL, editor. Utrecht, VSP.

Toro EF. (1997). *Riemann Solvers and Numerical Methods for Fluid Dynamics: A Practical Introduction*, Berlin, Springer-Verlag.

Tsuji Y, Kawaguchi T, Tanaka T. (1993). Discrete particle simulation of two-dimensional fluidized bed, *Powder Technol.*, 77:79–87.

Tsuji Y, Tanaka T, Ishida T. (1992). Lagrangian numerical simulation of plug flow of cohesionless particles in a horizontal pipe, *Powder Technol.*, 71:239–250.

Tu JY. (1997). Computation of turbulent two-phase flow on overlapped grids, *Num. Heat. Transfer, Part B, Fund.* 32:175–195.

Tu JY, Abu-Hijleh B, Xue C, Li CG. (2004a). CFD simulation of air/particle flow in the human nasal cavity, *Proc. 5th Int. Conf. Multiphase Flow*, Yokohama, Japan.

Tu JY, Fletcher CAJ. (1995). Numerical computation of turbulent gas-solid particle flow in a 90° bend, *AIChE Journal*, 41:2187–2197.

Tu JY, Fuchs L. (1992). Overlapping grids and multigrid methods for three-dimensional unsteady flow calculations in IC engines, *Int. J. Numer. Methods Fluids*, 15:693–714.

Tu JY, Yeoh GH, Morsi YS, Yang W. (2004b). A study of particle rebounding characteristics of gas-particle flow over curved surface, *Aerosol Sci. & Tech.*, 38:739–755.

Tucker P. (2003). Differential equation based length scales to improve DES and RANS simulations, *AIAA Paper 2003-3968*.

Tucker PG, Davidson L. (2004). Zonal *k-l* based large eddy simulation, *Comp, Fluids*, 33:267–287.

Turkel E. (1999). Preconditioning techniques in computational fluid dynamics, *Ann. Rev. Fluid Mech*, 31:385–416.

Unverdi S, Tryggvason G. (1992). A front-tracking method for viscous, incompressible, multifluid flows, *J. Comp. Phys.*, 100:25–42.

Valino L. (1998). A field Monte-Carlo formulation for calculating the probability density function of a single scalar in a turbulent flow, *Flow, Turb. Comb.*, 60:151–172.

Van Doormal JP, Raithby GD. (1984). Enhancements of the SIMPLE method for predicting incompressible fluid flows, *Numer, Heat Transfer*, 7:147–163.

Van Doormal JP, Raithby GD. (1985). An evaluation of the segregated approach for predicting incompressible fluid flows, ASME Paper 85-HT-9, *National Heat Transfer Conference*, Denver, CO.

Van Driest ER. (1952). Investigation of laminar boundary layer in compressible fluids using the Crocco method, *NACA Tech. Note 2597*, NTIS, U.S. Department of Commerce.

Van Driest ER. (1956). On turbulent flow near a wall, *J. Aero. Sci.*, 23:1007–1011.

Van Leer B. (1974). Towards the ultimate conservative difference scheme. II. Monotonicity and conservation combined in a second-order scheme, *J. Comp. Phys.*, 14:361–370.

Van Leer B. (1977a). Towards the ultimate conservative difference scheme. III. Upstream-centered finite difference schemes for ideal compressible flow, *J. Comp. Phys.*, 23:263–275.

Van Leer B. (1977b). Towards the ultimate conservative difference scheme. IV. A second order sequel to Godunov's method, *J. Comp. Phys.*, 23:276–299.

Van Leer B. (1979). Towards the ultimate conservative difference scheme. V. A new approach to numerical convection, *J. Comp. Phys.*, 32:101–136.

Verlet L. (1967). Computer experiments on classical fluids, I. Thermodynamical properties of Lennard-Jones molecules, *Phys. Rev. E*, 55:3546–3554.

Versteeg HK, Malalasekera W. (1995). *An Introduction to Computational Fluid Dynamics–The Finite Volume Method*, London, Pearson Education Ltd.

Vino G, Watkins S, Mousley P, Watmuff J, Prasad S. (2004). The unsteady near-wake of a simplified passenger car, *Proc. 15th Australasian Fluid Mech Conf.*, Sydney, Australia.

Vu-Quoc L, Zhang X. (1999). An elastoplastic force-displacement model in the normal direction: displacement-driven version, *Proc. R. Soc. London, Ser. A.*, 455:4013–4044.

Walton OR, Braun RL. (1986). Computer-simulation of the mechanical sorting of grains, *Powder Technol.*, 48:239–245.

Wang M, Moin P. (2002). Dynamic wall modeling for large-eddy simulation of complex turbulent flows, *Phys, Fluids*, 14:2043–2051.

Wang YQ, Jackson P, Phaneu TJ. et al. (2006). Turbulent flow through a staggered tube bank, *J. Thermophysics & Heat Transfer*, 20:738–747.

Watkins S, Vino G. (2004). On vehicle spacing and its effect on drag and lift, *Proc. 5th Int. Colloq. Bluff Body Aero. Appl. (BBAA5)*, Ottawa, Ontario.

Wesseling P. (1995). Introduction to multi-grid methods, CR-195045, ICASE 95-11, NASA.

Wilcox DC. (1998). *Turbulence Modeling for CFD*, DCW Industries Inc., La Canada, CA.

Wilson NM. (2002). Geometric algorithms and software architecture for computational prototyping: applications in vascular surgery and MEMS, Ph.D. Thesis, Stanford University, Stanford, CA.

Witry A, Al-Hajeri MH, Bondok AA. (2005). Thermal performance of automotive aluminium plate radiator, *Appl. Thermal Eng.*, 25:1207–1218.

Wolfshtein MW. (1969). The velocity and temperature distribution in a one-dimensional flow with turbulence augmentation and pressure gradient, *Int. J. Heat Mass Transfer*, 12:301–312.

Yakhot V, Orszag SA. (1986). Renormalization group analysis of turbulence. I. Basic theory, *J. Sci. Comp.*, 1:1–15.

Yakhot V, Orszag SA, Tangham S, Gatski TB, Speciale CG. (1992). Development of turbulence models for shear flows by a double expansion technique, *Phys Fluids A: Fluid Dynamics*, 4:1510–1520.

Yanenko NN. (1971). *The Method of Fractional Steps*, Berlin, Springer-Verlag.

Yang YX, Zhou B, Post JR, Scheepers E, Reuter MA, Boom R. (2006). Computational fluid dynamics simulation of pyrometallurgical processes, *Proc. 5th Int. Conf. CFD in the Minerals and Process Industries*, CSIRO, Melbourne, Australia.

Yang ZL, Dinh TN, Nourgaliev RR, Sehgal BR. (2000). Numerical investigation of bubble coalescence characteristics under nucleate boiling condition by a lattice Boltzmann model, *Int. J. Therm. Sci.*, 389:1–17.

Yeoh GH, Behnia M, de Vahl Davis G, Leonardi E. (1990). A numerical study of three-dimensional natural convection during freezing of water, *Int. J. Num. Meth. Eng.*, 30:899–914.

Yeoh GH, Lee EWM, Yuen RKK, Kwok WK. (2002). Fire and smoke distribution in a two-room compartment structure, *Int. J. Num. Meth. Heat Fluid Flow*, 12:178–194.

Yeoh GH, Tu JY. (2005). Thermal-hydrodynamics modelling of bubbly flows with heat and mass transfer, *AIChE J.*, 51:8–27.

Yeoh GH, Tu JY. (2006a). Numerical modelling of gas–liquid with and without heat and mass transfer, *Appl. Math. Modelling*, 30:1067–1095.

Yeoh GH, Tu JY. (2006b). Two-fluid and population balance models for subcooled boiling flow, *Appl. Math. Modelling*, 30:1370–1391.

Yerry MA, Shepard MS. (1984). Three-dimensional mesh generation by modified Octree technique, *Int. J. Num. Meth. Eng.*, 20:1965–1990.

Yu G, Xhang Z, Lessmann R. (1998). Fluid flow and particle diffusion in the human upper respiratory system, *Aerosol Sci, Tech.*, 28:146–158.

Yun G, Kim D, Choi H. (2006). Vortical structures behind a sphere at subcritical Reynolds numbers, *Phys. Fluid*, 18:1–14.

Zang Y, Street RL, Koseff JR. (1993). A dynamic mixed subgrid-scale model and its application to turbulent recirculating flows, *Phys, Fluids A*, 5:3186–3196.

Žukauskas A, Ulinskas R. (1988). *Heat Transfer in Tube Banks in Crossflow*, Washington, DC, Hemisphere.

Zukoski EE, Cetegen BM, Kubota T. (1984). Visible structures of buoyant diffusion flames, *Proc. 20th Symp. (Int.) Comb.*, The Combustion Institute, 20(1):361–366.

Zwartz GJ, Guilmette RA. (2001). Effect of flow rate on particle deposition in a replica of a human nasal airway, *Inhalation Toxicology*, 13:109–127.

Index

Note: Page numbers followed by *b* indicate boxes, *f* indicate figures and *t* indicate tables.

A

Accuracy, 37, 195–205, 269, 355–356, 366–367
Adaptive meshing, 357–358
Advection, 63, 73, 75, 77, 83–84, 93–94
Adverse pressure gradient, 55–56
Ahmed model, 308–309, 311
  aim, 408–409
  drafting configuration, 410
  drag coefficient, 410*f*
  lift coefficient, 411*f*
  single car configuration, 409–410
  two-dimensional geometry, 409*f*
Algebraic equations, 45–46, 123–125, 136–159, 177, 188–189
Analytical solution, 90, 127
Animation, 58
ANSYS-CFX graphical user interface, 32–34, 34*f*, 46–47, 46*f*, 49–50
ANSYS-FLUENT graphical user interface, 32–34, 35*f*, 46–47, 47*f*, 49–50
Applications
  aerospace, 8–9
  airfoil, 332
  automotive, 9–11
  biomedical, 11–14
  civil, 16–18
  combustion, 38–39, 377–379
  design tool, 8
  educational tool, 7–8
  environmental, 16–18
  fires, 39–40, 103, 303–308
  gas-particle flow, 283–288, 375
  heat exchangers, 288–296
  industrial processing, 14–16
  metallurgy, 18–19
  nasal airways, 13, 313–319
  nuclear safety, 19–21
  numerical experiments, 6–7
  power generation, 22–23
  research tool, 6–7
  sports, 23–26
  vehicle platoon, 308–313
  ventilation systems, 276, 280–282, 280*f*, 281*f*
Artificial compressibility, 351
Artificial dissipation, 353, 354

B

Back substitution process, 152–153, 154, 155
Backward-facing step flow, 244–245, 245*f*, 246*f*, 252–253, 253*f*, 255*f*, 256–257
Balsius calculation, 321–322
Bernoulli's equation, 74, 75
Body forces, 70–71, 104, 324, 397, 399–400
Boundary conditions, 40–43
  cyclic, 42, 115
  inflow, 40, 329–330
  inlet, 219–220
  no-slip condition, 65, 114, 247, 258, 325–326, 334
  outflow, 40, 41–42, 114, 242, 245–246
  outlet, 245–246
  periodic, 248–249
  symmetry, 42, 248–249
  wall, 246–247
Boundary layer
  hydrodynamic, 80–83
  instantaneous vorticity, iso-surfaces of, 338, 339, 339*f*
  laminar, 65, 326
  local, 322–323
  local Mach numbel, 327, 329*f*
  normalized temperature and velocity profiles, 327, 328*f*
  Reynolds number, 260
  pressure-surface, 267, 267*f*, 268*f*
  supersonic flow over flat plate, 321
  temperature contours and velocity vectors, 326, 327*f*
  thermal, 90–91, 92–93
  turbulent, 258–259, 259*f*, 374
Boundedness, 147

Printed in the United States
By Bookmasters